Science & Technology Policy and Indicators for Development

Perspectives from Developing Countries

Science & Technology Policy and Indicators for Development

Perspectives from Developing Countries

Editors
Seetha I. Wickremasinghe
V.K. Gupta

CENTRE FOR SCIENCE & TECHNOLOGY OF THE NON-ALIGNED AND OTHER DEVELOPING COUNTRIES (NAM S&T CENTRE)

2008
DAYA PUBLISHING HOUSE
Delhi - 110 035

ISBN10 81-7035-545-1
ISBN13 978-81-7035-545-8

Centre for Science and Technology of the Non-Aligned and Other Developing Countries (NAM S&T Centre)
Core-6A, 2nd Floor, India Habitat Centre, Lodhi Road,
New Delhi-110 003 (India)
Phone: +91-11-24644974, 24645134, Fax: +91-11-24644973
E-mail: namstct@gmail.com
Website: www.namstct.org

Published by : **Daya Publishing House**
1123/74, Deva Ram Park, Tri Nagar, Delhi - 110 035
Phone: 27383999, Fax: (011) 23260116
e-mail: dayabooks@vsnl.com
website: www.dayabooks.com
Showroom : 4760-61/23, Ansari Road, Darya Ganj
New Delhi - 110 002
Phone: 23245578, 23244987

Laser Typesetting : **Classic Computer Services**
Delhi - 110 035

Printed at : **Chawla Offset Printers**
Delhi - 110 052

Printed in India

Foreword

The developing countries should recognize science and technology as essential elements in the development process, and also develop capacities to use its full potential for the benefit of their people. The national S&T policies and their implementation should adjust to the new global realities and satisfy aspirations of providing food, shelter, health, education, energy, and higher levels of employment, quality of life and economic vitality for their societies without adversely affecting the environment. Given the limited financial resources, the key question before the policy-makers in non-aligned and other developing countries is how to tackle the challenge of globalization and get technology to produce good quality and low cost products to compete in local and export markets. The question of using traditional knowledge and skills by adding value through modern science and technology is also significant. The national S&T systems would need to address these and related issues and harness opportunities being brought out by advancements in science and technology.

I am glad that the Centre for Science and Technology of the Non-Aligned and other Developing Countries (NAM S&T Centre) organised a 3 day International Conference on "S&T Policy Research and Statistical Indicators" at Colombo, Sri Lanka during 8-10 November, 2006 jointly with the National Science Foundation (NSF) of Sri Lanka and National Science and Technology Commission (NASTEC) of Sri Lanka. This workshop provided an opportunity to focus upon many of the issues related to the formulation and implementation of S&T policies, and removing shortcomings and improving the national systems for developing policy relevant S&T statistics and indicators.

In Sri Lanka, we have taken several initiatives to bring science and technology to impact upon the critical problems facing the society including poverty and unemployment. We have to change the mindset of scientists and technologists and pursue a policy that encourages the flow of resources to those projects, which are indeed needed for our development. As developing nations, we have to pool our resources and capacities in order to maximize our outputs and competitiveness. We cannot be divided but work together to meet the tremendous challenge before us. This publication, which includes experiences from non-aligned and other developing countries in synergising science for development will be of immense value in spreading the message of cooperation amongst them and should be widely disseminated.

Prof. Tissa Vitarana
Minister of Science & Technology
Government of Sri Lanka

Preface

In the wake of globalisation, science and technology have gradually moved to the centre stage of the policy-making process in any government. Most developing countries have now started responding to this change by evolving policies and practices as well as reforming the mechanisms of policy-making and decision-making with a view to accelerate the process of synergising science and technology with the development. The investments in science and technology are also growing, which have emphasised the need for evolving short-term and long-term vision for science and technology and its directions of change. S&T policies can no longer be considered in isolation from macroeconomic policies and those in socio-economic sectors like food, agriculture, water, health, education, energy, trade and industry, which indeed are to be interlinked. This requires leveraging of science and technology for innovation. The nations need to understand the implications of S&T policy in socio-economic sectors and development as well as share each other's experiences and learn from them. The decision makers in developing countries are in need of better concepts, tools, techniques, data and other information for the formulation and implementation of S&T polices. This may include innovative approaches to collect S&T statistics and develop new and relevant indicators for measurement of technological innovation, technological balance of payments, share in high tech-trade, and in terms of S&T inputs (*i.e.* expenditures, human resources devoted to R&D, education level of R&D personnel) and outputs (*i.e.* publications, patents, technologies developed and commercialized) etc. There are several issues in formulation and implementation of S&T policies and creating S&T statistics, analytical capabilities and indicators to provide policy relevant insights. Recognizing the importance of these issues, the Centre for Science and Technology of the Non-Aligned and other Developing Countries (NAM S&T Centre) in New Delhi, India organised a 3-day International

Conference on "S&T Policy Research and Statistical Indicators" at Colombo, Sri Lanka during 8-10 November, 2006 jointly with the National Science Foundation (NSF), Sri Lanka and National Science and Technology Commission (NASTEC), Sri Lanka. Several eminent policy-makers, scientists, technologists, scholars and experts in the field of science and technology policy, S&T statistics and indicators attended the Conference. This book is an extended and updated version of the papers presented at the Conference and also includes specially invited papers.

The focus of the book is to share perspectives and experiences of developing countries on S&T policy, and on S&T statistics and indicators amongst the developing countries. The subject matter of the book is divided into three parts. Section 1 includes thirteen papers, which provide varied perspectives on S&T policy and indicators from the developing countries. Section 2 has five contributions, which describe the concepts and mechanisms for leveraging S&T policy for innovation. Section 3 reviews the implications of S&T policy in socio-economic sectors, which has six papers highlighting case studies and other experiences in sectors of agriculture, education, food, industry, shelter, and society.

In Section 1, the first paper by Doutor Eng. Venâncio Massingue, the Minister of Science and Technology, Government of Mozambique, points out the science and technology policy challenges and the vision of the Government of Mozambique. The vision integrates science and technology with developmental issues in the socio-economic sectors like agriculture, health, energy, marine sciences and fishing, construction, water, mineral resources, and environmental sustainability. It provides significant practical insights into science and technology policy formulation and implementation by indicating the lines of research and priorities for action. The case study of Cuba by Vito N. Quevedo Rodríguez, provides an excellent example of successful implementation of science policy, S&T statistics and indicators that have addressed the Cuban developmental concerns over a period of time. It demonstrates the importance of the Statistical Indicator System in the organisation and implementation of its scientific policy that resulted in the achievements in science, technology and innovation fields. Hendrik Christoffel Marais reflects upon the South African experience with a core argument that the S&T indicators can add considerable value to policy monitoring, evaluation and steering provided that the approach is linked to an underlying conceptual and methodological framework. It presents valuable insights based on a review of the South African National System of Innovation and the future role of quantitative measures in policy making. The paper by Seetha I. Wickremasinghe presents data relevant to the S&T policy framework, S&T structure, S&T statistical system in Sri Lanka. The paper highlights the work undertaken by the National Science Foundation (NSF), Sri Lanka during the last ten years and shares practical information, while pointing out the crucial bottlenecks in strengthening the S&T statistics and indicators in developing policy related insights. Ettu Obassi delves on the evolution of S&T policy in Nigeria and how the Federal Ministry of Science and Technology is articulating several initiatives to realize the vision and mission of science and technology enunciated in the 2003 S&T policy of the country. It shares the new economic initiatives of the Government, S&T reforms, and development of indicators in Nigeria.

India has a rich experience in formulation and implementation of S&T policies and developing a practical approach towards policy relevant indicators. Praveen Arora shares critical features of this experience, particularly, in bringing out R&D statistics, indicating issues and problems faced in developing indicators, and the need of new indicators. He reflects upon the role of the S&T statistical system, its relevance and policy implications at the national level. Naresh Kumar complements this experience by tracing the evolution of R&D planning and management in the Council of Scientific and Industrial Research (CSIR) in India and reflects on how S&T indicators help in planning and managing network and large R&D projects, making it dynamically responsive and a knowledge driven research organisation. Dudi Hidayat takes a look of S&T policies in Indonesia, its experience in making R&D and innovation surveys for collection and compilation of relevant S&T data, and the need for improving the S&T data collection. Seyed Mohsen Masoumzadeh from Iran provides the basic data on the status of research base in the country and points out the need for taking decisions on R&D priorities and innovation activities based on demand for such activities. Kamaruhzaman Mat Zin highlights the indicators to measure and map science and technology performance in Malaysia. It makes notable contribution in developing firm level indicators for technological innovations in Malaysian manufacturing sector. Tariq Bashir gives a broad coverage of the status of science and technology in Pakistan. He highlights the structure and mechanisms of the governance of science and technology including the statistics relating to science and technology manpower and R&D expenditure. More importantly, the paper indicates the current science and technology policy initiatives in strengthening R&D infrastructure and R&D programmes including those in the fields of information technology and biotechnology. Ministry of Science and Technology (MOST) in Myanmar is the main body for developing human resources and promoting research and development. Thidar from Myanmar details out the salient features of the institutions in implementing these programmes and activities in the country. Dilli Raj Joshi gives the basic information about the science and technology policy making system and identifies the issues that need to be addressed for making a more visible impact of science and technology in national development in Nepal.

The papers in Section 1 indicate the diversity of experiences from developing countries in making science and technology indicators to work for an effective formulation and implementation of science and technology policy. The contributions reflect that some countries have advanced in developing science and technology infrastructure, shaping policies and programmes, and developing science and technology statistics and indicators leading to synergies between science, technology and development while there are others that are still in the early stages of establishing and shaping such an effective system, policies and practices.

In Section 2, the paper by Ernesto Fernandez Polcuch of the UNESCO Institute for Statistics (UIS) elaborates its strategy for implementing renewed science and technology (S&T) statistics programme including concepts and methodologies for data collection taking into account specificities of developing countries. It captures several data concepts relating to human and financial resources, science education, innovation, the outputs and impact etc., that provide significant insights towards the benefit most developing countries. Rajeswari S. Raina argues that S&T policy makers use policies and policy research inputs that are inadequate to enable innovation for poverty reduction in developing countries. The policy relevant information for innovation can be generated through processes that entail interactive learning. In order to enable rural innovation, they should acknowledge that S&T was only one among several other sources of knowledge. S&T policy has to go through processes of institutional learning and change. M. Bandyopadhyay highlights the importance of collaborative links between academic, research institutions and industry in stimulating technological innovation for economic development and overall technological and industrial competitiveness. A network of institutions and technology transfer professionals in the NAM and other developing countries should be created for exchanging information and sharing of knowledge and expertise for establishing such collaborative linkages and create suitable infrastructure for technology transfer. In the emerging era of globalization and liberalization of investment climate, the issues of intellectual property rights have become of critical importance for developing countries. Vinod Kumar Gupta describes the basic concepts of intellectual property (IP) and delves upon the salient aspects of intellectual property protection in the specific context of technology acquisition from external sources, internal sources and in-house R&D and technology development. The critical aspects of negotiating intellectual property in finalizing the technology licenses have been discussed. Janet Ibrahim Youseef points out that innovation and invention are considered to be an essential factor in the advancement of social and economic status of Egypt. She points out that the Innovation and Invention Development Agency (IIDA) has been established, which has an effective role in developing technology and taking the distinctive Egyptian inventions into a commercialisation stage.

In Section 3, Lutta Muhammad from Kenya presents findings of a study conducted in South Western Kenya with the objectives of assessing the impact of small-scale irrigation and construction of terraces and analyzing factors influencing in its adoption. Factors affecting adoption and impact, and managing micro-finance have been indicated. A suggestion has been made for a policy on the Village Bank development. In the field of education, Dilrukshi Ranathunge and Seetha Wickremasinghe from Sri Lanka share results of a survey that was conducted to study the state of employment of S&T graduates passed out from the Sri Lankan universities during the period 1998-2003. The fluctuation in employment rates of S&T graduates does not appear to be demand driven, but showed a clear relationship to the Government policies adapted from time to time, to remedy the graduate unemployment in the country. The study indicates the importance of proper long term planning by policy planners in the country, in finding solutions to the unemployment of S&T graduates in such a way that the qualified manpower is utilized to the maximum extent. Sylvester Gyamfi from Ghana takes up the issues of food security in Ghana and indicates several initiatives and activities taken by the Government of Ghana for poverty reduction such as, youth empowerment and skills training programme, and micro-finance programmes to ensure food security and economic and financial wealth. Abu Kasim Bin Ali presents empirical results from a survey of sauce processor firms spread all over the Peninsular Malaysia. He undertakes technical profile analysis and discusses major problems of the industry. Inefficient resource use such as under utilisation of labour and machines are common in this industry. According to him, the productive efficiency of firms can be improved by narrowing the efficiency gap between the 'average' and the 'best practiced' firms. Shelter is a basic need of the people. Almost each country of the developing world is facing the problem of providing shelter to its people. Leelaratne from Sri Lanka reflects upon the significant achievements made in introducing new technologies and the challenges in good governance in providing shelter in Sri Lanka, which indeed is a valuable input for policy making by other developing countries. Finally, Juma Hamsini Seph from Tanzania shares results from a UNESCO funded project on the popularization of science and technology and its effective use to the society in Tanzania. The findings indicated that the contribution of S&T has had more positive impacts than negative. It makes suggestions to its Department of Science and Technology to strengthen popularization programmes on S&T to the general public.

In the Inaugural address to the Conference, Professor Tissa Vitaratna, Sri Lankan Minister for Science and Technology had put forth the daunting challenge before the developing countries. He stressed that the developing countries were required to be constantly aware of the new technological changes that were taking place in the emerging globalized economic systems. The priority consideration for them was to transfer the benefits of scientific advances to the poorest segment of the country. The papers in the book indeed address to the concerns raised by Professor Vitaratna and touch upon several dimensions of S&T policy and indicators for development that reflect diversity of experience from developing countries, which provide practical and useful insights for policy makers to meet such a challenge. The initiative of the Centre for Science and Technology of the Non-Aligned and other Developing Countries to bring out this publication for wider dissemination is indeed laudable.

Seetha I. Wickremasinghe

V.K. Gupta

Introduction

Knowledge has become a source of economic might and power in the 21st Century. Emergence of new technologies in the areas such as Information and Communications Technology, Biotechnology, New Materials and Nanotechnology has made unprecedented impact on economic growth and social development and we are witnessing overwhelming influence of science and technology over the advancement of human civilisation. Use of traditional knowledge and skills by adding value through modern science and technology has also become significant.

The S&T system of a country must be infused with new vitality if it has to play a decisive and beneficial role in advancing the well being of all sections of society. The country must recognise the central role played by S&T in raising the quality of life of the people, particularly of the disadvantaged sections of society, in creating wealth for all, in making the country globally competitive, in utilising natural resources in a sustainable manner, in protecting the environment and in ensuring national security. The national S&T policies and their implementation must adjust to the new global realities and also satisfy social and economic goals. The issue is how to facilitate innovation-led growth to develop relevant technologies, or enable industry to acquire and adapt technology from abroad.

Developing countries greatly suffer from the absence of a meaningful S&T policy that may stimulate the process of economic development by appropriate use of the S&T resources. Most of these are at crossroads in understanding the implication of new policy contexts and create S&T policies that are interlinked with production and service sectors, competition and other national public policies. They also lack appropriate human resources with adequate expertise in S&T policy formulation and implementation. In order to correct these imbalances, it is necessary to determine the S&T policy in a developing country through consultation between various stakeholders, *viz.* institutions, industry and the governments, by taking into account the specific technological needs and available resources. The decision makers in developing countries are in need of more and better information for the formulation and implementation of S&T polices including data on S&T statistics and indicators related to expenditures and human resources devoted to R&D, education of R&D personnel, measurement of technological innovation, and S&T outputs like publications, patents, technologies developed and commercialised technological balance of payments, and high-tech trade.

With the aforementioned in view, the Centre for Science and Technology of the Non-aligned and other Developing Countries (NAM S&T Centre) organised an International Conference on 'S&T Policy Research and Statistical Indicators' in Colombo, Sri Lanka during 8-10 November 2006 in association with the National Science Foundation (NSF) and the National Science and Technology Commission (NASTEC) of Sri Lanka. Sixty nine scientists, experts and professionals from 16 countries, *viz.* Cuba, Egypt, Ghana, India, Indonesia, Iran, Kenya, Malaysia, Myanmar, Nepal, Nigeria, Pakistan, South Africa, Sri Lanka, Tanzania and Thailand, and 2 international organisations (UNESCO and the NAM S&T Centre) participated in this event. The Conference was conducted in five technical sessions spread over 2½ days and concluded with a proposal to create an on-line network to exchange the information related to S&T Policy and Statistical Indicators and to continue the similar kind of events at regular intervals.

On behalf of the NAM S&T Centre, I express sincere gratitude to His Excellency Prof. Tissa Vitarana, Minister of Science and Technology, Government of Sri Lanka for his absolute support and patronage for this Conference and to Prof. Nalini Ratnasiri, Chairperson, NASTEC, Prof. Sirimali Fernando, Chairperson NSF and Dr. M.C.N. Jayasuriya, Director, NSF for hosting and making splendid arrangements for holding the event in Sri Lanka. I have a very special word of appreciation for Dr. Seetha I. Wickremasinghe of NSF who had been the real force behind the successful management of the event, and also because she toiled hard along with Dr. V. K. Gupta of the National Institute of Science, Technology and Development Studies (NISTADS), India, to whom also I wish to express my deepest gratitude, in jointly editing this publication. My grateful thanks are also due to the Council of Scientific and Industrial Research (CSIR), India for providing generous financial support to bring out this publication. Last, but not the least, I would like to appreciate the valuable services provided by the entire team of the NAM S&T Centre under the guidance of Mr. M. Bandyopadhyay, specifically by Mr. Gaurav Gaur and Mr. Pankaj Buttan, who coordinated the work on the organisation of the Conference and subsequently compiled the material for bringing out this publication.

I am confident that the material presented in this volume will be highly useful to the government officials and policy makers in assessing strengths and weaknesses of the S&T policies being implemented in their respective countries and finding suitable corrective measures in the same, and will serve as a valuable reference material for the specialists, professionals and people related with programmes in the areas of S&T policy, S&T statistics and information.

Arun P. Kulshreshtha

Director, NAM S&T Centre

Contents

Part II: Leveraging S&T Policy for Innovation

Part I

Perspectives on S&T Policy and Indicators

Chapter 1

Strengthening science, technology and innovation for economic growth and poverty reduction in Mozambique

Venâncio Massingue

Abstract

In spite of gaining a sustained economic growth in the past few decades, Mozambique is still one of the poorest countries in the world. The major reason behind is, the high illiteracy rate among its countrymen most of whom are still living in the rural areas of the country, and are outlying from the benefits of science and technology.

Therefore, this paper immensely concentrates on the need to strengthen science, technology and innovation in Mozambique, and explain the possible role of Ministry of Science and Technology (MST) of Mozambique in wealth creation, poverty eradication and improvement of the socio-economic welfare of the nation. As the Government of Mozambique has already been encouraging collaborative inter-disciplinary research, greater involvement of end-users in setting the research agendas, and participating in research and technology development, the paper further highlights the role of the national research and development system in knowledge, technology development and adoption for socio-economic development through the new approaches of Ministry and other relevant entities in order to ensure more improvement in the nation's economy.

Keywords: Agriculture, Construction, Education, Energy, Health, Human resource development, Marine science, Mozambique, National S&T funding, Poverty reduction, ST&I, Water

Introduction

In the last decade Mozambique has achieved sustained economic growth and reduction in poverty levels in the context of a stable political and economic system and environment.

Figure 1.1 shows that real GDP growth since 1993 averaged 8.1 per cent, but decreased to 7.1 per cent in 2003, primarily due to the floods of 2000. The real GDP growth was 6.2 per cent for 2005. However, most of this growth can be attributed to a few mega-projects, such as the Mozal aluminium smelter and the Sasol gas project. There is very little impact of this growth on the wider economy, particularly the small and medium enterprise (SME) sector, which has the potential of becoming the dominant engine of sustainable economic growth. There are few linkages between the mega-projects and the SMEs. A consequence of this imbalance is that most of the foreign direct investment into Mozambique is going into the mega-projects, with very little being used to stimulate the broader economy. Efforts are needed for the empowerment of Mozambicans through their involvement in similar projects.

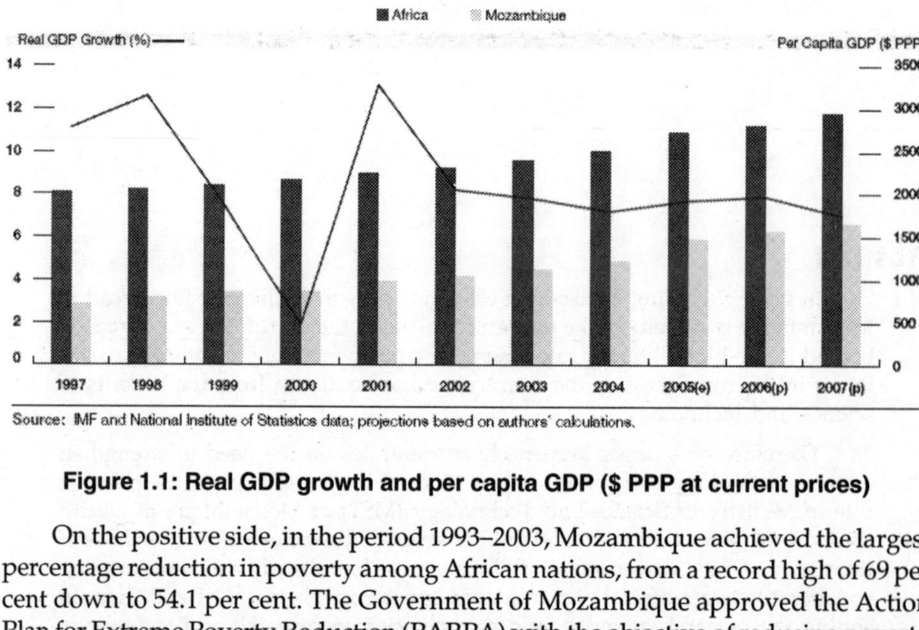

Source: IMF and National Institute of Statistics data; projections based on authors' calculations.

Figure 1.1: Real GDP growth and per capita GDP ($ PPP at current prices)

On the positive side, in the period 1993–2003, Mozambique achieved the largest percentage reduction in poverty among African nations, from a record high of 69 per cent down to 54.1 per cent. The Government of Mozambique approved the Action Plan for Extreme Poverty Reduction (PARPA) with the objective of reducing poverty up to 45 per cent by the end of the decade. The role of science and technology is crucial for the achievement of PARPA objectives, which is one of the crosscutter pillars.

Despite the major progress that Mozambique has made to reduce poverty, it remains one of the poorest countries in the world, with a GNP per capita of $347 (in the mid-1980s it was just $120) and with over 70 per cent of the population still living in rural areas. Endemic diseases such as malaria, tuberculosis, and HIV/AIDS are constant threats to all. A major barrier to socio-economic development is high illiteracy rate of 53.6 per cent, which is 68 per cent among women. In order to revert the current situation the country should greatly invest in science, technology and innovation (S, T&I) to promote research and technology transfer, thus, increase its contribution to socio-economic development and poverty reduction.

Mozambique is also committed with the global efforts to build S, T&I capacity through involvement in initiatives at national, regional, continental and international levels. For example, it is involved in the establishment of an entity that deals with science and technology at Southern African Development Community (SADC), implementation of the consolidated New Partnership for Africa's Development (NEPAD) action plan for science and technology, and in the implementation of UNESCO science and technology programs, with particular incidence to basic sciences.

This paper intends, therefore, to present the view of the Ministry of Science and Technology (MCT) on the need for strengthening science, technology and innovation in Mozambique as crosscutter sector as well as its role in wealth creation, poverty eradication and improvement of the socio-economic well being of Mozambicans. Specifically, the paper will highlight the need for science and technology in Mozambique, define the role of the national research and development system in knowledge, technology development and adoption for socio-economic development, and finally, define the role of the Ministry and other relevant entities in this process.

Science and technology challenges for Mozambique

The Ministry of Science and Technology of Mozambique (MCT) has the responsibility to guarantee equitable access to research and the associated results, which is to be reflected by the technological and social solutions to problems in rural areas, in particular, and in the country, in general. It is also recognised that promotion of equitable access as a right should be accompanied by appropriate strategies to encourage all society members to contribute to the development of science and technology in accordance with their different capacities.

Scientific knowledge is the result of a disciplined process of inquiry, seeking explanation through laws of logic and conceptual and empirical validation. Scientific knowledge should be socially verifiable and accessible. The credibility of science is derived from the fact that the validity of the results is always questioned. This has led to the necessity of a continuous evolution, in which past solutions are substituted by new ones, which provides a better clarification of the phenomena in question. This process, which is dynamic and iterative, differs from other forms of knowledge generation. To be sure that the definition is inclusive of knowledge about man, culture and the society, we can think of science as an investigative process trying to understand the universe in whole or part. Therefore, science as viewed in this document covers all fields in natural sciences, engineering, medical sciences, agricultural sciences, social sciences and humanities.

Figure 1.2 illustrates the vision of the Government of Mozambique as far as science and technology is concerned. It indicates that the link between science and technology is bi-directional, in which there is a need for regular renewal of technologies to confront new problems, which in turn require the renewal of knowledge through further research. The approach and priority areas as well as different levels and types of research and technology development are also indicated in this figure.

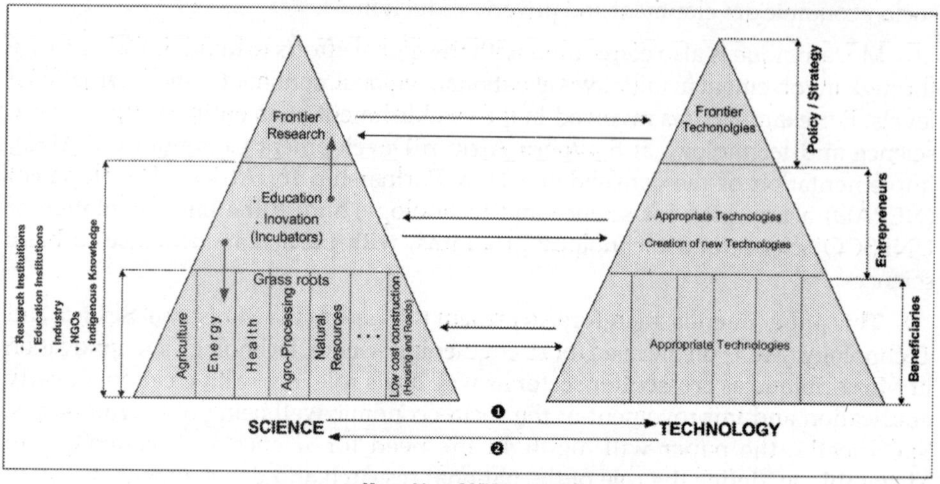

Mozambique S&T Development Concept

Figure 1.2: The vision of GoM on S&T

In order to promote equitable access to science and technology, the Government of Mozambique, recently, approved the Mozambique Science, Technology and Innovation Strategy (MOSTIS) that will enable science, technology and innovation to be harnessed, thereby enhancing their contribution to poverty reduction, to economic growth and to the wealth creation in the country. In the MOSTIS, the human resource capacity building is a key issue in order to have knowledge and expertise in the sciences, engineering and technology. This is essential for the successful application of science and technology to the challenge of poverty reduction and socio-economic development. A strategic plan has been approved for the development of human resources for ST&I, with a view to address the severe need for additional properly-trained researchers. In 2002, there were about 470 researchers in the research institutions, which is clearly inadequate. The plan calls for a stepped approach to meeting its full target of 6595 researchers by 2025.

The Government of Mozambique is also aware that there is need for simplifying and increasing access to scientific information and technological solutions for the society. We encourage and support a more collaborative inter-disciplinary research and greater involvement of end-users in setting the research agendas, and participating in research and technology development. It will be, therefore, more proactive in ensuring that the research that takes place at the community level informs science and technology development to make it more relevant and responsive to the needs of the society.

The strategic areas and strategic crosscutting areas

In the context of pressing national priorities and limited resources, and in order to break out of a spiral of dependence, it is important that Mozambique should focus on investments in STI within strategic areas. Following an analysis of the strengths and weaknesses of the Mozambican economy and society and of Mozambican science and technology, the potential threats that must be faced, along with the opportunities and challenges that exist for developing the potential of the country and its people, several strategic areas have been identified, along with several strategic crosscutting areas. The thrusts of the MOSTIS towards poverty reduction, and also towards the economic development of the country, will focus primarily on these areas.

The strategic areas include human resource development, education, agriculture, health, energy, marine sciences and fishing, construction, water, and mineral resources. The strategic crosscutting areas include social and human sciences and culture, gender equity, HIV/AIDS, environmental sustainability, and ethno-botany. The enabling areas acting as catalysts of the development of strategic areas are Information and Communication Technology (ICT), and biotechnology.

The strategic issues and research lines that are listed below for each strategic area have been identified through a process of interaction with representatives of each particular area. Programmes and research lines for addressing priority issues are also identified for each strategic area and indicated as follows:

Human resource development

In 2002 there were about 470 researchers in the research institutions (excluding Higher Education Institutes (HEIs)), which is quite inadequate when compared to the levels in countries that are showing significant progress in S&T. Taking sub-Saharan Africa as a baseline, in which there are a little more than 200 researchers per million inhabitants, there would need to currently be more than 4000 researchers in Mozambique.

Year	*Number of researchers*	*Proportion of 2025 total*
2010	660	10 per cent
2015	2638	40 per cent
2020	5276	80 per cent
2025	6595	100 per cent

The plan calls for a stepped approach to meeting its full target of 6595 researchers by 2025. It is intended that the researchers be trained in the following areas:

1. Natural sciences (15 per cent)
2. Engineering and technology (20 per cent)
3. Medical sciences (20 per cent)
4. Agronomical sciences (20 per cent)
5. Social sciences (12.5 per cent), and
6. Humanities (12.5 per cent).

The plan will be achieved through four programmes for graduate and post-graduate training both inside and outside Mozambique. Provision is also made for support that includes infrastructure, teaching staff, research publications, a research fund and scholarships.

Education

This strategic area has to do with education in the school system (primary, secondary and intermediate) and the need to develop human resources with S&T skills in the long term. The foundation of any strategy to increase the number of S&T graduates is the basic science and ICT education that is provided by the primary, secondary and intermediate education system. The basic sciences include mathematics, physics, chemistry and biology. In this area dramatic improvements are needed to prepare high-school students for pursuing science or engineering studies at a tertiary level.

It is needed to well equip science laboratories to prepare high-school students for pursuing science or engineering at a tertiary level. Access to computers and the Internet should be improved. It also is intended to increase the numbers of science and mathematics teachers as well as capacity building in order to improve their quality. Furthermore, children both in primary and secondary schools should be more exposed to science and technology to attract them to consider careers in science and technology.

It has been, therefore, planned to undertake initiatives to improve the laboratory infrastructure in schools, both for science and for ICT, throughout the country.

Agriculture

The agriculture strategy is founded on the promotion of research interventions leading to a more active and participative role of farmers in production to meet basic needs, to create self-sufficiency and well-being and ultimately for creating individual wealth as well as improving the country's wealth, on the basis of sustainable use and management of agricultural and natural resources. Agricultural research should prioritise applied research for the short term and medium term, based on the study of systems of production and socio-economic studies. It should emphasise participative and multi-disciplinary research, recognising agro-ecological differences, the importance of the regional advantages, the impact of value addition to agricultural products, and the opportunities presented by favourable environments for innovation and technological change, with the objective in particular to develop the family-oriented small-scale segment.

The immediate objective is to institute mechanisms for local prioritisation and responsibility so that the research agenda is more responsive, representative and is adapted to the main opportunities and limitations related to regional socio-economic development issues. In this context, three main pillars to be addressed have been identified: industrial farming, small-holder farming, and agro-ecological sustainability.

Agricultural research lines

1. Inventory, sustainable use, management and preservation of agricultural and natural resources (crops, livestock, soil, water, forestry, ethno-botanical resources)

2. Data collation and sharing systems for natural resources and biodiversity
3. Characterisation of production systems and their potential impact on agro-ecological and socio-economic issues
4. Inventory and preservation of genetic material of local and adapted resources
5. Post-harvest preservation and processing technologies, including the link to the agro-industry
6. Application of enabling technologies such as biotechnology and breeding for the improvement of production and productivity of local resources (plants, animals, forestry) and exotic species
7. Development and adaptation of irrigation technologies for different production systems (small-holder and agro-industrial farming)
8. Food quality control and safety
9. Control of diseases, vectors of diseases and pests of plants and animals, as well as zoonotic diseases
10. Ethno-botanical studies, to harness the potential of different plant species and promote their use in health and nutrition
11. Domestication and management of wildlife in captivity

Health

Poor health has a negative impact on the quality of life of Mozambican individuals, communities and society as a whole, as well as draining the economy due to reduced productivity, and the multiple costs related to all aspects of health-care. Effective health care is much more than the prevention and treatment of diseases and trauma. Therefore, mechanisms and processes will be established to promote scientific research in health in the public sector, the private sector, and amongst Non-Government Organizations (NGOs). While significant progress has been made in health research, the diffusion of the resulting S&T needs to be improved. Steps will be taken to strengthen diffusion channels.

Health research lines

1. Research leading to the enhancement of health policies and the health system
2. Diseases, and determinants of diseases, that have the largest impact on the health of the Mozambican population, such as HIV/AIDS, TB, malaria, intestinal parasites and other endemic diseases
3. Nutrition-related health issues, such as ways to improve nutrition, the health impact of poor nutrition; nutrient balance; alternative nutrients
4. Indigenous Knowledge Systems directed towards traditional medicine, including validation and dissemination of information concerning indigenous plants for medicinal and nutritional purposes

Energy

Reliable energy sources are fundamental to the aspirations for a good quality of life of all the people of Mozambique, as well as being essential for any sustainable economic development, including the emergence and growth of entrepreneurial ventures. Over 70 per cent of Mozambicans live in rural areas and depend on traditional fuels (especially wood), using inefficient and environmentally unsustainable technologies to try to satisfy their basic needs of nutrition, warmth and light. Efficient and effective energy systems are primary drivers of quality of life as well as sustainable economic activity and growth.

The traditional fuels will continue to be used for many years to come as a main energy source for a significant part of the Mozambican population. The efforts will be focused towards the promotion of efficient use of traditional energy sources by means of ST&I. In our strategy we intend to develop new and renewable energy sources, particularly based on wind, solar, hydro, biogas and other sources.

Energy research lines

1. The means of energy production, including bio-fuels, biomass gasification, biogas production, the production of briquettes, pellets and charcoal, and alternative energy sources, such as wind, photovoltaic, hydric, among others
2. The means of efficient use of energy, related to water decontamination, lighting, domestic and industrial ovens, and food preservation
3. Policies and programmes in the area of energy, including pricing, subsidies and fiscal incentives, as well as regulation on the efficiency of the technologies adopted in the production system

Marine sciences and fisheries

Mozambique has a surface of approximately 800 000 Km^2 and a coast line of about 2780 Km. The territorial marine water covers an area of more than 100 000 Km^2, and the inland waters an area of 20 000 Km^2. There are, however, numerous threats to the sustainable development of marine resources, such as pollution, imbalanced coastal development and over-exploitation of the ecosystems and resources. Therefore, there is need for management practices grounded on scientific principles, and of viable technologies for a sustainable use of the sea and the coast, to ensure the preservation of resources for today's and coming generations.

Strengthening the capacity and capabilities of Mozambican research institutions (including world class expertise and equipment, as well as adequate number of researchers) to create a national centre of excellence to stimulate and coordinate a wide range of research activities related to the Mozambican fishing sector will be the mean to improve research capacity and impact in this sector.

Marine sciences and fishing research lines

1. Evaluation of the fishing resources and fishing management measures (differentiating between small-scale, semi-industrial and industrial fishing)

2. Research on marine and aquatic environment (oceanography, limnology), and its relation to fishing resources
3. Research and selection of species with better genetic potential, for the improvement of germ line quality
4. Research to address the issue of reduction in populations of species being fished
5. Technical testing for commercial and aquatic species production adapted to the Mozambican environmental conditions (marine and fresh water species production, including crustaceans, molluscs, sea weed, and fish)
6. Research on fishing technologies that ensure sustainability of the fishing resources
7. Research on adequate fishing technologies to improve handling, processing and preservation, leading to added value of fishing products
8. Socio-economic and anthropologic aspects of the fishing resource utilisation, particularly in small-scale farming
9. Microbiological fishing research (main factors of contamination, and measures to conserve fish)
10. Maintenance of good health and productivity of the marine and coastal ecosystems
11. Sustainable management of marine and coastal ecosystems
12. Studies on the interaction land ocean-atmosphere, for identification of better management practises
13. Marine biotechnology as means of adding value to marine-derived products
14. Studies on marine pollution, its sources, and conception of systems to reverse the effects
15. Establishment of an effective and efficient surveillance of the sea and coast management
16. Use of ICT in diverse aspects of marine and fishery sciences, including remote sensing technologies for surveillance

Construction

Adequate housing is a primary need for everyone. Adequate housing implies the physical integrity of the building, the protection of the occupants from adverse environmental conditions, privacy, as well as access to clean water and other basic services in the context of a clean environment. A key input to adequate housing is the use of appropriate construction materials and techniques.

In Mozambique, the efforts will be made to improve the quality of housing, in particular to reduce vulnerability to natural disasters such as flooding. While a number of factors need to be addressed in this regard, including regulatory and legal, a key issue is the need to improve the quality of materials used in construction, along with the application of appropriate construction techniques. Therefore, research on local, low-cost construction materials and associated labour-intensive techniques, for both housing and roads will be carried out.

Construction research lines

1. Research on house construction technologies and techniques, construction equipment suitable for different cultural and geographical contexts, and the adoption of related construction standards and regulations
2. Research into safety matters, reformulation and adoption of construction standards and regulations based on current local construction practices and conditions
3. Development of basic quality control criteria in the production and application of construction materials
4. Research into reducing the production costs of construction materials, and into low-cost materials
5. Research to identify the potential of locally-available construction materials, and to evaluate the socio-economic viability of the use of such materials
6. Research into construction materials that reduce energy consumption

Water

Water is essential for life and health, but it has also brought death, destruction and despair to many Mozambicans through floods, storms, disease and pollution. A wider appreciation of water and the many roles it plays in the lives of humans, animals, plants and the biosphere will empower Mozambicans both to value and to better manage this unique resource. Mozambique needs to dramatically reduce the incidence and impact of water related disasters, and also to better utilise water as a renewable resource for irrigation, hydro-electric power, human consumption, manufacturing, waste management, transportation, and many forms of human leisure.

Water research lines

1. Legal and sovereignty issues related to water
2. Inventory of subterranean and surface water resources, their quality, quantity and potential use (subterranean water, rivers, lakes)
3. Effective use of water resources, including rivers
4. The water cycle and its management
5. Purification and recycling technologies for water
6. Technologies and management systems for water in the industrial, domestic and agricultural spheres
7. Water conservation, including financial and political incentives

The operationalisation of the referred research programs will be through the implementation of the following activities.

Thrust area 1: Promoting research production through funding of research proposals within selected strategic areas of water resources management. These include hydrological and hydro geological studies, water quantity, quality and spatial variations, hydrological modelling, climate change related events, and oceanographic studies.

Thrust area 2: Promoting technological innovation through testing and development of new technologies. These include rainwater harvest, ground water management, water recycling, irrigation and drainage systems, and water conservation technologies.

Thrust area 3: Promotion of internship programmes in "field laboratories" through testing and application of knowledge and technological innovation, water in agriculture–*e.g.* Chókwé irrigation system, water and fisheries–*e.g.* Aquaculture fields, water and energy–*e.g.* Hexachlorobenzene (HCB), water for industry and urban supply–*e.g.* Maputo/Matola, and water in the environment–*e.g.* Sofala Bank/Zambezi Delta.

Thrust area 4: Promotion of short-term training programmes such as on the job training, professional attachments, and visiting professors.

Thrust area 5: Promotion of training programmes in strategic areas.

Thrust area 6: Promotion of scientific symposiums and conferences.

Thrust area 7: Scientific magazine/newsletter.

Mineral resources

Mineral resources, including hydrocarbons, existing in the country can contribute to economic growth and development through sustainable production, processing, local use and exports. Some minerals can be of use for agriculture, for example, in improvement of soil quality and in pest control, and therefore, contribute to agriculture production and development. The innovative aspects of the applied technologies should be combined with, and draw from, local experiences, knowledge and skills. It is relevant to investigate techniques and technologies that are environment-friendly, along with providing increased production and productivity. Techniques and technologies adding value to mineral products will be investigated for a better use of these resources for the benefit of Mozambicans.

Mineral resources research lines

1. Appropriate technology for small-scale and low-cost mining
2. Local processing of mineral products and hydrocarbons, in order to provide for national demands and increase the opportunities for export
3. Use of metallic and non-metallic minerals
4. Use of mineral products in construction materials
5. Exploitation of hydrocarbons for energy production and other utilities
6. Applied geology and the environment
7. Application of geophysics in the mapping of water resources and in seismology research
8. The use of minerals for agriculture, pest control and health

Strategic enabling and crosscutting areas

The social and human sciences and culture

When dealing with communities there is a need to take into account several aspects such as their culture, habits and traditions. Thus, the social and human sciences have an important role to play in community development, in general, and the application of S&T for poverty reduction, in particular. Scientific knowledge concerning a community's *modus vivendi* will facilitate the introduction of new technologies, especially as these relate to attitudes and behaviour.

Research lines

Social studies on the impact of socio-economic development programmes and of potential conflicts between modern and traditional approaches

1. Studies on the adoption and impact of the new technologies that are introduced within communities
2. Studies concerning community involvement in S&T-based measures related to diseases, water and sanitation

Gender equity

Women are significantly under-represented in the S&T system. While the causes of this may be several and varied, specific interventions are necessary to redress this imbalance.

An important component of the solution will be to encourage more girls to take basic science courses in secondary school, enrol more women students for science-related degrees at the tertiary level, so that there is a larger pool of women with S&T expertise that may work within the S&T system, implement specially-designed programmes to enable women to gain the necessary S&T expertise, and fast-track women into playing significant roles within the S&T system.

HIV/AIDS

HIV / AIDS is taking a big toll on the lives of millions of Mozambicans and it is a socio-economic problem of growing concern. The disease has harsh effects in vulnerable sectors such as education and agriculture, where reduction of the numbers of teaching staff and in the workforce for an economic activity threatens the availability of the human resources that provide the means of fighting poverty. The effects of the pandemic in other productive and social spheres are equally damaging. Therefore, the challenge for S&T in this area lies in addressing the important issues relating to the prevention and control of HIV / AIDS, while devising interventions leading to preserving the human resource base which is crucial for implementing the plans to eradicate poverty and for economic development.

The overall aim of the MOSTIS in connection with HIV / AIDS is to minimise the setbacks to the quality of life of Mozambicans, to the economy, and to the pursuit of the development goals. Within this scope there are two inter-linked and complementary thrusts for S&T to address the HIV / AIDS pandemic. These are (i) strengthening the existing national programme for the prevention and control of HIV/AIDS, and (ii) creation of new research opportunities in strategic issues for control. Examples of such programmes would be evaluation and monitoring of retroviral therapy, involving step-by-step monitoring of the effects of therapy, alternative therapies that are affordable by the poor, evaluation of vaccines and adjustments of vaccine design to protect against locally-occurring strains of HIV, developing a vaccine that is most suited to local HIV strains, where Mozambique has the role of contributing to the characterisation of local strains and their diversity, and generation of knowledge leading to the prevention and alleviation of HIV/AIDS, and related problems.

Environmental sustainability

Environmental sustainability refers to the meeting of human needs without reducing the ability of the environment to provide for those needs and support life in the long term. Sustaining the environment is a crosscutting issue that should be incorporated into all policies and strategies. In particular, we will pay attention to the management of natural resources, of waste, and of biodiversity.

There are several global and regional initiatives that are building the capability to address these types of problems such as Global Earth Observation System of Systems (GEOSS) and Strategic Analysis and Knowledge Support System (SAKSS). Mozambique (though MCT) can partner and benefit from these initiatives. It should also begin to build its own national, provincial and local capabilities so that they are fully compatible with, can benefit from and can support these multi-national applied S&T initiatives. It is also important that all communities be empowered with knowledge concerning the risks of natural disasters and measures that can be taken to reduce or nullify their impact.

Research lines

1. Coastal area management
2. Urban environment management
3. Conservation of natural resources
4. Climatic changes
5. Environment and poverty
6. Environment and economic impact

Ethno-botany

Ethno-botany is a highly inter-disciplinary area of research, combining both the natural and social sciences, and overlapping with several existing disciplines and areas such as botany, biochemistry, health and traditional medicine. It investigates the traditional interaction between people and plants, with a view to applying such knowledge to many areas of Mozambique's society and economy. Such research focuses on the identification, naming and classification of plants (and related standardization issues), their place in indigenous knowledge, their value, and their use and management. This approach may present an opportunity for commercial, industrial and scientific development of new products and knowledge from the existing but unexplored indigenous knowledge and experiences. Ethno-botany can be applied to two areas of knowledge, the traditional and the modern scientific.

To allow an effective implementation of the planned actions in this area, a research centre for ethno-botany will be established to promote and coordinate research activities, deal with conservation of natural resources (plants), interact with other institutions, define priorities and application of legislation. The institution will promote the production of and market assessment for plant-based products by small, medium and large-scale enterprises.

Research lines

The lines of research that should be focused on for ethno-botany include research into the characterisation of traditional practices, including social aspects, research to validate traditional knowledge, and research into the use, production and commercialisation of products based on ethno-botanical knowledge.

Information and communication technologies

There is a need of further strengthening ICTs so that these may be able to act as booster of the country's development. To enable Mozambique to become an information-literate nation, and the development of Mozambique as an Information Society, an extensive capacity development programme needs to be developed covering a wide range of ICT skills and skill levels. This programme is key to a mindset of government and it will be addressed through interventions in the following dimensions:

1. ICT Capacity Development
2. ICT Infrastructure and Access
3. Common Architectures and Platforms
4. ICT Entrepreneurship
5. ICT Industry development

Biotechnology

Biotechnology is a prime example of a crosscutting technology that has enormous potential to add value to a range of products and services in different sectors. It is a technological platform that combines different scientific disciplines to generate alternative solutions to, amongst others, food production, human health, and industrial and environmental problems. It contributes to the current rapid developments in sectors as agriculture, medicine, the environment, marine biology, the processing industry and biomaterials production worldwide. For example, biotechnology has the potential to provide practical solutions to needs such as: (*a*) increasing production efficiency, (*b*) improving the nutritional qualities of food crops, (*c*) developing crops resistant to diseases, pests and adverse conditions, (*d*) simplifying rapid diagnosis of diseases, (*e*) new medicines and vaccines for the treatment and control of disease, or (*f*) biological processing of industrial products, and (*g*) methods for managing polluted environments and for reversing environmental degradation.

Mozambique's current capacity in biotechnology is extremely limited. The change in this scenario requires strategic commitment of resources for this area, so that its potential can contribute to development. In addition, Mozambique can contribute to institutions and networks in the SADC region, to better address long term goals related to food security, health, environmental sustainability, economic stability and development. Therefore, we are intending to develop a specific biotechnology strategy for Mozambique that will be in line with Cluster 1 of NEPAD's flagship research and development programmes on Biotechnology.

Research lines

1. Adaptation and application of current knowledge and technologies generated in developing countries for improvement of the productivity of subsistence crops and livestock breeds
2. Alternative technologies of land management for agriculture and livestock production
3. Characterization of genetic markers of relevant traits of local food crops and animal breeds, for selection and breeding purposes
4. Identification of determinants, major pathogens and vectors of high-impact diseases of humans, animals, plants and marine species
5. Application of existing (and development of new) molecular detection and molecular epidemiology methods to support the control of disease
6. Development and testing of new generation diagnostic assays, therapeutic and prophylactic products, derived from gene technology
7. Evaluation of the diversity of Mozambique's biological resources
8. Research on mechanisms of disease infection and protection in humans and animals, and mechanisms of resistance and adaptation in crops and plants that are important locally
9. Risk analysis of biotechnology-related solutions

The role of MCT and other Ministries in S, T&I

The MCT undertakes its functions as described in Presidential Decree no. 17/2005 of 31 March 2005, covering three primary roles: (1) policy making for S&T, (2) monitoring of the performance of all scientific research by the public sector, and (3) coordination of all S&T-related activities within the public sector. Thus, MCT establishes an enabling environment for development of science and technology, while the line ministries are responsible for the implementation of research activities and S&T-based interventions. In fulfilling its role in the implementation of the strategy, MCT will exercise an overall managing function, and will be appropriately involved in the planning of programmes, follow-up, monitoring, and in evaluation. At a programme level, MCT will be directly responsible for research activities, while line ministries will be responsible for roll-out and implementation. It will be important to avoid competition between ministries during implementation. Instead, a team approach should be followed, in which the baton of responsibility is passed from one ministry to the next at the appropriate time during the lifecycle of an intervention.

By its nature, S&T is a crosscutting issue that is relevant for all sectors (depicted in Figure 1.3). This strengthens the need to ensure that S&T is mainstreamed within each vertical sector. MCT should encourage such mainstreaming efforts in partnership with each applicable line ministry.

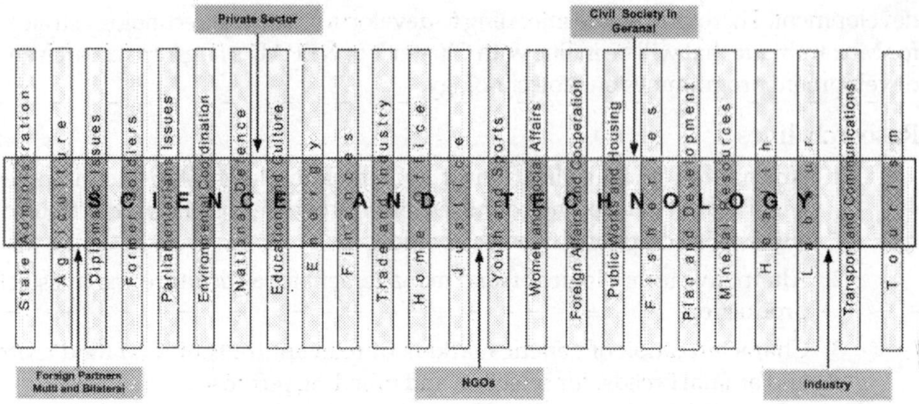

Figure 1.3: The role of MCT as a crosscutter

The crosscutting nature of MCT's role requires that careful attention be paid to the areas of intersection between MCT and each relevant line ministry. Line ministry S&T activities will be dealt with through memoranda of understanding between the concerned parties. To ensure the greatest possible clarity of roles concerning S&T activities in a particular sector, the respective roles of the two ministries should be clearly understood in relation to the full lifecycle of an S&T-based intervention. In general, MCT is responsible for research and development, while the line ministries are responsible for implementation in their respective sectors. Thus, an intervention may commence with research performed under the direction of MCT, which leads to a potential solution to a problem in a particular sector. The solution may then be tested through a pilot implementation within the sector, with the cooperation of the applicable line ministry. If the pilot programme is successful, then the next phase is adoption of the solution by the community. Once adoption has taken place, the intervention will be fully in the hands of the line ministry. To ensure alignment between the two ministries, and a smooth and effective handover between MCT and the line ministry, staff from both ministries should be involved in the intervention from the start. As the most difficult phase is the transfer of responsibility from MCT to the line ministry, it should be the subject of careful planning and effective management.

Current actions for science and technology support and promotion in the country

A number of actions aiming at promoting the creation of institutions (*e.g.* Thematic Scientific Councils) and professional associations to support and promote science and technology activities in the country are being undertaken by the MCT. These Regional Centres of Science and Technology (RCST) were created and scientific expeditions had been taking place. On the other hand, the promotion of the commercialisation of products generated by research and technology activities and job creation through incubators for manufacturing and services is also being implemented. There is also a need for technical assistance towards scientific induction for improvement of S, T and Innovation contribution towards our goals. Some of the actions are listed below.

The thematic scientific councils

The objectives of the Councils include:

1. Identification of the key research areas for the development of Mozambique,
2. Identification of the short, medium and long term strategic objectives for each key area, so that the sustainable development objectives are attained for the reduction of poverty and economic growth,
3. Promotion of high quality scientific research in the areas of strategic development, according to the identified strategic objectives and ethical, social and environmental principles, and
4. Advising the government concerning its sectoral interventions, based on the research results, and on the principles of independence, competence and professionalism of the members of the Councils.

Five Thematic Scientific Councils have been created, namely, Energy Scientific Council and Water Scientific Council in 2005, Ethno-botanic Scientific Council and Agriculture Scientific Council in 2006, and Health Scientific Council in 2007, respectively.

The regional centres for science and technology

With the establishment of the three RCSTs, the MCT will have the means of implementing programmes and projects all over the country, and will also have an effective presence close to the main beneficiaries.

The RCSTs are empowered to:

1. Evaluate socio-economical potential of technological innovations and identify the kind of support needed so that the potential may be realised,
2. Carry out, coordinate and monitor scientific research and innovation, and support appropriate technology transfer for the main socio-economic activities in their region,
3. Promote and support scientific activities, technological development and innovation, and the popularisation of S&T,
4. Carry out capacity building activities for workers, technicians and graduates in general, in terms of new technologies, literacy and technology adoption,
5. Promote the establishment of experimental research laboratories with a local focus,
6. Promote the publication of appropriate S&T, together with the central functions of MCT, by organizing exhibitions, fairs, bazaars and other programmes,

7. Evaluate the efficacy development activities and make recommendations as to their enhancement, and
8. Mobilise scientific and private sector partners, civil society, NGOs and international institutions, in order to support the RCSTs activities.

The Mozambican industry-Government group for high technology (MIGHT)

The role of the Government as a major player for the development of an economy has shifted to the private sector. Mozambique as a nation that wants to advance its economy has to adjust to this challenge. The government needs to stimulate the private sector if it intends to push the economy forward. The MCT is coming with a concept called "Mozambican Industry-Government Group for High Technology (MIGHT)" to link the government with the private sector. MIGHT is a partnership between the Government and private sector working in synergy to identify business and investment opportunities for industry and to provide inputs for Government's policy dissemination. MIGHT is to promote and coordinate the partnership, and it is aimed to address issues in High Technology development such as research priorities in areas of competitive advantage, innovation, human capital, and funding through its various consensus-based programs.

The objectives of the initiative are to:

1. Link the researchers, industry and the government by taking the knowledge from the academic institutions and research centres to be applied in the industry,
2. Create more opportunities for innovation to stimulate the Mozambican economy growth,
3. Establish a platform to convert the ideas (knowledge) from the researchers into marketable product/wealth,
4. Establish a dependent/sequential platform which links the generators of knowledge (researchers),
5. Develop and implement operational plans to support the partnership between the government-industry-researchers,
6. Create ways to assist the private sector, and
7. Build skills for growth and competitiveness.

Approval of researcher's statutes

One of our strategies is centred on valorisation of Mozambican researchers. For this purpose a specific instrument for researchers approved by the government aiming at stimulating them through a number of incentives through additional payment to their wages, such as:

1. Exclusivity to do research
2. Risk subside linked to research activities
3. Creativity subside related to scientists who create

5. Innovation subside related to scientists who innovate
6. Payment to scientists involved in research programs that generate income (Royalties).

Funding the S&T system

Funding organisations and instruments

Funding organisations and instruments have a very important task in integrating the S&T system, and in ensuring that activities are appropriately focused. A key source of funding for S&T in Mozambique is through international funding partners, agencies and instruments. Currently such funding is essential for S&T to be used effectively in reducing poverty and enabling the creation of wealth. The funds for many of the funding instruments that need to be established in Mozambique will be sourced from such funding organisations. About 0.21 per cent of the GDP is currently being allocated for research and the challenge, as per our plan, is to increase the levels investment in research to about 0.8 per cent by 2010.

Public-private partnerships can be enhanced by appropriate funding instruments, for example, tax payment by private companies and its allocation to research entities such as in case of commercial crops, namely cotton, tobacco, sugar, cashew, among others. The institutional funding that covers the basic costs of universities and research institutes needs to be complemented by competitive funding. Competitive funding involves open competition by research teams for funds allocated to targeted research domains, and the allocation is normally based on policy objectives, but these need to be sufficiently broad.

Funding and coordination

The scarcity of resources, however, demands that science and technology activities be directed to areas and sectors with largest potential impact on poverty reduction, wealth creation and improvement of socio-economic well-being. Hence, funding instruments have to be tailored to project and programme objectives. To keep the government and the international financing agencies fully appraised concerning the use of funds for S&T, and to promote accountability on behalf of the research institutions and other S&T institutions involved in managing the funds, a meeting of all these parties is arranged on an annual basis to assess the results of the previous year, and to consider the plans for the following year.

As response to this challenge, the Government of Mozambique recently established the National Research Fund. Research and related activities will be funded through the National Research Fund (see below) and through funds allocated to a particular sector. The latter funds will be provided by a funding agency to the relevant research institutions in the sector on the basis of the research plans that they have submitted.

The national research fund

A National Research Fund (NRF) has been established and key mechanism for providing funding streams for S&T are currently being developed. The NRF is a national, independent institution, established by and operating under a mandate provided by the Minister of Science and Technology. It invites proposals for funding, evaluates them, awards funding, and monitors and evaluates the implementation and the results of the funded proposals, as well as fund on its own initiative programmes and projects that promote and enhance S&T in the country.

The fund will make use of several funding instruments, with each instrument being designed to achieve a specific purpose. Examples of such instruments include:

Research projects, on a competitive basis

Researchers will be invited to submit research proposals for funding. These will be adjudicated and funding will be awarded on a competitive basis, using published criteria.

Institutional development

In order to enhance Mozambique's research capacity, some funds will be available for use for additions and improvements to the research infrastructure of research institutions. These funds will also be made available on the basis of successful proposals.

Government-commissioned research project

It must be anticipated that from time to time the Government will identify specific research projects that need urgent attention to address national priorities. This funding instrument will be used to fund such projects.

Innovation and technology transfer

A portion of the NRF's resources will be allocated for funding the high-risk phases of the innovation and commercialisation of new products and services, as well as for the transfer of technology. Funding will be awarded on the basis of successful proposals.

S&T development

This fund will be used to assist in creating a culture of S&T, to build awareness, to enhance S&T capabilities and capacities, and for related initiatives. Both solicited and unsolicited proposals for funding will be considered.

Some funds are allocated by the MCT specifically for each funding instrument on an annual basis, and the NRF should mobilize resources directly elsewhere (from donors, credits, revenues, and other funding sources) to complement the funds allocated by the MCT. In order to guarantee good management, funds designated for one funding instrument may not be awarded to a recipient through another instrument without the prior approval of the Minister of Science and Technology.

The government and its privileged partners in building science, technology and innovation capacity in the country, namely the World Bank, Department for Research Co-operation (SIDA)/Swedish International Development Agency (SAREC), Danish International Development Agency (DANIDA), International Development Research Centre (IDRC) and Ford Foundation are currently contributing to the National

Research Fund, and the negotiation with other partners such as Department for International Development (DFID) and L'Institut de recherché pour le développement *i.e.* Institute for Research and Development (IRD) are in progress. Therefore, the country would like to exhort other partners to join the team in promoting and strengthening scientific research and technology development capacity in Mozambique through support to the NRF.

Final remarks

The Ministry of Science and Technology (MCT) as a coordinating institution is geared towards serving the different actors and beneficiaries in science and technology. The ministry strives to be action oriented and always preoccupied with the final impact of science and technology on the economic and social well being of Mozambicans. The MCT strives also to promote research and technology development, which are geared towards resolving specific problems, wealth creation and generation of knowledge. For efficiency and cost effectiveness, the MCT will promote horizontal duplication of successful cases in science and technology. That is, successful science and technological solutions (acquired or developed) will be adapted in locations with similar conditions. This approach will reduce cost and speed up development. Figure 1.4 gives the organigram of the Ministry of Science and Technology.

The MCT counts on the collaboration of all actors and beneficiaries as well as development and cooperation partners in fulfilling its mission. The Ministry believes

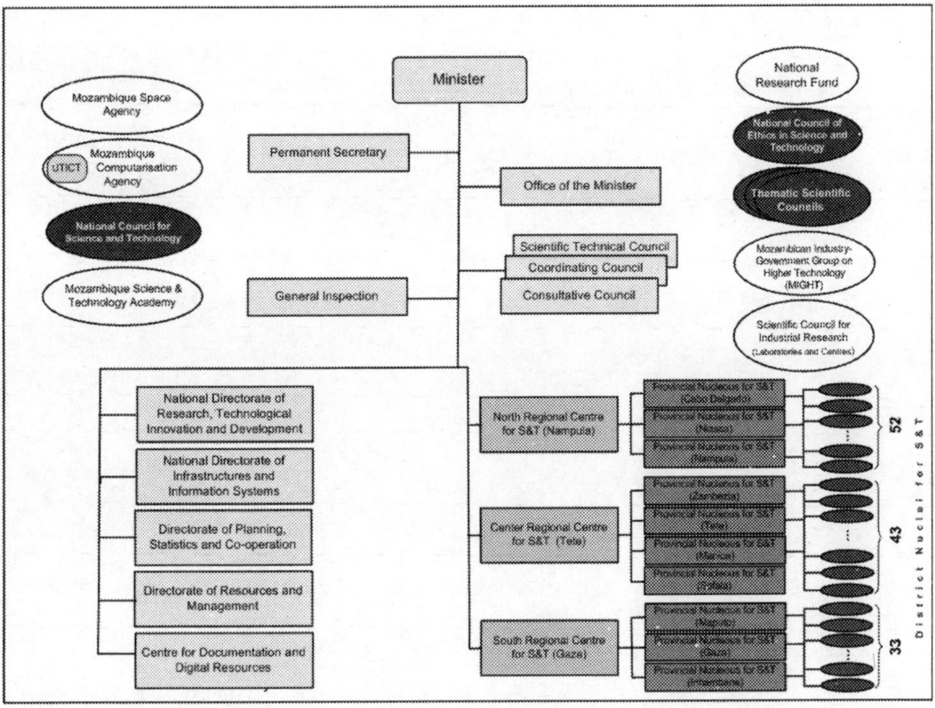

Figure 1.4: Organigram of the Ministry of Science and Technology

that, it is only through support and articulated actions with same objectives of social and economic development that science and technology will have an accelerated impact on poverty alleviation and improvement of well being of all Mozambicans. These actions are to be guided by the science and technology policy, and the strategic plan recently approved by the Government of Mozambique and for an effective implementation of the approved S, T&I strategy. It is intended to make an appeal to the development partners a strong support through adequate resource allocation.

Finally, it must be emphasised that building science, technology and innovation capacity is a serious challenge that deserves serious attention from all, if the current situation in Mozambique is to be reverted, as far as the economic growth and poverty reduction is concerned.

Chapter 2

Evolution and current state of the policy and the statistical indicators of science and technology in the Republic of Cuba

Vito N. Quevedo Rodríguez

Abstract

The present work exhibits results from a research study on the evolution of the scientific policy over the last 45 years that have been brought about in the Republic of Cuba and the main concerns that have been defined every time, in order to develop the country. The science and technology status in the entire nation has been assessed through a Statistical Indicator System. It is demonstrated how excellent results have been achieved in science, technology and innovation fields, in spite of being under globalization and threatening situation for more than 45 years. Some of these results are capable to be rivalled with those being obtained in developed countries. The historical succession of the statistical indicators on science and technology, have been illustrated in this paper. It shows the attentiveness, priority and fresh impetus that the Cuban state has rendered in the development of science in the country. The outcomes have been shared, to a large extent, prominently and selflessly with other people.

The paper also presents the mainstays of the Cuban scientific policy, its implementation, ways of organization, planning, funding and encouragement as well as the assessment of its performance from the defined indicators in the statistical system of Science and Technology. Experiences on evaluation of the scientific activities through the effect of their outcomes in the economical, social and environmental activities are also discussed. Finally, several offers on the forthcoming development issues have been made.

Keywords: Capabilities, Impact, Indicators, Innovation, Policy, Science and technology, Strategy, System.

Introduction

Cuba is a small island located in the Caribbean Sea and belongs to the Latin-American nations community, with a little more than eleven million inhabitants. Cuba's shows an interesting and attractive experience of what can be done in the field of science and technology for the development of the country, which has limited availability of financial resources. This development has been possible in Cuba since the State prioritizes, and encourages the science and technology activities by understanding the mind set of people, and their interests. The state promotes science and technology for the benefit of the development of the country as well as for the development of their human capital.

In 1959, the Cuban Revolution achieved its definitive victory in the country. It inhabited one million illiterate people, more than 30 per cent of the population in those days. There were scarcely any research centres. There were four experimental stations in which less than 100 persons used to work. Practically, the funds, from the Government, did not come for the development of research programs. There were only 3 Universities. This unfavourable situation for the development of science and technology activity did not restrain the rise of scientific personalities, bestowed with the universal knowledge, in the national milieu. Among them, it is worthy to recall a few like Dr. Carlos J. Finlay, Dr. Pedro Kourí and Dr. Tomás Romay.

The present work, thoroughly, looks into the evolution of the scientific policy of the nation, supported with enough data and figures. It, particularly, focuses on its role in the development of the country, preparation of human capital and creation of own knowledge, without ignoring the ones furnished by the rest of the world. The work shows the way how scientific policy has, at every time, been keeping with country's priorities, the international situation and demands, and existing trends in scientific fields. How the scientific policy had always been under the perspectives of the national interest, and in extreme accordance with the economic and social project that country had adopted.

Approximately 10 years back, Cuba established its current and in force, Cuban System for Science and Technological Innovation. This was established as a result and consequence of the developments on science and technology issues and was aimed at upgrading the effectiveness, efficacy and excellence of the country's economic activities. The system, among its core targets, aims to achieve a larger linkage between science and economy, and to keep improving in the social, economic and environmental impact of its achievements. Nowadays, these achievements are momentous, and one of the aims of this system is to endeavour to reach, at a shortest span, to a major contribution of Science, Technology and Innovation to the development of the country and its priorities. In this way, it is at present beginning to give definite shape to the national scientific and technological policy, which has settled on a pyramid down in the orderliness of research whose origin, formation, and development are demonstrated in this paper.

Cuba has made significant national contributions in many scientific fields, in terms of publications, patents, their implementation and impact on the economy and the society. The meaning of the IMPACT of the scientific, technological and innovation achievements is to measure these activities and results because of their tangible contributions to the economy, knowledge, and level of quality of life of Cubans people. There are two core aspects in the current national scientific policy in respect of measuring processes, *viz.* (i) measuring by impact, and (ii) measuring along with the integration of cooperation, linkage, and interface in the science and technology activity. To accomplish the full development of the devised scientific policy is an issue being currently focused by many entities and centres that have been closely related to this activity throughout the country. The measuring of the technological innovation is of key interest. It is something where we have made progress but not to the satisfactory level.

The Cuban scientific policy

The Cuban scientific policy was first laid down in 1960, with the comments of the Commander in Chief Fidel Castro, to quote, "the future of Cuba must be a future of men of sciences, of men of thought ...". His vision constituted a long-term strategy that guided the creation of highly qualified and committed human resources for the development of the nation. This was achieved as a result of the literacy campaign, carried out in 1961, until present days, giving place to the creation of schools, universities, investigation centres and other important institutions where a sustained formation process was materialized. This would continue to advance. This activity has been strengthened with the recent creation of the Municipal University Headquarters that facilitated the access, and in consequence, increased registration in teaching levels. At the moment, hundreds of these headquarters are functioning in the municipalities all around the country, and there, thousands of Cuban students are studying, irrespective of their age. It demonstrates the principle defended by Cuba, which considers its main treasure as the human being, and its main resource as its the human capital.

There are at present 65 universities in Cuba, and about 200 investigation entities– of which 115 are big and well-established scientific centres of excellence, constituting a wide and well-structured system of national education. Above-mentioned municipal universities headquarters play a significant role in the execution of the traced strategy, in the important sphere of preparation and formation of high quality human resources.

In the country, the student degree letter embraces 100 per cent of the population and it doesn't present limitations of health to consent to her. There are more than 700000 students graduated from universities and they receive an intense postgraduate programme, an activity that allows them to develop themselves in a regular manner. Presently, there are 1.8 engineers and investigators per thousand inhabitants. There are 8500 Doctors and about 5500 categorized investigators. The country is planning to have about 80 thousand workers in science and technology spheres. This will allow us to be sure that they will contribute successful results in the whole system, and in the necessary and enough quantities to face the challenges related to the national development and international solidarity. Starting from this strategic lesson, the Cuban scientific policy has been developed and organized, and at present, it is sustained in five pillars, namely, own scientific potential, support to the development of the country, generation of own technologies, assimilation of knowledge and international technologies, and integration.

The evolution and implementation of this policy has lead to elaborate science and technology strategies, with support from and contributing to development of territories, production and services sectors. It included several enterprises, entities and communities, and promotion of the local development in harmony with the development of their economic, social and environmental sectors, with the aim of reaching sustainable development. It has also lead to strengthening of science and technology activities in respective territories, achieved by a wide and popular movement around science, technology and innovation, and creation of the system of science and technological innovation. It substituted the previous science, and technique system in the country. The execution measures have been taken such as, menstruation for the impact of the results and not for the process by which they are obtained, and pyramidal organization of the investigations (Figure 2.1), starting from the PRIORITIES defined in the strategies, those which become PROGRAMS according to their relevance and contain PROJECTS whose execution leads to the IMPACT in the economic, social and environmental activities that the nation needed. Project is the basic cell of the organization. The financing and the execution of the investigations and innovations contribute to the development of the country. Projects are organized in national, branch, territorial and institutional levels.

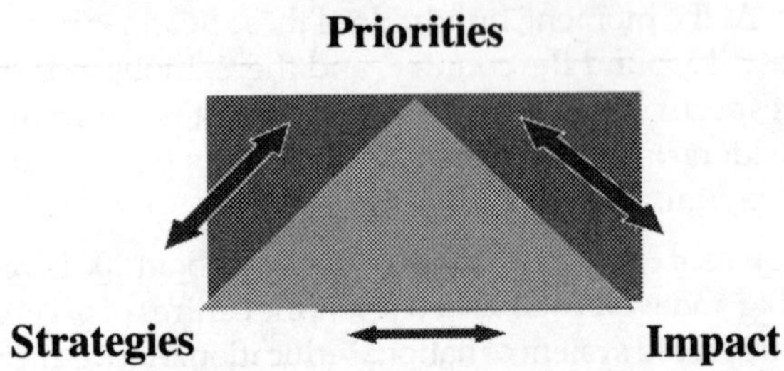

Figure 2.1: Pyramidal organization of the investigations

The policy also contributed in the establishment of the science and technological innovation plans in all the pertinent organizational and hierarchical levels, financing for projects, and elaboration and setting off the regulatory frame.

At the beginning of 2005, the Cuban System of Science and Technological Innovation was executing, without considering the institutional projects, about 3400 investigation and innovation projects–1082 in the territorial level, 2058 in the productive or service branches, and 275 at the national level (Figure 2.2). During the

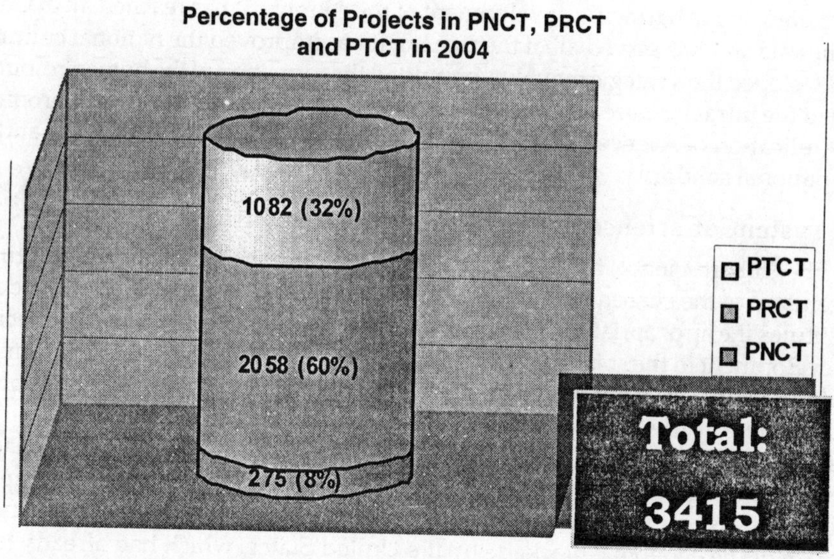

Figure 2.2: Structures of the system of programs and projects Inside the system of science and technological innovation

10 years the current system had existed, 585 national projects with a financing of 179 million pesos, were executed and approved by properly organized experts groups. Of these, 30 per cent of the projects have been consequently applied, which is not satisfactory. This is so because, even when the advances of Cuba are, broadly, well known in topics like biotechnology, health, computer science, meteorology, struggle against disasters, sports, education and many other branches, we are convinced that there are organizational reserves, which with a non excessive financing could increase the introduction of such results and improve the impact. Therefore, presently, the measures, which could contribute to get more results in this field, were being examined. In these years, the priorities of science and the technology in Cuba have been placed in areas like production of foods, health, education, energy, environment, biotechnology, technologies of the information, and society.

Many results have been reached by Cuba with indicators of health, which could be compared with the ones of developed countries. Cuba occupies a prominent place in the human development index (HDI). 95 per cent of its population has access to energy services and drinkable water. It has developed vaccines, which have only been achieved in rich countries. It presents high security levels against disasters and other risks. It has elevated its international authority in the sphere of the environment. It has put to the disposition of many countries, especially the poorest ones, a quick and efficient method of literacy, a technology proven to cure sight illnesses which is being implemented under the denominated "Miracle Operation" and many others where science, technology and innovation have had an important role.

The elements that have been analysed up to now, demonstrate that the scientific policy traced in Cuba has contributed in a decisive way, although we are still unsatisfied, to the harmonic development of the country. It has reached all of Cuban inhabitants and has served all of them. Also, it has improved the national culture. It has developed the strategic thoughts, permitted the formation of the human resources, created the infrastructure of institutions and necessary organizations, and promoted the application of science in the interest of human development, the peace and the international solidarity.

The system of science and technological innovation

The Cuban science, technological and innovation system (STIS) is a legitimate descendant of the consequent application of national scientific policy, and today, it constitutes the appropriate model to continue its harmonic development, especially, its development in the sphere of science, technology and innovation. The system was created, approximately 10 years ago, as a consequence of the development accumulated in the sphere, the status of art, and the demands of national and international environment. It was just the period when Cuba was forced to establish an economic regime of severe material limitations, widely-known as the "special period". It was recovering of the strike that the collapse of the socialist field and the increase of the economic blockade by the United States, which had already lasted more than 45 years. It was also the time when globalization started to increase the rhythm to win land quickly. The change that occurred in the society of knowledge was evident. Under these conditions, the change of the former science and technology system took place to the current system of science and technological innovation. It was necessary to increase the generation of knowledge and its application and effectiveness at the time to do it, because it was not a simple change of name but a change of concept. It aimed to place the new system in the main place in the management of knowledge and its application in the service of the already created systems of health, education, sports and to increase the role of the enterprises. At this moment, this projection has found an important field in the energy sphere in which Cuba has already reached results that are likely to grow in a very short period of time as part of the program, namely, "The Energetic Revolution".

The STIS is an organizational form through which the scientific and technological policy has to be implemented. It covers a wide spectrum ranging from the generation and knowledge accumulation, its diffusion, application and generalization until the materialization of new production of goods and services, and their marketing. The components of the STIS (Figure 2.3) are, entities that participate directly in innovation and development in different stages of the innovation process, organs that participate in its direction and organization, integration elements and methodological legal base.

One essential element of the stage has been the integration reached among the different organizational forms (*e.g.* scientific and productive poles, fronts, science and technical forum, national association of innovators, youth technical brigades, science academy of Cuba, scientific societies) so that the science becomes an important element, *i.e.* a catalyst element for the development of the country, increasing the efficiency of the economy and the quality of life of the population. These forms are described below.

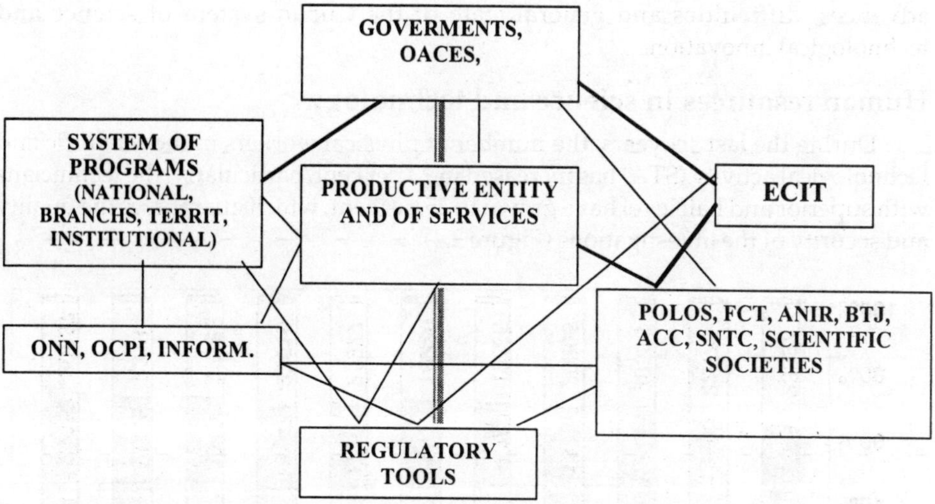

Figure 2.3: Components of the system of science and technological innovation

1. Scientific–productive poles: space or organizational form where, science, goods and services, and production combine in order to facilitate obtaining of the results and enhancement of the human, financial and infrastructure capacities to obtain concrete and priorities objectives. They stimulate the integration and training of cooperation nets.
2. Science and technical forum: popular and massive movement that contributes to the integration of science and technology entities with enterprises, hospitals, hotels, schools, and others, by principles of integration, cooperation and over crowding.
3. National innovators association and rationalizations and youth techniques brigades: guarantees the participation of workers and youths in the innovative process.

The integration elements have guaranteed the biggest participation of our people in the economic and social development and their own welfare, which formed a platform for the integration of scientific development and the participative democracy that materialized in different systems in Cuba, *viz.* science and technological innovation, health and education.

Results and discussion

The discussion about the scientific policy results, apart from some of the elements advanced in the text, will be carried out, starting from the historical series of the statistical indicators of science and technology that cover 10 years of work. It constitutes one of the tools to evaluate the Cuban system of science and technological innovation. We will also consider other national and international sources as the national surveys of innovation that reflect with eloquence the results and they demonstrate the advances, difficulties and general state of the Cuban system of science and technological innovation.

Human resources in science and technology

During the last six years, the number of physical workers in the scientific and technological activity (STA) has increased in 23 per cent, particularly, the technicians with superior and half level have grown in 25 per cent, which guarantees the quality and security of the investigations (Figure 2.4).

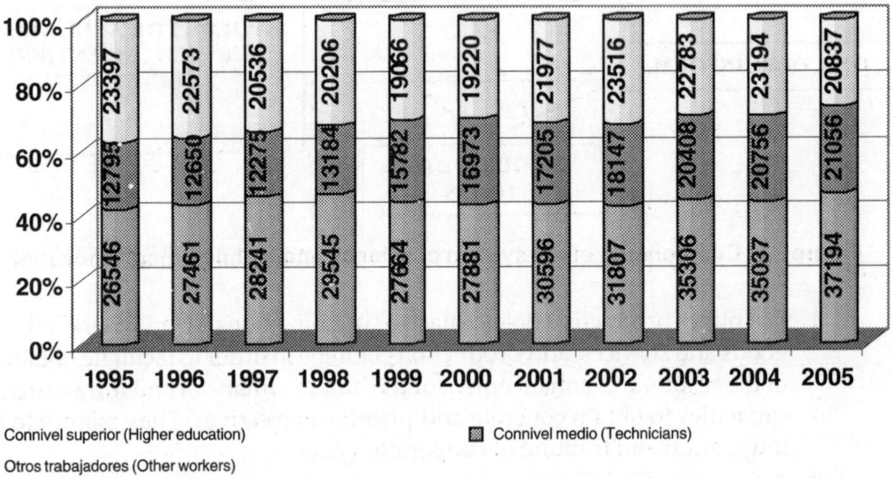

Figure 2.4: Physical workers in STA according to educational level

The categorized investigators indicate a reduction during 2000 and 2002 while showing an increase since 2003 to present day. They have surpassed the biggest number they ever reached *i.e.* 5526. It means Cuba has 500 investigators with scientific category per million inhabitants (Figure 2.5).

The increment of Cubans women participation in the STA is shown in Figure 2.6. One observes in this indicator that the number of women who intervene in this activity has overcome the masculine participation, from the year 2000 so far, in a stable and upward form (Figure 2.6).

Figure 2.7 presents how the indicator related with the graduate students from universities has evolved, between 1999 and 2005, according to different branches of science. The values of this indicator point out a substantial increase in each of the branch, in correspondence with the necessities and priorities determined by the country. The grant of scientific grades, expression of the elevation of the quality of graduates, increased 40 per cent between 2000 and 2005 (Figure 2.8). This increase, observed in all the branches of science, is a demonstration of the harmonic growth achieved in this activity. In relation with the social sciences, the increase has been significant.

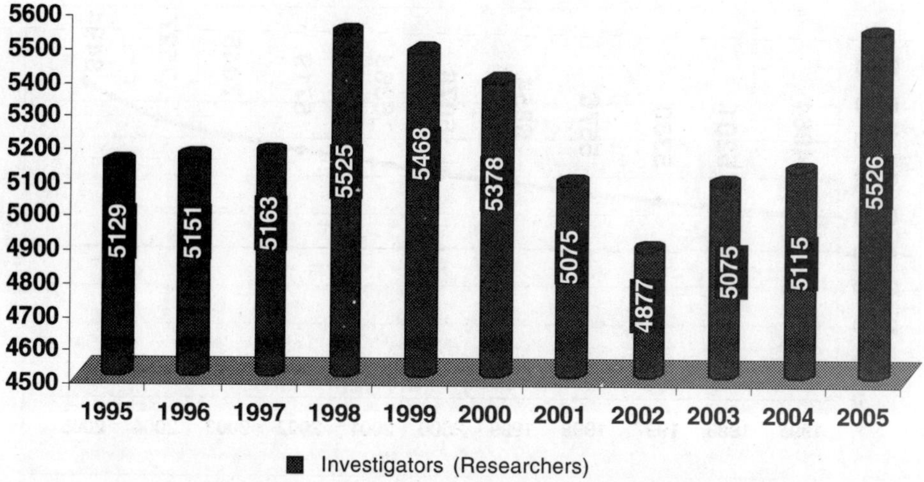

Figure 2.5: Investigators with scientific category

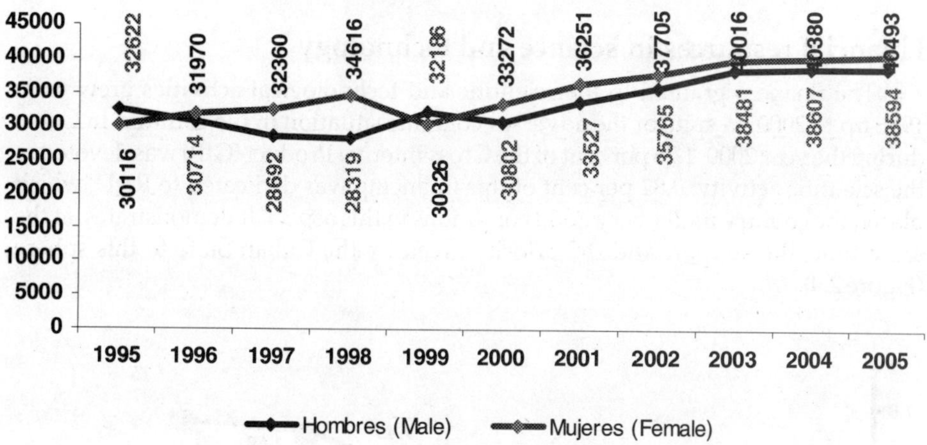

Figure 2.6: Physical workers in STA according to sex

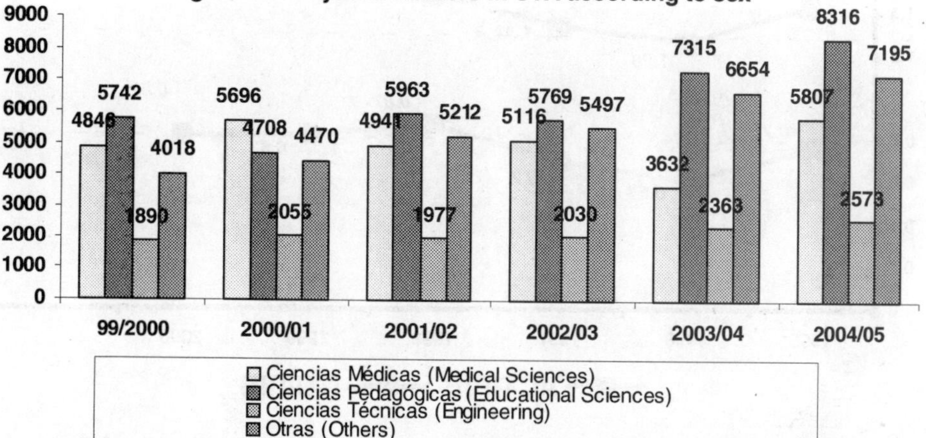

Figure 2.7: Graduate university students for branch of the science

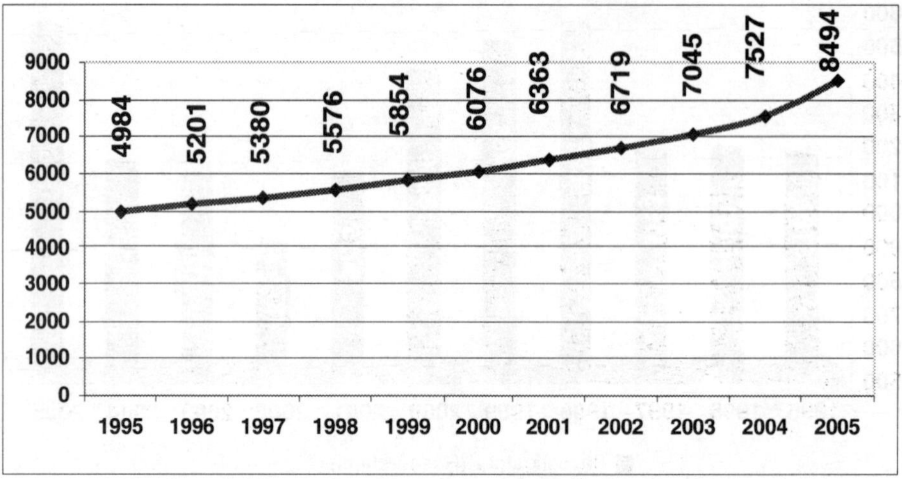

Figure 2.8: Granted scientific grades

Financial resources in science and technology

The finances granted to the scientific and technological activities grew from 1996 up to 2000, in spite of the adverse economic situation in the country. In Cuba, during the year 2000, 1.75 per cent of the Gross Internal Product (GIP) was devoted to the scientific activity; 0.82 per cent of this financing was dedicated to R&D, which placed the country under very good conditions in this aspect. It demonstrates, at the same time, the support and the priority given by the Cuban State to this sphere (Figure 2.9).

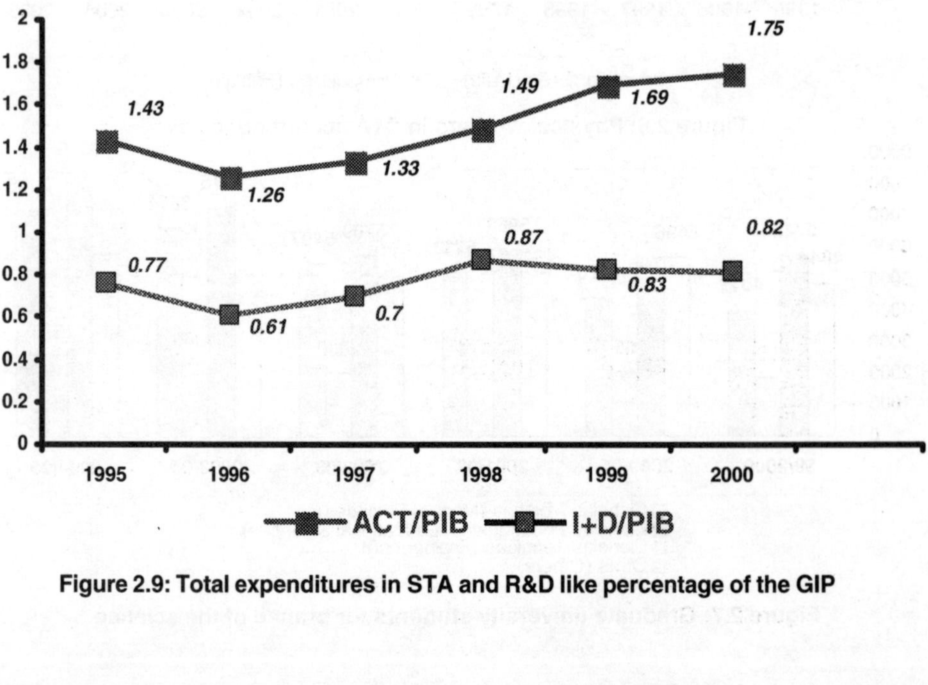

Figure 2.9: Total expenditures in STA and R&D like percentage of the GIP

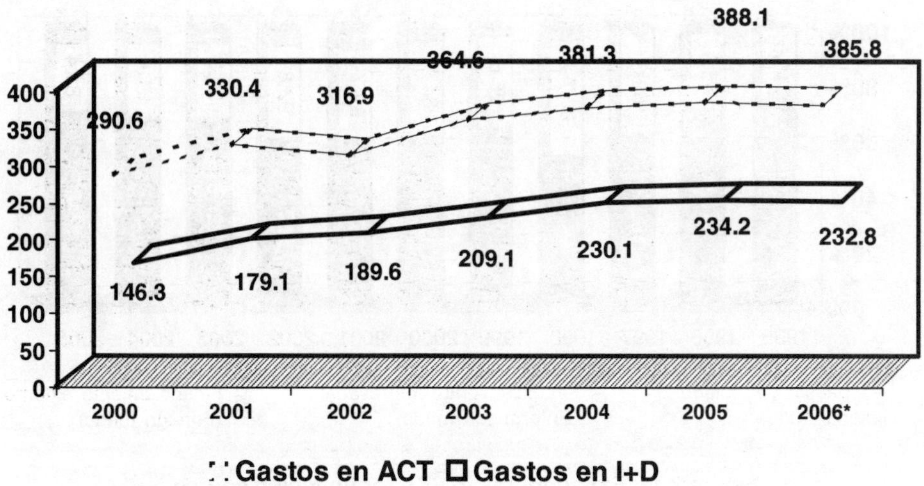

Figure 2.10: Total expenditures in STA and R&D (to average prices)

The mensuration used in the historical series of the GIP starting from 2000, considers a revaluation of the social services, when applying rates that incorporate elements of profitability. They reflect the added value of these activities better, which takes in to consideration the characteristics of the economy of Cuba and its social system. The services of health, education, cultural, sports and others have a very superior final socio-economic effect to the registered one for the expenses that it is the method of traditional calculation for these activities.

The country has made a great effort in regard to the investments for the STA, which is not appreciated easily in the methodology of calculation of the GIP. In the Figure 2.8, it is not shown as the expenses in STA measured with relationship to the per cent of the GIP, as previously explained. As indicated in absolute terms, the STA would receive 2 per cent of the GDP, practically, an aspect that cannot get lost sight of when evaluating the SCIT in Cuba.

In correspondence with that exposed work previously, presently, the science technology and innovation system located in the enterprises is the centre of its attention, and in consequence, there contribution to the STA has increased more than twice (Figure 2.11). It has rebounded in the obtaining of new products, services and processes with added value of the science in the enterprises, favouring the level of exports and the substitution of imports, which has placed the STA as the most dynamic sector in the country in the sphere of the external trade. This is termed as the impact of the scientific results and of innovation.

Industrial property

With the development of the STIS, the activity of industrial property has reached a great peak in our country. It has been possible to enhance the culture in this sphere, systems of industrial property have settled down in the enterprises. It has resulted in

Figure 2.11: Expenditures in STA for financing source

increase in the transfer of technology and the grants of patents in other countries to Cuba, mainly in biotechnical branches. The Figures 2.12 and 2.13 show the applications to get registration, presented by Cuba, as well as the applications of inventions granted up to 2005. In a similar way, it points out the evolution in matters of registration of industrial models and marks. The activity will grow up and will improve progressively.

Besides the analysis of these statistical indicators, it is useful to evaluate the activity of science, technology and innovation system in Cuba, to make a valuation of other sources of information that contribute in the evaluation of the System. One of them is constituted by the RAND report about the capacities of science and technology published in March 2001. The index of capacity in science and technology elaborated

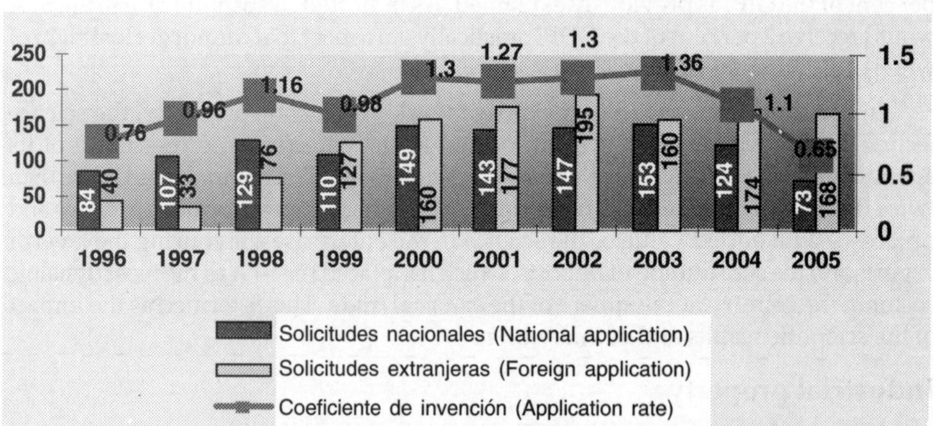

Figure 2.12: Applications of registrations presented in Cuba

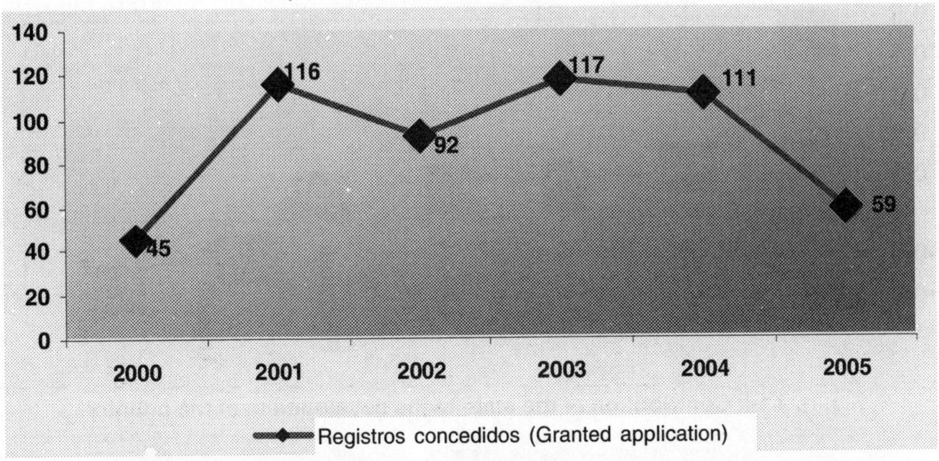

Figure 2.13: Application of registration of granted application

by RAND report is the base that this work team used for the classification of the countries in connection with its scientific development. This compound index was built starting from the available data in the countries for a group of indicators that were selected to the effect. Starting from the index of capacity in science and technology, the RAND report considers four categories of countries according to this position in connection with the scientific development. These are, countries scientifically advanced, countries scientifically precocious, countries scientifically in development and countries scientifically delayed.

Cuba, with its general index of 0.11 got the second category, above all the Countries of Latin American Region and the Caribbean countries.

A summary of the results of the RAND report for seven countries of different continents is shown in the Table 2.1 and in Figure 2.14, where some of them show superior indicators to the Cubans ones, and others below them. That is the reason why we can emphasize the standing assigned by this RAND report to our country.

Finally, it is significant to appreciate as science, technology and innovation have contributed to the development of the Country.

The results which generated better impact during the last 10 years include, vaccines (anti-meningocóccica BC, anti–hepatitis B, against the Haemophilus, Influenzae, vaccines polyvalent), antibiotics, medical teams, automation meteorological radars, systems of early alert to the drought, effectiveness of the presage of the time, improvement of varieties to urban agriculture, treatment and improvement of soils, vaccinates and veterinary medications (B. microplus), studies of the mountain ecosystems, dynamics of the beaches, and the production of crude oil and natural gas, guaranteeing 100 per cent of their use in the industry. There were outstanding results in the system of education (educational software, change in secondary education programs, introduction of tools of teaching of more effectiveness) as well as significant

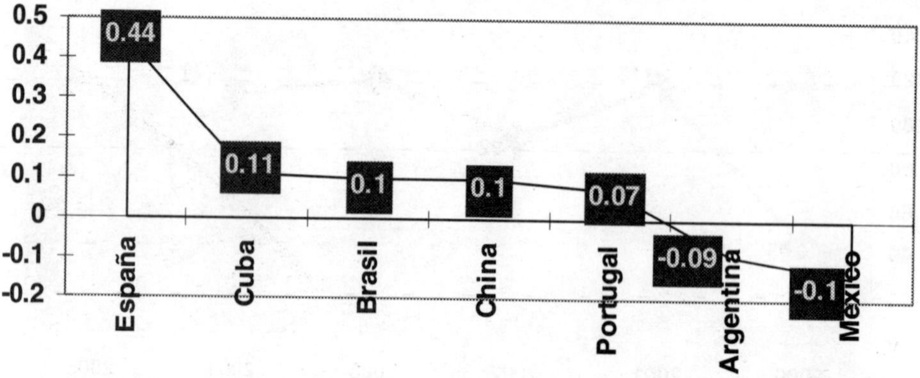

Figure 2.14: Contribution of the state to the development of the country

contribution of the products and services with added value of the science and the innovation to the Cuban external trade.

Table 2.1: Indicators expressed by the RAND report of the World Bank (March 2001)

Country	*GIP per capita*	*Scientific and technical personnel per million inhabitants*	*Published scientific articles*	*Expenditures in STA (% PIB)*	*Institutions, universities per million people*	*Patents in USA or EU*	*General index*
Cuba	2194	1612 Total: 17, 732 1.6 por c/1000 hab.	147	0.8	2.16	12	0.11
Brazil	4630	168	3511	0.8	0.65	440	0.1
México	3840	214	1758	0.3	0.56	262	-0.14
Argentina	8030	660	1944	0.4	2.22	176	-0.09
Portugal	10670	1182	968	0.6	4.2	77	0.07
Spain	14100	1305	10557	0.9	3.07	1904	0.44
China	750	454	7763	0.7	0.3	466	0.1
Canada	19170	2719	20989	1.7	91.75	12559	2.08

Conclusions

As we proposed at the beginning of the work, the hypothesis to be demonstrated was the favourable evolution of science and technology in Cuba, the practical value and the grade of execution of the policies traced for the sphere and the contribution level from the activity to the national development.

The analysis we have done, allows us to demonstrate that in Cuba the policies of science and technology are not merely limited to documents. It is really possible to appreciate the recognition and support given by the State to the activities when prioritizing, financing and promoting them. The results have contributed in a significant way to the economic, social and environmental activities, which could be demonstrated in a lot of examples in the majority of the branches, territories, communities, enterprises and other social and productive entities of the nation. In Cuba, it has been possible to establish a wide and comprehensive education system that assures the continuous preparation of a mass of high quality human resources, even in municipality level, in an organized, legalized and controlled manner. This is a sure way for the local development. In our country, science, technology and innovation activities are popular. It covers gender and number expectations. The country has the basic financing and it has been able to create a scientific culture attracting more people to science, technology and innovation. The statistical system of indicators has been developed and it advances more every moment. It is being used to improve the mensuration of the contributions of science to the well-being of people.

It will be necessary to achieve a bigger level in the application of the results, to give a better form to the system of indicators, to improve the measuring of the impacts, and to complete the statistical system of indicators of science and technology. As it has been demonstrated in the work, the science in Cuba is a genuine work of the revolution. The country would necessarily work better in order to achieve the target expressed by the Commander- in- Chief, Fidel Castro, *viz.* "in the future the country will have to live off its intellectual productions." We are absolutely sure of this target because we have much better and vast experiences, and we have prepared a large number of human resources for the purpose.

In conclusion, Cuba has been able to show the advantages in the field of science and technology, not only by its practical results and direct contribution to the economy, society and environment, but also for its organization, promotion and control which can be seen in the establishment of policies and strategies and an ordinate system of statistical indicators of science and technology.

The development, during these years, of such tools for taking decisions and proper guidance of the work has allowed us to install some capacities in our country, which must be considered and used. Therefore, we would like to express our disposition to offer such tools to all of you and to anyone who requests for them.

Propositions

1. To create and to increase the net of indicators of science and technology of the Non-aligned (NOAL) and to direct them to guarantee the exchange of information, the conformation of databases, the elaboration of statistical indicators, according to the characteristics of the member countries as well as their technological and scientific demands.
2. To create a working group as a result of the conference to carry out and to present a conclusive study, before the beginning of the second semester of the year 2007, about the convenience of creating the office of transfer of

technology of the NOAL which if accepted would work with their own norms and in benefit of all the member countries.

3. To develop during the year 2007, an international conference on "Energy, Innovation and Sustainability" among the member countries of the NOAL, the enterprises which could be considered for their contribution to the theme directed to exchange technologies of energy saving, elevation of the energy efficiency, development of renewable energy sources and the alternative fuels.
4. To propose, or in its effect, for inclusion of the current centre (NAM S&T Centre), a centre for the innovation which will have to be in charge of developing the idea to employ the technologically cooperated parks among the member countries, the incubators of enterprises and other structures to energize the innovation.

Acknowledgement

To my work partners in the present investigation, for the demonstrated effort and the borrowed interest to the realization of this work.

References

Brisolla, Sandra, 2001. Producción y difusión de indicadores de ciencia y tecnología: su utilización en las esferas nacional y regional. V Taller de indicadores de Ciencia y tecnología, RICYT.

Documentos Rectores de la Ciencia y la Innovación Tecnológica, La Habana, 2001.

Documento "Prioridades de la ciencia y la Tecnología cubanas". CITMA, 2004. Dirección de Tecnología e Innovación, (2000) Documento "Tabla estadística para la medición de impacto".

Indicadores de Ciencia y Tecnología, CITMA, 1996–2005.

Informe RAND. Banco Mundial. Marzo, 2001.

Licha, Isabel, 1994. Indicadores endógenos de desarrollo científico y tecnológico, y de gestión de la investigación, en Martínez, Eduardo (ed.), Ciencia, tecnología y desarrollo: interrelaciones teóricas y metodológicas, Nueva Sociedad, Caracas.

OCDE, 1997. Revisión de las clasificaciones de los sectores y de los productos de alta tecnología.

Quevedo, Vito, Chía, Jesús, y Rodríguez Batista, Armando, 2002. Midiendo el impacto. Ciencia, Innovación y Desarrollo. 7(1).

Quevedo, Vito, *et al.*, 2003. *Bases para el Perfeccionamiento y Desarrollo de la Innovación. Ediciones Academia, CITMA, Cuba.*

Resolución Económica del V Congreso del PCC, 1995.

Trueba, Gerardo, 2003. Principales características del financiamiento de la ciencia y la tecnología en la República de Cuba. Documento de trabajo.

UNESCO-ICSU, Declaración de Budapest. Declaración sobre la ciencia y el uso del saber científico. (http://www.campus-oei.org/budapestdec.htm)

UNESCO, Instituto de Estadísticas, 2002. Reseña internacional de estadísticas e indicadores de ciencia y tecnología. Estadísticas e indicadores de ciencia y tecnología en países en desarrollo: perspectivas y desafíos.

Chapter 3

The South African STI policy as an indicator driven one: Approach, nature, size, performance and challenges

Hendrik Christoffel Marais and Simon Mpele

Abstract

The core argument of the paper is that indicators can add considerable value to policy monitoring, evaluation and steering, provided that the approach is linked to an underlying conceptual and methodological framework. The paper is mainly based on a background report for an OECD country review of the South African National System of Innovation and on the ongoing policy related research undertaken by the National Advisory Council on Innovation and other research centres in the country.

The paper starts with two introductory notes on, firstly, the historical context of South Africa's science, technology and innovation (STI) system, and secondly, the basic methodological points of departure on measurement, in general, and indicators, in particular. With regard to the latter it is proposed that methodological criteria such as reliability and validity have to be explicated in research and policy development. The paper lists the major policies, strategies, institutions and instruments, and further shows that STI system is, in principle, an indicator driven one, as can clearly be seen from the national R&D strategy, the importance attached to national R&D surveys that have been declared official statistics, and governance provisions for the regular collecting and updating of statistical information on different components of the NSI, *e.g.* key performance indicator reports submitted annually by the public research entities to the Department of S&T.

Taking the ratio of GERD to GDP of 0.87 per cent (for 2004/2005) as point of reference, a concise comparative, albeit selective, overview is given of the most recent standard performance indicators (inputs, outputs as well as outcomes) for the main components of the system. On the basis of the selected indicators presented, it is concluded, on the one hand, that within the developing world context, South Africa has a relatively productive system of innovation, but that it has slipped backwards over the past thirty odd years relative to its peer countries. Against this background, the paper next summarises some of the STI system's strengths and weaknesses.

The paper concludes with the future perspectives and the role of quantitative measures in future.

Keywords: Science, Technology and innovation indicators (STI), R&D strategy, S&T policy, S&T indicators, South Africa.

Background

The National Advisory Council on Innovation (NACI) is a statutory body that advises the Minister of Science and Technology (S&T) on all matters pertaining to innovation. In this capacity it has commissioned several studies in recent years on the key aspects of the South African national system of innovation (NSI), including various aspects of human capital, physical infrastructure, return on public investment in research and development (R&D), innovation and competitiveness, regional profiling of innovation systems (for an overview, see NACI, 2005) and a background report for an OECD country review of the South African NSI requested by the Department of Science and Technology (DST). The paper is partially based on the said report but further discusses the ongoing policy related research undertaken by NACI and other research centres in the country.

Although the paper explicitly argues for science, technology and innovation (STI) indicators to serve as at least one of the sets of guides to steering policy making in a national system of innovation–also of developing countries–it by no means elevates an indicator orientation to the status of representing the methodological apex in evidence-based policy development. As will be shown briefly in the paper, the proper use of STI indicators requires an understanding of the theory underlying the development of indicators. Given this more critical orientation to an indicator approach, the thesis of the paper can be summarised as follows:

1. Indicators can add considerable value to policy monitoring, evaluation and steering of national systems such as health, economics, education, social development, and in the present case, STI, provided that the approach is linked to an underlying conceptual framework and not seen as being *sui generis*.

The themes presented in the paper include conceptual notes on STI measurement, followed by a selected overview of current South African STI policies, strategies and instruments, and relating these to the use of indicators in the government's monitoring, evaluating and steering of the system. The fourth theme entails a comparative, albeit selective, overview of the most recent standard performance indicators (inputs, outputs as well as outcomes) for the main components of the system, leading to a brief summary of strengths and weaknesses of the South African STI system. The paper concludes with methodological and policy reflections on the South African experience and suggests a possible agenda for the further development of quantitative and qualitative policy research approach.

Conceptual considerations

This conference focuses on indicators and it would, therefore, be appropriate to introduce the paper with a minimum of conceptual considerations on the indicator philosophy, approach and methodology. A quick review of Research Methodology would produce a definition of indicators, *viz.* an indicator is an operationalisation of an attribute or property (*i.e.* approximately an operational definition) of an abstractly defined concept or construct (*i.e.* as constitutively defined). To put it differently, an indicator is a measurement of part of an abstract concept–that is, the representation in the empirical meaning space–that is to be found in the theoretical meaning space.

Three corollaries of this definition should be mentioned. Firstly, in terms of the above very crude definition, an indicator is just what the name suggests, *viz.* something that points towards something else–hence, the common description of an indicator as a proxy for something else. In practice, this means that the person using indicators should always refer back to the underlying concept or construct. Secondly, it is generally accepted in most scholarly disciplines that most concepts are characterised by more than one property/attribute. This assumption implies that no single given indicator should simply be looked upon as the one and only measure of a concept, unless theoretical or empirical support is offered. Thirdly, measurement can, for the purpose of this presentation, be defined as the attribution of numbers to a phenomenon in such as way that the properties of the number construct correspond to the properties of the phenomenon being measured. An obvious implication is that indicators are normally seen as quantitative indices. It follows that various methodological criteria have to be explicated and considered when using quantitative/indicator-type measures and data in policy research. Some of the more obvious criteria include reliability and internal validity of the instrument that produced the indicator (*e.g.* Frascati-type R&D questionnaire), external validity (*i.e.* the representativity of the sample), level of measurement (*e.g.* rank-order, interval or ration scale), and appropriateness of statistical techniques applied to the data. 'Quantitative' or 'indicator' does not necessarily imply 'good'.

The South African STI system: Size, policies, strategies and instruments

This section offers a concise overview of the STI policy scene in South Africa (more information on a selection of policies, strategies and instruments can be found at the NACI web address: http://www.naci.org.za/Innovation_gateway/Policies.htm).

By way of introduction, an historical note should be made. The transformation of South Africa in 1994 from the apartheid state, it was since colonial days to an inclusive constitutional democracy, led to a wide spectrum of policy changes in virtually every sector of the country, all aimed at reconstructing and developing the country in such a way as to become a just and equitable place in which to live and work (cf. Marais, 2000). STI was no exception. The policies briefly referred to below, are all part and parcel of this transformation process. The following list offers a bird's eye view of the policy landscape (see http://www.naci.org.za/OECD/OECD%20Report%2021Aug.pdf).

It was thought useful to introduce this section by offering a simple overview of the size of the South African STI system, answering the expected question as to what kind of system is being served by the policies and strategies.

Size

The size of the South African STI system can be inferred from the following selected indicators.

Table 3.1: Size of the NSI[0]

Sector	*Number of institutions*	*R&D expenditure[1]*	*Human capital*		
			Total R&D staff[1]	*Under-graduates[2]*	*Post-graduates[2]*
Business	651[3]	45 per cent	14 337	NA	NA
Higher education	23	21 per cent	23 793[4]	597 609	123 704
Science councils	9	17 per cent	6 170	NA	NA

Notes:

0: Table copied from NACI 2006. *Annual Report 2005/2006*. Pretoria: NACI.

1: Data from the 2004/05 R&D Survey (CeSTII)

2: Data supplied by the CHE

3: Businesses that provided data for the 2004/05 R&D Survey.

4: Excluding doctoral students and postdoctoral fellows

Policies

The first major initiative was to upgrade the S&T function to the ministry and the departmental level with the formation of the Department of Arts, Culture, Science and Technology in 1994. This department later split to become the Department of Science and Technology (DST) and the Department of Arts and Culture. The next major policy initiative was the publication in 1996 of a new consolidated policy, the *White Paper on Science and Technology* (DST, 1996), which was followed in 2002 by the *National R&D Strategy* (DST, 2002a). The equally important *White Paper on Education* was published in 1997 (DoE, 1997). The period since 1996 has seen the approval of at least nine major S&T sectoral/thematic policies and strategies and the establishment of at least 19 new institutions/programmes, a selection of which are surveyed below.

Strategies

Flowing from the first policy document, a number of specific strategies have been approved by the South African government. Some of these have already been implemented and others are being implemented. The following list provides a selective overview of strategies approved since 2000 (the relevant department and the year is listed):

1. National R&D strategy (DST, 2002): This is the most important strategic document to have emerged over the past decade. Its main objectives were (i) to promote innovation and new national technology missions (biotechnology, information technology, technology for advanced manufacturing, technology for and from natural resource sectors and technology for poverty reduction), (ii) to improve and diversify human resources, (iii) to promote a new set of science missions, and (iv) to create an effective government S&T system. Most of the strategic objectives are already being implemented. It further proposes a set of performance measures, *i.e.* indicators, according to which the progress of the NSI should be monitored. This specification is revisited in a later section of the paper (Kaplan, 2004).
2. National biotechnology strategy (DST, 2001) supports the development of a bio-economy, including the formation of Biotechnology Regional Innovation Centres and extensive funding for the close-to-market biotechnology R&D.
3. The accelerated and shared growth initiative for South Africa (ASGISA) (South African Presidency, 2006) aims at addressing constraints to higher sustainable economic growth, including scarce skills, and is expected to have an impact on various aspects of the NSI.
4. A number of other relevant strategies developed by different government departments, include the national skills development strategy for South Africa, open source software and standards strategy, towards a framework for the monitoring and evaluation of South African higher education, advanced manufacturing technology strategy, indigenous knowledge systems policy, nanotechnology strategy and the higher education qualifications framework.
5. In addition to the above, a framework for intellectual property from publicly financed research, a technology transfer strategy, the hydrogen economy initiative, advanced metals initiative, and gender equity strategies are being developed or have recently been completed.
6. Earlier this year the government announced a 150 per cent tax deduction for R&D expenditure by companies.

Institutional landscape, programmes and instruments

The past decade saw significant changes in the institutional landscape. The major changes are summarised below:

1. The S&T line function was elevated to a full-fledged central government department.

2. The National Research Foundation was established as a funding agency to serve all the disciplines excluding the Medical sciences,
3. New STI points of gravity were established, *e.g.* for energy, innovation, and automotive development,
4. Higher education was rationalised to consolidate the 36 institutions to 23,
5. Six centres of excellence were established at higher education institutions,
6. A number of public research institutions were significantly strengthened,
7. A number of STI promotion trusts, with centres under their control, were established, including the Innovation Fund,
8. Other new programmes include annual R&D surveys, an Antarctic programme, the square kilometre array and large SA telescope, and the pebble bed modular reactor, and
9. Finally, it is necessary to refer to the country's evaluation tradition given the focus of this conference, *viz.* evaluation and measurement of the STI system. As far back as 1997, the White Paper on S&T specified the intention of regular institutional reviews. Since then, an outstanding characteristic of the NSI has been the fact that almost all aspects of the STI system have been reviewed on a regular basis and following standard international practice. At least 21 reviews have taken place ranging from the progress in implementing the White Paper on S&T, through two cross-institutional reviews, to four-yearly international reviews of all the public research institutions (PRIs), annual key performance indicator reviews, and programme reviews of all significant government funded programmes. In addition, the Council on Higher Education manages an ongoing programme of evaluating various components of higher education institutions, ranging from the execution of an institutions R&D function to specific graduate programmes, such as MBA programmes.

Summary

The South African government has introduced a broad spectrum of new STI policies, strategies and programmes. In this way, the government reorganised the institutional landscape in order to bring it up to par with what had been happening elsewhere in the world, on the one hand and to raise its productivity, competitiveness and equity, on the other. All of these are in some stage of implementation, some fully, while others only recently announced. It is obviously too early to assess the outcomes and impact of most of these initiatives. What is shown in the next section is a commitment to monitor and evaluate the development of the new policy configuration in general and the individual initiatives in particular.

The role attributed to indicators in the South African system

Reference was, firstly, made in the preceding section to the fact that the *National R&D Strategy* (NRDS) specifies that the strategy- and subsequent strategies–was an indicator driven one. Secondly, reference was also made to the pervasiveness of reviews of various components of the South African STI system. The following two paragraphs elaborate somewhat more on the argument that the South African STI system could be described as an indicator guided one.

The NRDS, a key director of the South African STI system, explicitly states the importance of indicators in mapping progress in the implementation of the strategy. To quote, "*In order to develop a robust, effective strategy, it is necessary to track indicators of the "health" of the science, technology and innovation systems through time and to set objectives for the future. (...) The indicators are given for the enablers and the investment and outcomes of the strategy up to 2012*" (DST, 2002).

Parallel to the drafting of the NRDS, the DST committed itself to strengthen capacity in the area of indicator measurement by contracting a PRI, namely the Human Science Research Council (HSRC) to undertake regular national R&D Surveys (Frascati) and national innovation surveys (Oslo). So far, the HSRC's Centre for science, technology and innovation indicators (CeSTII) has completed three R&D surveys and is busy processing the data of the first national innovation survey (DST, 2006). This initiative should be seen against the background of the country's relatively long tradition of national R&D surveys, dating back to at least 1956 (Blankley and Kahn, 2005). Two important developments in 2005 signified the quality of the national R&D survey. Firstly, Statistics South Africa announced that the data of the national R&D survey would become part of the country's official statistics. Secondly, the OECD's national experts on S&T indicators (NESTI) agreed to include the South African data on R&D in its main S&T indicators publication. It is also worth mentioning that at least two other research centres are intensively involved in indicator research, especially bibliometric and patenting indicators, *viz.* the Institute for Technological Innovation, University of Pretoria, and the Centre for Research on Science and Technology, University of Stellenbosch.

Given a reasonably vibrant indicator research-for-policy community in the country, the next question that springs to mind is whether indicators are indeed utilised for planning and tracking of progress in the implementation of policies, strategies and programmes at national and institutional levels. No systematic study has so far been done in South Africa on the utilisation of indicators in the STI policy field. The following unsystematised notes can be offered as pointers in finding an answer to this question:

1. Indicator type surveys and analyses across a broad range of topics have become part of the programme of a number of research agencies, and the contracts with government departments are also common. On the one hand, this shows the quality of indicator research capacity in the country and the acknowledgement of government by the potential value of indicators, on the other.
2. At departmental level, indicator data are commonly used as part of arguments to obtain government approval of budgets, new strategies, instruments and programmes.
3. Advisory bodies, such as NACI and the Academy of Science of South Africa (ASSaF), use indicators as a matter of course in the development of policy recommendations to government.

4. At institutional level, PRIs use indicator frameworks as key instruments in their strategic planning, corporate business plans and annual performance reports.
5. Ministers and senior officials increasingly quote information generated by national indicator studies.

At this point, it would be fair to conclude that the notion of measurement of various aspects of STI policy, in general, and the use of STI indicators, in particular, have become an important part of policy planning, development, implementation, resourcing and monitoring.

Status of STI in South Africa

This section offers a highly selective overview of the most recent standard performance indicators (inputs, outputs as well as outcomes) for the main components of the system. These were taken from the recent background document to an OECD peer review of the country[1] (http://www.naci.org.za/OECD/OECD%20Report%2021Aug.pdf). A short discussion accompanies the figures and tables, since most of them are self-explanatory.

Financial inputs

The most recent R&D Survey (2004) showed that the country spent SAR12.0 billion (SAR 1=US$ 0.13) on R&D, which represents 0.87 per cent of GDP. This represents a steady increase from 0.69 per cent in 1997. The official target is to reach 1 per cent by 2008, which is the level attained by countries such as Brazil and the Czech Republic.

An interesting specification of the gross expenditure on R&D can be found in Figure 3.2, which shows that:

1. The business sector is the major financer and performer of R&D in the country, *viz.* 45 per cent and 58 per cent, respectively.
2. Government is responsible for 33 per cent of the financing and 21 per cent of R&D performance. The former is comparable to the OECD average, and China but higher than South Korea.
3. The higher education sector performs 21 per cent of all R&D.
4. It is interesting to note that international funders account for approximately 15 per cent of total funding.

The national R&D survey 2004/05 (DST, 2006) also reported (not shown in Figure 3.1) that 18.6 per cent of GERD is allocated to basic and strategic basic research.

---

[1] The report was compiled by Hermi Boraine, Steven Gelb, Hendrik Marais, Simon Mpele, Charles Mokonoto, Kate Morris, Adi Paterson and David Walwyn (editor), under a NACI oversight committee consisting of Dhesigen Naidoo, Adi Paterson, Johannes Potgieter, Bethuel Sehlapelo and John Stewart (convenor). The background report was approved by the National Advisory Council on Innovation. The direct and indirect contributions of these individuals to the data reflected in this report are gratefully appreciated.

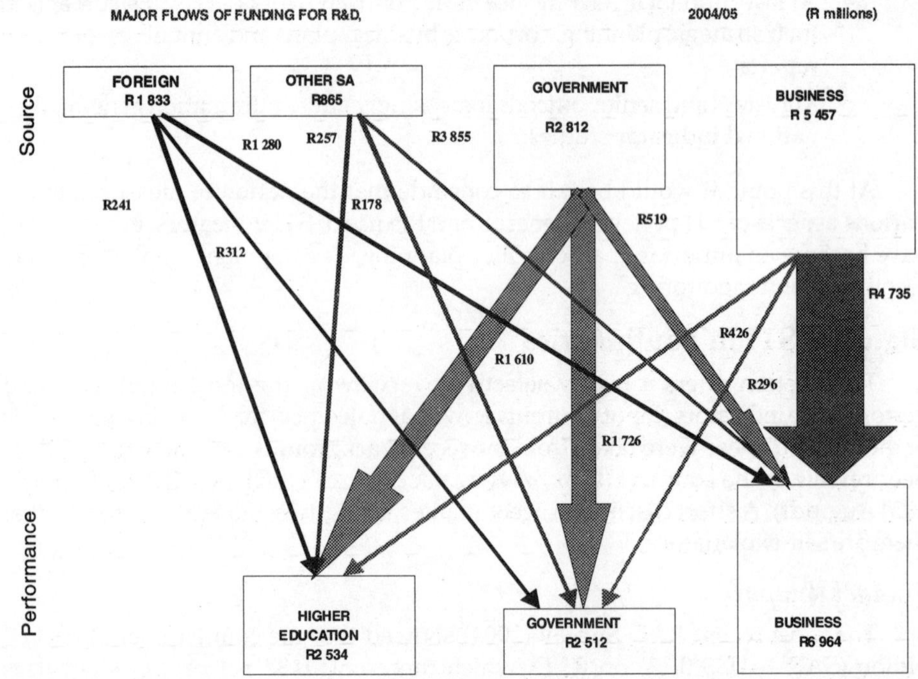

Figure 3.1: Major flows of funding for R&D (2004/05) (R million; SAR 1=US$ 0.13[2])
***Source*: 2004/05 National R&D Survey.**

Analysis of expenditure per research field show that engineering sciences account for approximately 25 per cent, the natural sciences for 20 per cent, social science 12 per cent and the agricultural sciences for approximately 7 per cent.

A last indication of financial investment in STI can be offered by referring to expenditure in the higher education sector. The higher education expenditure accounts for 2.7 per cent of the State budget and 20.6 per cent of the education budget, and 0.72 per cent of GDP, while the government expenditure per FTE student in 2002 was approximately SAR14 500 per annum.

Human capital inputs

Table 3.1 showed that the country has a total R&D personnel of just over 44 000. This figure translates into nearly 30 000 FTE personnel. From a different perspective, the country has 2.7 researchers per 1000 employees as compared to the highly industrialised countries and lower ratio of peer countries in South America. An estimated 4.5 per cent of the estimated 20 million South Africans between the ages of 25 and 64 enjoyed a tertiary education.

---

[2] Current exchange rate.

Publication performance

Publications are normally accepted as one of the standard categories of outputs of STI in a country. The following paragraphs reflect a sample of the country's publication production and are largely based on a recent report by the Academy of Science of South Africa (ASSaF, 2006). The report that emanated from this comprehensive ASSaF project is currently being discussed with various government role players. It would be reasonable to expect a rethink of relevant policies and/or strategies to follow in due course.

Figure 3.2 provides comparative information on peer-reviewed South African publications over two decades. At least two conclusions spring to mind from even a superficial interpretation of the table. Firstly, while the absolute number of publications grew more or less steadily from just over 2000 to over 4000, the country's share of the world output declined from about 0.7 per cent to 0.48 per cent. Some countries, which had started from a much lower base than South Africa such as Brazil and South Korea, overtook us a decade later. Analysis, reported elsewhere, showed a skewed distribution across institutions, with the most productive one accounting for 18 per cent of the national output and the lowest five together accounting for less than 1 per cent of the total (ASSaF, 2006).

As far as quality is concerned, Figure 3.3 shows that South Africa's citation rate per article has improved in recent years. The country was actually doing better than countries with which we often compare ourselves. South Africa publishes more than

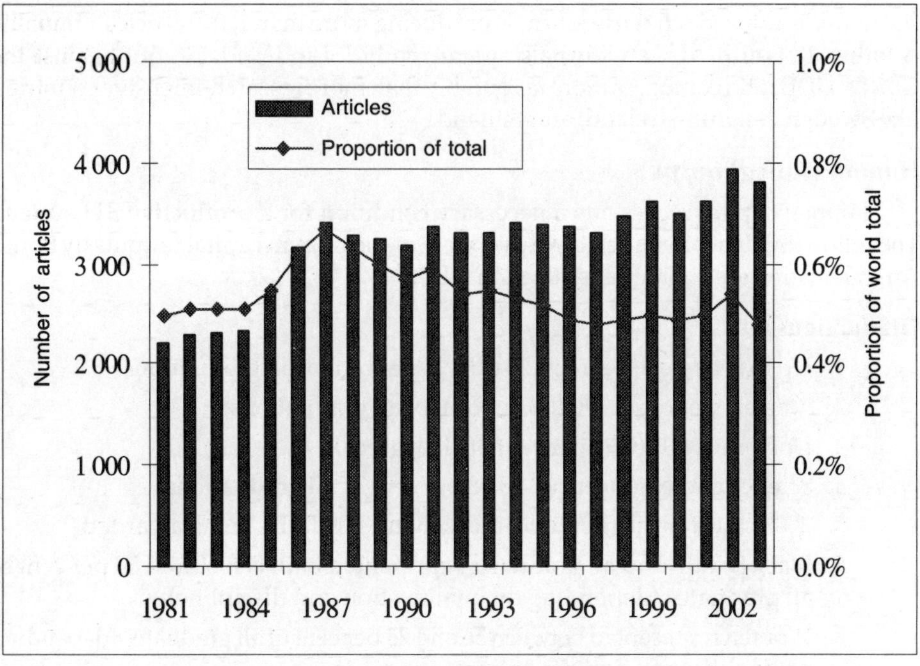

Figure 3.2: Number and proportion of South African publications in the ISI

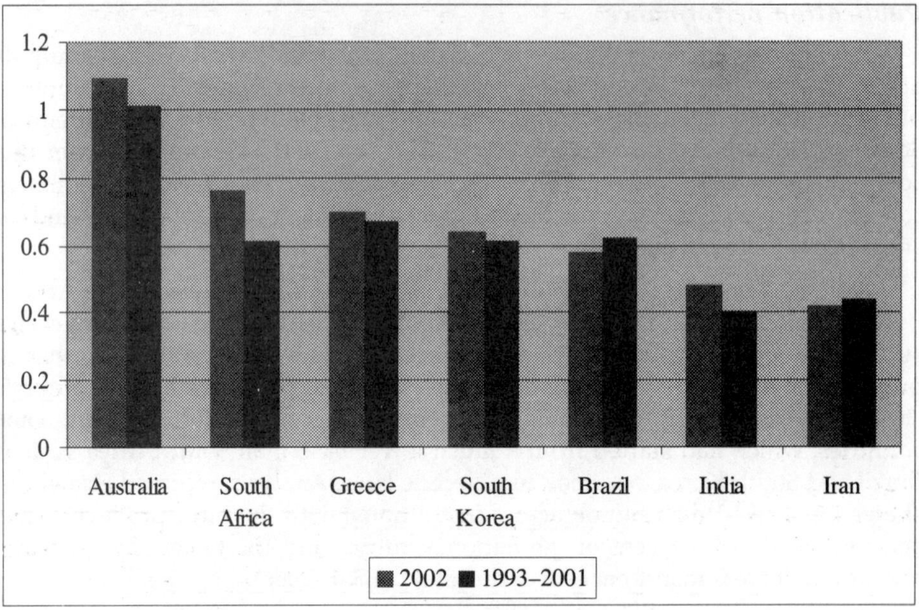

Figure 3.3: Normalised citation rates per article (1993–2002)
***Source*: King, 2004**

200 non-ISI-indexed scholarly journals producing more than 1, 000 articles, annually. Another 17 South African journals are currently ISI-indexed. Set out against the GERD/GDP ratio compares more favourably than more research-intensive countries, like Sweden, Belgium, Ireland, and Finland.

Human capital output

Human capital represents a necessary condition for a productive STI system. The following data offer a selective overview of the human capital outputs by South African institutions of higher education (HE).

Graduations (2001)

1. Up to four-year degrees, 73 615 (91 per cent of all graduations)
2. Masters degree, 6 296 ((.8 per cent of all graduations)
3. Doctorate, 801 (0.99 per cent of all degrees)
4. Science and engineering, 20.38 per cent of all graduations
5. 2005 saw, for the first time, more than 1 000 PhDs being awarded
6. Black South African graduates represented between 60 and 30 per cent of all graduates (depending on qualification and discipline)
7. Women represented between 56 and 23 per cent of all graduates (depending on qualification and discipline)
8. One study has shown that 88 per cent of graduates find employment within six months after graduation.

It is generally accepted that the subject of human capital outputs, in general, and the future projections, in particular, are subject to the potential influence of HIV/ AIDS, given that the proportion of South Africans so affected is one of the highest in the world. One study estimated the infection rate amongst undergraduate students to be of the order of 33 per cent and that of postgraduate students to range between approximately 22 and 46 per cent, depending on the type of institution.

Downstream outputs, outcomes and impact

In 1996, South Africa became the first country to officially approve an innovation rather than S&T policy, and to label the STI system as a national system of innovation. It follows that an overview of the system needs to include innovation as the downstream phase and component of a national overview. The following paragraphs briefly offer reference to indicators of innovation such as patents and exports of technological products.

South Africa has a notable history in technological innovation in fields such as deep and hard rock mining, petrochemicals, nuclear technology, and high-technology armaments development and production. However, in absolute terms, South Africa's performance in patenting has remained stable over the past decade at a modest, approximately, 100 USPTO granted patents, although the country has slipped back significantly in comparative terms (Pouris, 2005). For example, in 1994 this country and India received 101 and 27 USPTO granted patents, respectively, while the figures for 2004 were 100 and 363, respectively. Income from royalties and licence fees has been negligible at an US$/person of 1.4, in comparison to Chile's 17.9 and Brazil's 0.7.

In conclusion, a few notes on indicators even further down the innovation chain, *viz.* exports of technology goods and trade balance of payments would be appropriate.

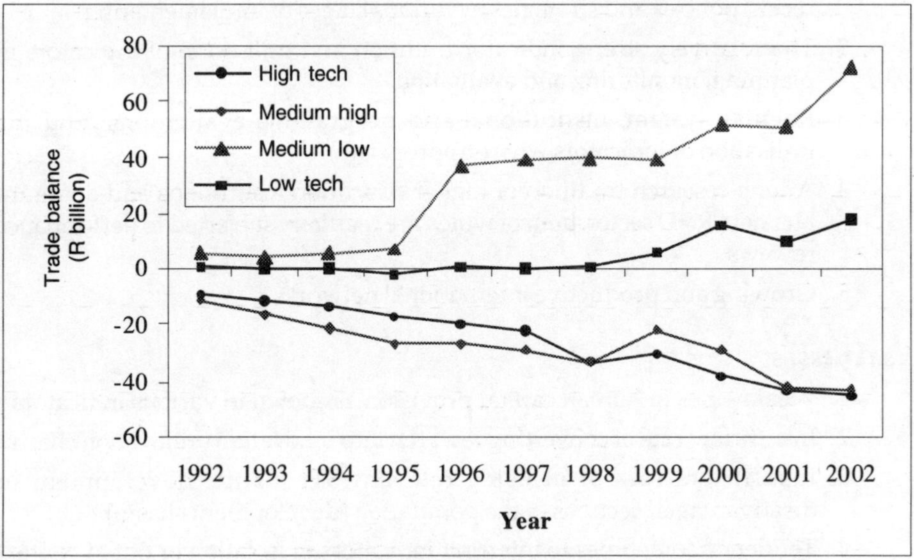

Figure 3.4: South Africa's trade balance of payments (1993-2003)

***Source*: NACI/Pouris, 2005**

Given that international trade is dominated (up to 64 per cent) by high and medium technology, it would be relevant to look at South Africa's performance in this regard; this information is reflected in Figure 3.4. The trend shown here is clear. In the case of high research intensity products, the country imports more than it exports. However, it exports more low to medium low technology than what it imports. Nevertheless, South Africa's comparative advantage in medium low and low technology is higher than that of some other resource based economies, such as Australia.

Conclusions

The overview of performance indicators offered in the preceding paragraphs suggests that South Africa has improved its absolute productivity marginally, if at all, but fallen behind countries with which it tends to compare itself. This applies to classical R&D type indicators such as publications, as well as innovation ones like patents. On the one hand, the strong accelerated performance of those countries clearly explains part of the fact that we are 'loosing' out on them. On the other hand, it can not be denied that South Africa's development in science, technology and innovation will have to be accelerated, if it wants to utilise STI as serious contributor to economic growth and improvement of the quality of life of all its people.

Selected strengths and weaknesses

This paper so far has offered an overview of the South African STI system–policies, inputs and outputs–against the background of conceptual notes on indicator methodology. This section offers a listing (without elaboration) of some strengths and weaknesses implied and reflected in this paper.

Strengths

1. The importance attached to STI by the government and a wide spectrum of recent policies and strategies in various stages of implementation
2. The relatively strong indicator tradition and utilisation of indicators in planning, monitoring and evaluation
3. Regular system, institutional and programme evaluations (and the utilisation of indicators where appropriate)
4. A long research tradition of higher education institutions and a vibrant business R&D sector, both of which are regularly subjected to performance reviews
5. Growing and productive international networks

Weaknesses

1. Weaknesses in human capital provision as shown in various indicators
2. Insufficient rate of converting research into innovation (innovation chasm)
3. Insufficient rate of utilising relevant STI for the development of disadvantaged sections of the population (development chasm)
4. Tendency sometimes to interpret indicators in isolation or out of context and the need for meta-methodological analyses of indicators.

Future perspectives

The country has already demonstrated its commitment to take serious cognizance of and participate in STI taking place elsewhere in the world–including Africa and other developing regions. It is likely that this trend will be strengthened in future with an array of consequences such as the internationalisation of research agendas, teams and resourcing. While this trend would bring its undoubted rewards, the challenges to R&D management at national, institutional and programme level should not be underestimated. A condition for and at the same time a consequence of such a trend would be the optimisation of communication. In this regard, new quantitative measures and indicators are in the process of development, *e.g.*, globalisation and mobility. Indicator approaches will grow in importance to the extent that countries wish to measure their own progress domestically and internationally, and it would require a better understanding of the methodology underlying the approach.

References

ASSAF, 2006. Report on Strategic Approach to Research Publishing in South Africa. Academy of Science of South Africa. Pretoria: ASSAf.

Blankley, W., and Kahn, M., 2005. The history of research and experimental development measurement in South Africa and some current perspectives, South African Journal of Science, 101, pp. 151–156.

DoE, 1997. White Paper on Education: A Programme for the Transformation of Higher Education. Pretoria: Department of Education.

DST, 1996. White Paper on Science and Technology: Preparing for the 21st Century. Pretoria: Department of Science and Technology.

DST, 2002. South Africa's National Research and Development Strategy. Pretoria: Department of Science and Technology.

DST, 2006. National Survey of Research and Experimental Development 2004/05. Pretoria: Department of Science and Technology.

Website of NACI's Innovation gateway (http://www.naci.org.za/Innovation_gateway/Policies.htm)

Kaplan, D., 2004. South Africa's national R&D strategy: a review, Science, Technology and Society, 9, pp. 273–294.

King, A.D., 2004. The scientific impact of nations, Nature, 430, pp. 311–316.

Marais, H.C., 2000. Perspectives on Science Policy in South Africa. Pretoria: Network Publishers.

NACI, 2006. Annual Report 2005/2006. Pretoria: National Advisory Council on Innovation.

NACI., 2006. The South African National System of Innovation: Background report to the OECD country review. Pretoria: Website of National Advisory Council on Innovation. (http://www.naci.org.za/OECD/OECD%20Report%2021Aug.pdf)

NACI and DST, 2002. South African Science and Technology: Key Facts and Figures–2002. Pretoria: National Advisory Council on Innovation and Department of Science and Technology.

NACI and DST, 2004. Facing the Facts: Women's Participation in Science, Engineering and Technology. Pretoria: National Advisory Council on Innovation and Department of Science and Technology.

Pouris, A., 2005. Technological performance judged by American patents awarded to South African inventors, South African Journal of Science, 101, pp. 1–4.

Chapter 4

Science and technology structure, statistical system and the current scenario in Sri Lanka

Seetha I. Wickremasinghe

Abstract

The role played by science and technology (S&T) in the socio-economic development of a country is widely recognized. The developed countries make a significant amount of investments for research and development (R&D) activities out of the Gross Development Product (GDP) of the country (more than 3 per cent), while many developing countries invest an amount of less than 1 per cent for the same. Since R&D activities contribute much to the improvement of national capabilities, the developing countries should also be encouraged to increase the amount of percentage GDP towards R&D activities. However, the outcomes and impacts of these investments are equally important in order to guide the governments for future investments on science and technology. Measuring science becomes important at this juncture. For this purpose, it is necessary for the countries to have the national science and technology statistical systems and the Management of Information Systems (MIS) well in place. Highlighting the above facts, this paper presents the science and technology structure, the statistical system and the current S&T scenario in Sri Lanka and also discusses the issues related to the development of relevant indicators in the country.

Keywords: Management information systems, Measuring science, Science and technology indicators, Science and technology structure, Science policy, Sri Lanka.

Introduction

Science and technology (S&T) influence the society as never before. The fact that science is the driving force behind not only the technological innovations, but also the socio economic growth of a particular country is now widely recognized and accepted among scientists and economists. Driven by this knowledge, the governments of the developed countries support basic or academic research in large scale. According to Mansfield (1991), 10 per cent of new products and processes would never have been developed in the absence of academic research. The governments of the developing countries have now realized this fact. They have slowly begun to increase allocation of funds for academic research. However, many developing countries still spend rather a small amount of funds for S&T, which in fact is a minute percentage of their Gross Development Product (GDP). Unlike the developed countries, many developing countries do not have a proper S&T structure identified and therefore, the attention of governments of such countries to this important aspect is rather low. To remedy this situation, it is important for a country to have an identified S&T structure well in place.

Many developing countries are at the cross roads in understanding the implications of policy contexts and creating science policies that are interlinked with other public policies towards the socio-economic development. They make considerable investments in research with the intention of developing national capabilities, improving technology capacity, and enhancing economic performance. On one hand, it is important to know the outcomes and impacts of these investments in order to guide the decisions taken by the governments about future investments on science and technology. On the other hand, it is essential to maintain a science and technology statistical system or a network of such systems in different scientific organizations. These statistical systems, commonly known as Management Information Systems (MIS) will readily provide information on the input and the output measures, human resources and other relevant socio-economic parameters. Therefore, setting up of Management Information Systems at the national level and also at the sectorial level has been the current interest of many stakeholders of the S&T sector in various countries.

It is interesting to note that when the developing countries start thinking of allocating more money to academic and/or hard sciences research, the developed countries are keen on allocating more funds for analysing the socio-economic issues of scientific research that will have a direct impact on the growth and development of the country. Measuring science, thus, becomes important at this juncture. By measuring science, one could develop science and technology indicators through which a critical analysis of science and its impact to the society has been possible. Analysis of comprehensive data with correct facts may greatly influence the science and technology policy of a particular government towards the betterment of its own people. But where do these indicators come from? It was about thirty years ago, the subject of S&T indicators was first discussed by Chris Freeman (1968) and Derek de Solla Price (1969). While they were developing the concept, theory and methodology of S&T indicators, the governments and other public policy organizations were beginning to take an interest in the measurement of science and the use of quantitative analysis for public policy in science and technology. Use of S&T statistics (input and output indicators) is of importance to the politicians and policy makers for their subject interests either to support or oppose the government policies. As observed by

Holbrook in 1992, these indicators also appear in public debates on science policy, at all levels of government; consequently, their use (and misuse) has become an important issue in the analysis of S&T policy options. In this context, let us now examine the science and technology scenario in Sri Lanka.

Some socio-economic parameters of Sri Lanka

Sri Lanka (formerly Ceylon) is an island in the Indian Ocean situated in the south and slightly to the east of the southernmost point of India. It is separated from India by the 23 km (14 miles) wide Palk Strait. Sri Lanka has a total land area of 65,610 sq km (25,332 sq m). The Gross Development Product (GDP) of the country in 2004 had been SL Rs. million 1,797,941. The services sector contributed to 55 per cent of the GDP while the contribution of the industry sector and the agriculture sector had been 27 per cent and 18 per cent, respectively. The country was achieving nearly 7 per cent growth rate by the latter part of 2006. Other key socio-economic parameters are as follows:

1. Population (in million)	19.2
2. Population growth (per cent)	1.3
3. Unemployment rate (per cent of labour force)	7.7
4. GDP per Capita 2002 (US$)	873
5. R&D expenditure for year 2004 (per cent of GDP)	0.21
6. Researchers in R&D (per million people)	191
7. Physicians (per million people)	430
8. Literacy rate (per cent age 15 and above)	92.1
9. HDI rank (out of 173 countries)	96

S&T policy framework

The cabinet Ministry of Science and Technology of the Government of Sri Lanka provides a framework for the science and technology structure in the country. According to the Science and Technology Development Act No. 11 of the government, the mission of this Ministry is to formulate programmes and projects based on the national policy in respect of science and technology and implementation of such programs and projects. There are two institutions namely, the National Science and Technology Commission (NASTEC) and the National Science Foundation (NSF) that operate under the S&T Ministry in this connection. The first is responsible for formulating S&T policies based on the recommendations and the data provided by the latter which is mandated for conducting S&T policy research. In addition, there are other institutions coming under various ministries (*i.e.* Council for Agricultural Research Policy of the Ministry of Agriculture) that are inter-connected with S&T Policy framework in the country. Figure 4.1 shows the various institutes coming under the S&T Ministry of Sri Lanka.

There are other Ministries in the government of Sri Lanka that handle the S&T related activities *i.e.* Ministries of Agriculture; Plantation Industries; Power and Energy;

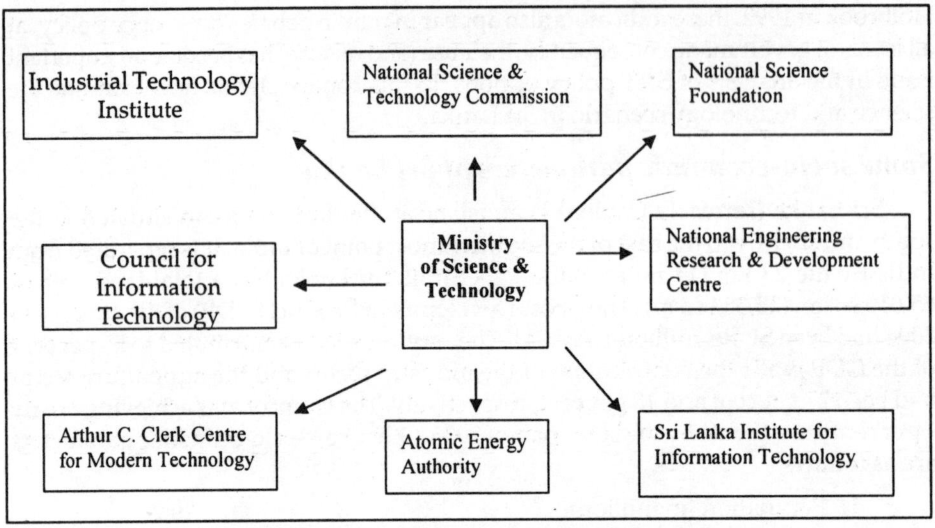

Figure 4.1: Institutes coming under the Ministry of Science and Technology in Sri Lanka

Education; Environment and Natural Resources; Highways; Small and Rural Industry; Land and Indigenous Medicine; Industry and Investment Promotion; Fisheries and Aquatic Resources etc.

S&T structure

At present, there are 111 different state organizations handling S&T activities which are coming under different Ministries. These include 18 government departments, 15 universities, 29 research institutions, 10 S&T institutions, 05 health related institutes, 07 corporations, 08 boards, 07 authorities, 04 bureau agencies, and 08 others. In addition, there are many industries in the private sector that are involved in R&D related activities along with a few international organizations. The non-governmental organizations (NGOs) operating in the country also conduct research, which is mainly related to social sciences. The local R&D institutes such as Tea Research Institute (TRI), Rubber Research Institute (RRI), Coconut Research Institute (CRI), Institute of Fundamental Studies (IFS) etc., are involved in 100 per cent R&D work while the universities are mainly involved in the higher education. Universities also concentrate on basic and academic research. S&T institutes mainly provide services to the public and other sectors, but some of them do undertake research, especially, in the field of social sciences. For example, the Science and Technology Policy Research Division (STPRD) of the NSF conducts S&T policy research. Many key industries that operate in Sri Lanka do not conduct research per se, but import technologies from their parent companies abroad, and adjust such technologies to suit into the local conditions.

S&T statistical system

There are eight (8) institutions in Sri Lanka that have S&T statistical systems or Management Information Systems in place now (Figure 4.2).

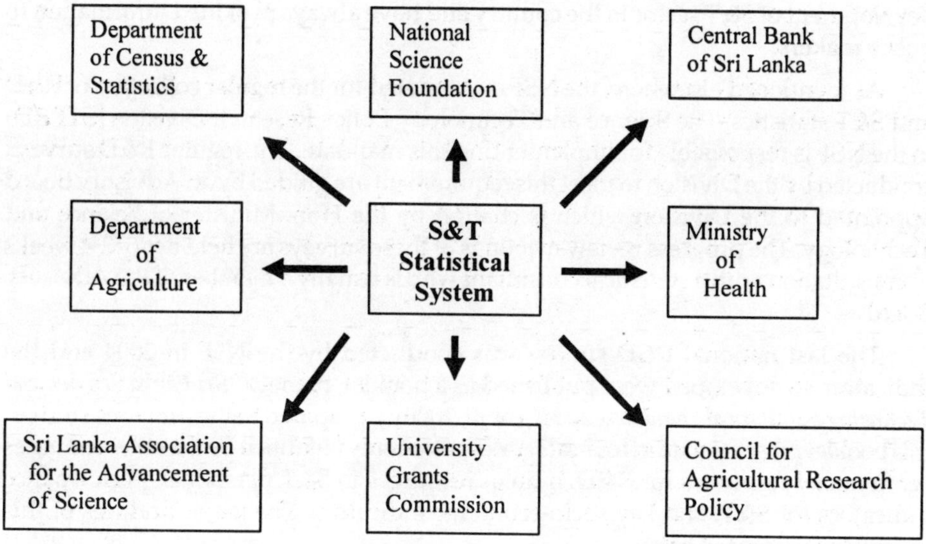

Figure 4.2: S&T Statistical System in Sri Lanka

Sri Lanka is one of the countries in South Asia, which has a long tradition in developing science and technology indicators. The status of S&T sector in Sri Lanka has been a subject of study for the National Science Foundation (NSF) and its predecessors, the Natural Resources, Energy and Science Authority (NARESA) and the National Science Council (NSC) since early 1970s.

Early studies on S&T indicators were carried out in 1970 to assess the scientific resources in Sri Lanka in terms of public expenditure on research and development by the Ceylon Institute for Scientific and Industrial Research (CISIR), presently known as the Industrial Technology Institute (ITI). This was followed by a survey on scientific and technical manpower in Sri Lanka carried out in 1972 by the Ceylon Association for the Advancement of Science, which is now known as the Sri Lanka Association for the Advancement of Science (SLAAS).

However, the first systematically designed survey on scientific and industrial manpower potential in the country was carried out in 1974 by the National Science Council following the guidelines of UNESCO to assess, monitor and forecast the scientific and technical manpower in the country. In 1977, the NSC carried out its second major survey which included an assessment of funding for basic and applied research in the country for the period 1956-1965 and the technical manpower during 1977. The third study in the series was carried out in 1984 by the Natural Resources, Energy and Science Authority, which was the successor to NSC. The fourth comprehensive survey on research and development in Sri Lanka was then carried out in 1996 by NARESA (Fernando and Amaradasa, 1998). This survey covered the higher education in the public and private sector institutions. A subsequent survey was then carried out by the NSF in 2000, which covered only the higher education sector and S&T public sector (NSF, 2003). These studies have shown the trend in the development of S&T sector in the country and have always provided information to policy makers.

As mentioned elsewhere, the NSF is mandated for the regular collection of R&D and S&T statistics. The Science and Technology Policy Research Division (STPRD) in the NSF is responsible for implementing this mandate. The regular R&D surveys conducted by the Division to meet this requirement are guided by an Advisory Board appointed to the Division which is chaired by the Hon. Minister of Science and Technology. The progress review meetings of these surveys are held every 2-4 weeks in consultation with a statistical consultant who is usually a member of the Advisory Board.

The last national R&D survey was conducted by the NSF in 2004 and the indicators so developed were published as a booklet, namely "*Sri Lanka Science and Technology Statistical Handbook 2004*" (NSF, 2006). Compared to the previous studies, this booklet presents data in four categories commonly identified by the other countries *i.e.* financial resources for R&D, human resources in S&T personnel, performance indicators for S&T, and key socio-economic indicators. The major findings of this survey are presented below.

S&T indicators

There are two most common types of indicators. The first type is the Input indicators, which measure resources that are required as inputs for the S&T activities, mainly, the funds and the personnel.

S&T input indicators

As indicated in the Table 4.1, the government has throughout been the major contributor to research expenditure, which is about 67.5 per cent at present, while the foreign funds contribute to 22.6 per cent of the total expenditure. The private sector contribution has been very low which is about 0.6 per cent. An amount of 9.3 per cent comes from the 'other' category which is usually from the funds generated by the institutes themselves.

Table 4.1: National Expenditure on R&D (GERD) by source of funding 2004 (Rs. in million)

Source of Funding	*Recurrent*	*Capital*	*Total*
Government	2,115.3 (55.6 per cent)	456.0 (12.0 per cent)	2,571.2 (67.5 per cent)
Private	17.1 (0.5 per cent)	4.7 (0.1 per cent)	21.9 (0.6 per cent)
Foreign	675.7 (17.7 per cent)	186.2 (4.9 per cent)	861.8 (22.6 per cent)
Other	276.4 (7.3 per cent)	76.1 (2.0 per cent)	352.5 (9.3 per cent)
Total	3,084.5 (81.0 per cent)	723.1 (19.0 per cent)	3,807.5 (100.0 per cent)

However, the Government expenditure on R&D or GERD expressed as a percentage of the national GDP has been below 0.2 per cent for many years. This is much lower than the level recommended by the Action Plan of the United Nations for the developing countries which is a minimum of 1 per cent of GDP. There has not been any considerable difference in the government expenditure on R&D during the past two decades, which varied between 0.18–0.2 per cent, indicating a relative stagnation of the country's R&D effort (Figure 4.3). The relevant figures calculated for the year 2004 is Rs million 3,807 (~US$ 40.9 million) indicating that the research expenditure as percentage of GDP for 2004 is 0.21 per cent.

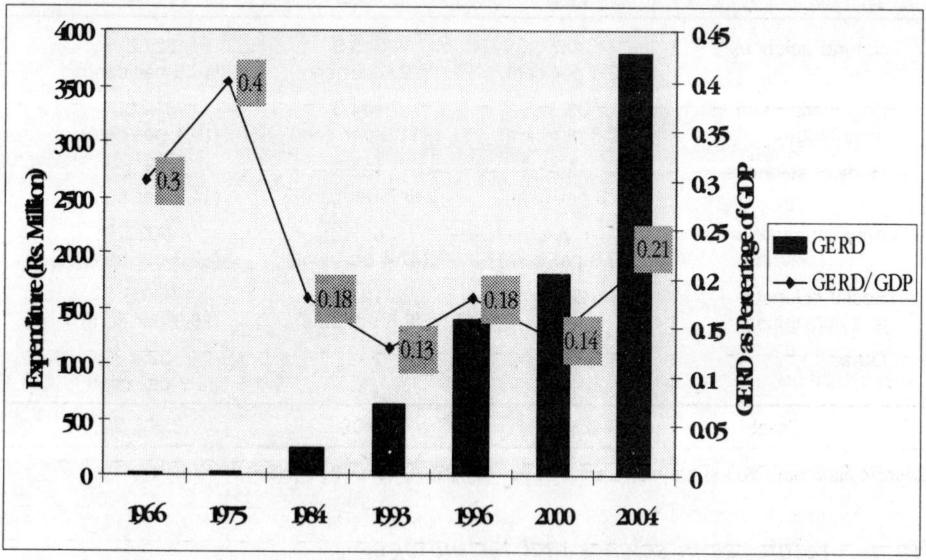

Figure 4.3: Growth of national R&D expenditure from 1966–2004

Out of the total R&D expenditure in 2004, more than 75 per cent has been spent on applied research (Table 4.2).

Table 4.2: National R&D expenditure by nature of research activity (Rs. in million)

Nature of research	*1984*		*1996*		*2004*	
	Amount	*%*	*Amount*	*%*	*Amount*	*%*
Basic research	24.8	10.0	446.3	32.0	519.60	13.6
Applied research	185.2	72.0	867.3	61.0	2,886.10	75.8
Experimental development	46.9	18.0	96.0	7.0	401.80	10.6
Total	256.8	100.0	1,409.6	100.0	3,807.50	100.0

Source: National R&D survey Sri Lanka (NARESA, 1996) and (NSF, 2004).

In terms of support to different disciplines, Agriculture comes first, which is 26.3 per cent understandable for an agricultural country like Sri Lanka. However, it is of interest to note that similar amount of funds has been expended in the field of social sciences and humanities in 2004, indicating steep increase in R&D work in the particular field over the years (Table 4.3). This is followed by the Natural sciences (16.5 per cent), Engineering and Technology (16.1 per cent), and Medical Sciences (14.0 per cent).

Table 4.3: National R&D expenditure by discipline 1984–2004 (Rs. in million)

Discipline	*Year*		
	1984	*1996*	*2004*
Natural sciences	30.7 (12.0 per cent)	318.3 (22.6 per cent)	627.6 (16.5 per cent)
Engineering and technology	32.8 (12.8 per cent)	164.3 (11.6 per cent)	614.0 (16.1 per cent)
Medical sciences	13.3 (5.2 per cent)	136.6 (9.7 per cent)	531.4 (14.0 per cent)
Agricultural sciences	153.4 (59.8 per cent)	669.2 (47.4 per cent)	1002.5 (26.3 per cent)
Social sciences and humanities	26.5	121.2 (8.6 per cent)	999.5 (26.3 per cent)
Other	–	–	32.4 (0.9 per cent)
Total	256.7	1,409.6	3,807.5

Source: National R&D survey Sri Lanka 1996 (NARESA) and 2004 (NSF).

Human resources in science and technology

According to the national research and development survey of 2004 conducted by the Science and Technology Policy Research Division of the National Science Foundation, Colombo, the "head count" of 'scientists and engineers' in Sri Lanka at present is revealed as 9,746. The total S&T personnel in the country in 2004 had been 28,432, which included the scientists, engineers, technicians and other supporting staff (Table 4.4).

The highest number of the science and technology personnel was attached to the government sector in the year 2004, which was nearly 45 per cent of the total (Table 4.5).

Even though the total number of scientists in the country is 9,746 in 2004 the total number of R&D scientists was 4,602 and out of this, the total number of Full Time Equivalent (FTE) researchers was only 2,679. When the distribution of FTE researchers among subject areas is considered, the natural sciences and the engineering and technology fields comprise the majority followed by the filed of agricultural sciences. However, this is way below the optimum level. Of the R&D force about a quarter are doctoral degree holders (22.9 per cent), and about 40 per cent are over the age of forty.

Table 4.4: Distribution of S&T personnel by category (2004)

STP category	*1996*			*2004*		
	Total number	*% of STP*	*Per million inhabitants*	*Total number*	*% of STP*	*Per million inhabitants*
Scientists and engineers	13286	47.8	726	9,746	34.3	502
Technicians	14,514*	52.2*	793*	12,302	43.3	634
Other Supporting Staff				6,384	22.5	329
STP	27,800	100.0	1,519	28,432	100.0	1,465

Source: National R&D survey Sri Lanka 1996 (NARESA) and 2004 (NSF)

Note: Total STP includes all scientists, engineers, technicians, and other supporting staff

* 1996 Survey had not counted technicians and supporting staff separately.

Table 4.5: Distribution of science and technology personnel (STP) by sector

Sector	*1996*		*2004*	
	STP	*Per cent*	*STP*	*Per cent*
Higher education	3,990	14.3	4,285	15.1
State	18,645	67.1	12,685	44.6
Private	5,165	18.6	11,462	40.3
Total	27,800	100.0	28,432	100.0

Source: National R&D survey Sri Lanka 1996 (NARESA) and 2004 (NSF).

It was also noted that only 10.3 per cent of the total scientific community (9,746) in Sri Lanka has PhDs although the number has been growing slowly over the years (Table 4.6). It is a known fact that the ability to conduct independent research needs research experience, preferably, PhD training. Therefore, increase of R&D budget is essential for allocation of funds for PhD level training.

Table 4.6: Educational qualifications of S&T scientists

Qualification	*1996*		*2004*	
	No.	*Per cent*	*No.*	*Per cent*
Ph.D.	1,473	14.0	1,001	10.3
M.Phil./M.Sc.	2,116	20.0	1,407	14.4
B.Sc. + P.G. Diploma	6966	66.0*	410	4.2
B.Sc. (Special)			6,377	65.4
B.Sc. (General)			551	5.7
Total	10,555	100.0	9,746	100.0

* Includes B.Sc. (Special) and B.Sc. (General) Degrees.

Out of the total R&D scientists, 32.1 per cent are females. In the case of engineering and technology sector, the female representation is only 25 per cent. However, the total requirement of scientific research personnel for a country with the size and population of Sri Lanka should be more than double this number to contribute significantly to the socio-economic development of the country through science and technology.

S&T output indicators

The output indicators attempt to measure the direct products of S&T activities such as publications and patents. The various NSF surveys have also revealed the signs of progress in S&T in Sri Lanka over the years. Up to the year 2000, the number of non-resident patents registered locally was higher than that of resident patents. But since then, the number of resident patents registered locally has increased, clearly showing the growing trend in local technology development (Figure 4.4).

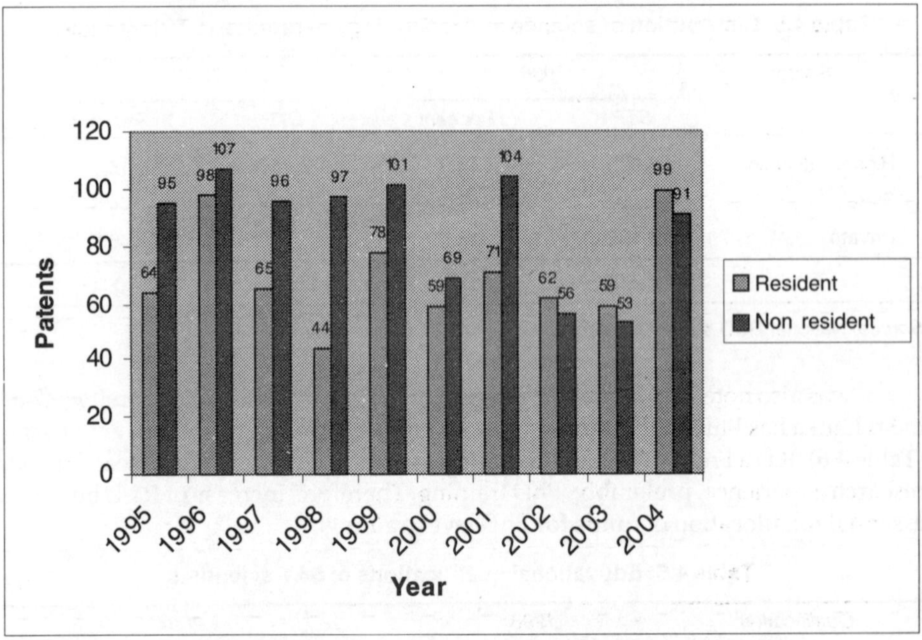

Figure 4.4: Number of patents registered from 1995–2004

Unfortunately, the important aspects such as the transfer of technologies and commercialization of research findings have still not taken off the ground, though our indicators show that there has been increase since 2001. As at 2004, the numbers of activities that have taken place in relation to the transfer of technologies were 2,324. The 2004 survey reveals that the individual inventors claimed 76 of patents, and private institutions claimed only 13. Of this only 9 belonged to public institutions (Table 4.7). The study also indicated that majority of patents were granted for small technologies in the areas of innovations in domestic appliances/utilities, product development and other miscellaneous categories.

Table 4.7: Number of patents registered locally, during the period 2001–2004 by sector

Sector	*2001*	*2002*	*2003*	*2004*
S&T institutes	5	5	8	5
Higher education inst.	1	1	4	4
Private	4	7	8	13
Individual	64	49	41	75
Total	71	62	61	99

Source: National R&D survey Sri Lanka 1996 (NARESA) and 2004 (NSF).

According to the International Science Citation Index data too, the number of research publications by Sri Lankans in all S&T fields has been increasing over the years. The field of medical and health sciences recorded the highest number of research articles followed by the fields of biological sciences, chemical sciences and agricultural sciences (Figure 4.5).

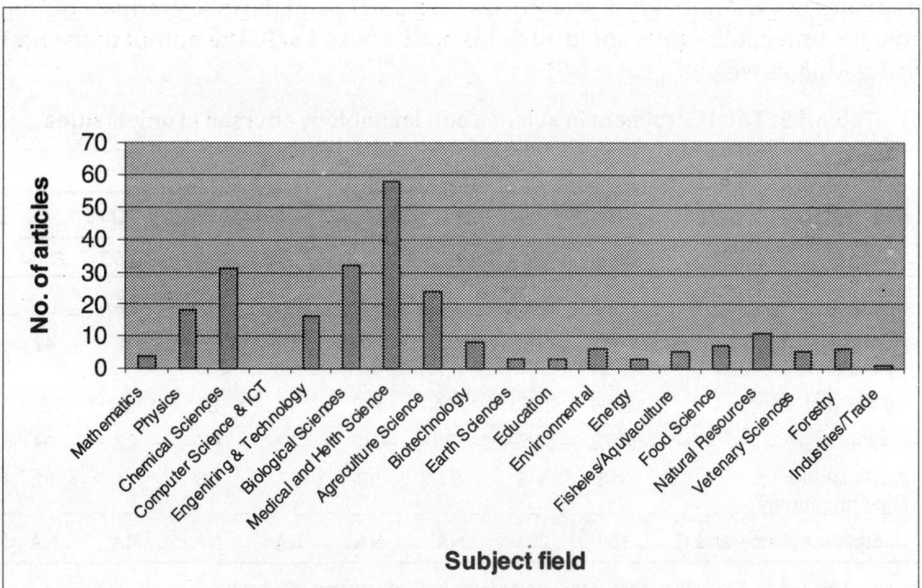

Figure 4.5: Number of research articles appeared in the science citation index in 2004

However, in the case of Sri Lanka, the highest number of patents has been granted to individuals over the years followed by the private sector (Table 4.7).

Publication of research results at the Sri Lanka Association for the Advancement of Science (SLAAS) Annual Sessions has also increased over the years (Table 4.8).

Table 4.8: Number of presentations at SLAAS annual sessions

Field	2000	2001	2002	2003	2004
Medical sciences	52	25	30	27	36
Agric. sciences	102	85	98	95	118
Engineering and technology	24	46	59	24	28
Life sciences	64	52	89	52	61
Physical sciences	43	21	43	24	27
Chemical sciences	49	28	57	44	23
Social sciences	31	28	34	31	41
Total	365	285	410	297	334

Source: SLAAS Annual Proceedings 2000-2004.

Higher education sector

In regard to higher education, out of the total graduates, the science and technology graduates accounted for 13.7 per cent. As at 2004, the total enrolment of under graduates in various fields of science has been 25,212 and the total graduate output from the universities relevant to all fields of S&T was 4,479. The output of the S&T postgraduates was 347 (Table 4.9).

Table 4.9: Total Enrolment in science and technology courses at universities

Course	Undergraduates				Post graduates			
	Enrolment		Output		Enrolment		Output	
	2003	2004	2003	2004	2003	2004	2003	2004
Science	8588	8652	1726	1875	493	359	98	147
Medicine/Dental/ Veterinary	7483	7425	1408	1427	45	71	8	42
Agricultural sciences	2658	2853	394	402	41	28	2	2
Engineering	5073	5269	629	683	32	117	22	94
Architecture/ Quantity survey	513	812	81	92	27	66	14	62
Computer science and IT	150	201	NA	NA	NA	NA	NA	NA

Source: University Statistics 2005, University Grants Commission, Colombo.

NA: Not Available.

It was also noted that the number of technical collages (36) has remained the same since 2001, although the expenditure on technical education has slightly increased over the years which amounts to Rs. million 515 in the year 2004.

Concluding remarks

However, the inescapable conclusion arising from the above S&T indicators is that the national R&D effort in Sri Lanka has remained at best at a standstill. Despite the importance of emerging areas such as biotechnology, molecular biology, nanotechnology etc., Sri Lanka has not managed to bolster institutionalization and professionalization towards such areas. The lack of an adequate science and innovation base for these frontier areas of science is likely to have serious repercussions for the relevance of such institutions to the Sri Lankan economy in future, which otherwise make her dependant on plantations and other connected industries.

The present government plans to dispense with overseas sponsored feasibility studies for local development projects. It plans to opt for local expertise at much reduced costs along with the move to ascertain the pool of Sri Lankan talent currently available overseas. The issue of Brain Gain is another important aspect gaining attraction in the current S&T scenario in the country. Our preliminary analysis in this direction shows that the United States leads in having many Sri Lankan scientific expatriates followed by Australia and Canada. The sample also reveals that majority is married (87 per cent), male (80 per cent), and falls between the age group of 45 to 55 years. More than 60 per cent of them have PhDs. Therefore, it is essential to workout a proper plan to obtain their services to the country with a view to increase the contribution of S&T to the socio-economic development of Sri Lanka.

In this context, a considerable increase of the percentage of GDP for the S&T sector and inclusion of Science and Technology in the next ten year development plan of the country become most significant.

It should also be mentioned that despite the high costs involved, there are many difficulties that one would encounter in developing the S&T statistical indicators in developing countries like Sri Lanka. These include the following issues:

1. No mechanism for regular updating of S&T data in the institutes
2. Many stakeholders in the S&T sector are not interested in R&D surveys and have their own reservations
3. Only a few S&T sector institutions collect data regularly for their particular interests but there is no network/link even among such institutes.
4. Many S&T sector institutes do not have regular data collection system
5. Some OECD definitions are not suitable for developing countries and hence need some amendments *i.e.* S&T fields–Social Sciences, and Law.
6. Individual funds received are not recorded in the financial reports of the Institutes
7. Funds awarded for collaborative research could be recorded twice or more
8. Standard questionnaire may not be perfectly applicable for the business enterprise sector
9. Some institutions have definitions such as researcher/engineer, but they do not have suitable educational qualifications
10. Many institutions do not keep their previous records or do not refer to the same

11. Some institutes do not like to change their old questionnaire format due to the difficulty in comparison with their own previous records
12. Basic and applied research–funds received for basic research are also used for infrastructure development and training. Hence, there is confusion as to whether the whole amount is counted under basic research

In conclusion, it may be emphasized that the development of up to date indicators for science and technology by the developing countries is of importance to guide the policy decisions taken by the changing governments. It is also necessary to make the local community aware of the importance of measuring science, in particular, the stakeholders to improve their response to data collection. Regular data collection from the S&T sector and maintenance of databases to cover such information in a comprehensive manner will help the development of S&T indicators based on the correct facts. This will encourage the governments of the developing countries for fair funding for S&T research towards the socio-economic development and meet the demands that arise towards globalization.

Acknowledgements

The cooperation extended by the Scientific Officers Dr Dilrukshi Ranathunge and Mr Ruwan Nagahawatte of the Science and Technology Policy Research Division of the National Science Foundation, Sri Lanka in preparation of this paper is thankfully acknowledged.

References

Fenando, C.M. and Amaradasa, W., 1996. National Survey of Research and Development in Sri Lanka 1996, Natural Resources, Energy and Science Authority, 47/5, Maitland Place, Colombo 07, Sri Lanka

Freeman, C., 1968. Science and economy at the national level. *Problems of Science Policy*, Organization for Economic Cooperation and Development, Paris, p. 58.

Holbrook, J.A.D., 1992. Why measure science? Editorial introduction to a special issue of Science and Public Policy on S&T Indicators, Science and Public Policy, 19 (5), October 1992.

Mansfield, E., 1991. Academic Research and Industrial Innovation. Research Policy, 20, pp1-12.

NSF, 2006. Sri Lanka Science and Technology Statistical Handbook 2004, National Science Foundation, 47/5, Maitland Place, Colombo 07, Sri Lanka

NSF, 2003. National Survey of Research and Development in Sri Lanka 2000, *Unpublished data*, National Science Foundation, 47/5, Maitland Place, Colombo07, Sri Lanka

Solla Price, D. J. de., 1969. Measuring the size of science, Proceedings of the Israel Academy of Science and Humanities, Vol.4.pp.98-111.

UGC, 2005. Sri Lanka University Statistics, University Grants Commission, 20, Ward Place, Colombo 07, Sri Lanka.

Chapter 5

Science and technology (S&T) development policy in Nigeria

Ettu Obassi

Abstract

The S&T sector in sub-Saharan Africa has not been able to create a dynamic and sustainable process of scientific production and reproduction of national scientific communities, over the past 30 years. Nigeria has made significant initiatives in this context, under the Federal Ministry of Science and Technology (FMST). The paper describes some of the significant features of the S&T policies in Nigeria and delves upon the S&T statistic. In particular, it gives details of the main scientific orientations of S&T, S&T and education and the major Nigerian initiatives. Science and technology are, today, more important than ever for Nigeria, if it is to raise the standards of living of its people, consolidate a modern economy and participate as a significant partner in the global arena.

Keywords: Nigeria, S&T policies, S&T statistics.

Introduction

Over the past 30 years, the S&T sector in sub-Saharan Africa has not been able to create a dynamic and sustainable process of scientific production and reproduction of national scientific communities. Informed public constituencies need to be formed to generate transparent and convincing S&T policies (Gaillard, 2002). Science institutions need to be better prepared to engage effectively in the economic sphere. The countries of Africa or the continent, in general, would otherwise, unlikely be able to respond to the challenges posed by new and advanced S&T areas such as biotechnology, or new and advanced materials as well as due to the current low status of agricultural biotechnology, and the corresponding low level of public funding on research in sub-Saharan Africa. R&D spending is low but it is increasing. Institutional arrangements are being made in some countries including Nigeria to stimulate research in biotechnology and other areas. Institutional arrangements for the development of regulatory instruments in some other countries are still to be created.

It might, however be argued that there were very hopeful signs since funding for various researches appeared to be increasing in Africa. If public constituencies can be created, S&T in agriculture, medicine, education, and other sectors may develop rapidly as it has, elsewhere, in developed countries and developing countries of Asia. Nigeria is making her own efforts towards this as will be indicated, in due course of our discussions on agencies and parastatal under the Federal Ministry of Science and Technology (FMST).

Background

Nigeria Republic in the western Africa is situated along the coast of the Atlantic Ocean on the Gulf of Guinea. Most of the land area of Nigeria consists of a low plateau cut by rivers, especially the Niger and its largest tributary, the Benue. The country takes its name from its chief river. Until 1991, Lagos, the capital, was the largest city, on the southwestern coast; afterwards, the city of Abuja, in the country's interior, became the capital. The government moved from Lagos to Abuja in 1991 in the hope of creating a national capital where none of the country's ethnic groups would be dominant. The country is bordered by Cameroon in East, Republic of Benin in the West, Niger Republic and the Chad in the North, and washed by the Atlantic Ocean in the South. Nigeria, for a long time, had an agricultural economy but now depends almost entirely on the production of petroleum, which lies in large reserves below the Niger Delta.

Nigeria is by far the most populated among the African countries, with more than one-seventh of the continent's population. The people belong to many different ethnic groups. These groups give the country a rich culture but they also pose major challenges to nation building. Ethnic strife has plagued Nigeria since it gained independence in 1960.

The Federal Ministry of Science and Technology, (FMST), Nigeria

Establishment of FMST

The Federal Ministry of Science and Technology, (FMST) was established as a full-fledged ministry following the apparent realization of the importance of S&T to the national development, with the emergence of the democratic government of Alhaji Shehu Shagari, through Act No. 1 of January, 1980 replacing the National Science and Technology Development Agency (NSTDA) established by Decree No. 5 of 1977.

The mandates of FMST

The mandates are designed to cultivate durable S&T culture that would address current problems and future challenges covering the areas of formulation, monitoring and reviewing of the national policy on S&T, promotion and coordination of scientific and technological research, promotion and administration of technology acquisition, adaptation and transfer, promotion and coordination of national production of scientific and technological goods and services, promotion and coordination of development of scientific and technological infrastructures, promotion and coordination of integrated, (*i.e.* human infrastructural and institutional) scientific and technological capacity building, and establishment and maintenance of relations with national and international, public and private scientific and technological organizations.

Table 5.1: General information

1.	Official name	Federal Republic of Nigeria
2.	Capital	ABUJA–FEDERAL CAPITAL TERRITORY [FCT]
3.	Independence	The country gained her independence from Britain in 1960
4.	Government	Federal Presidential System with 3 tiers of government; the Federal, the States (36) and the Local Government Areas (about 772). The Country operates bicameral legislation (the Senate- upper house and the House of Representative–lower house)
5.	Land Mass	About 923,768 sq km (356,669 sq miles)
6.	Population	In 2005 Nigeria's estimated population was 128,771,988, yielding an average density of 141 persons per sq km (366 per sq miles)
7.	Official and other languages	English is common among educated people. About 400 native Nigerian languages have been identified. The most common of the native languages are Hausa, Yoruba, and Igbo. Other major languages include Fulfulde, Kanuri, Ibibio, Tiv, Efik, Edo, Ijo, and Nupe.
8.	Ethnicity/tribes	Nigeria's three largest ethnic groups–the Hausa-Fulani, Yoruba, and Igbo–represent about 70 percent of the population. About 10 percent of the total population consists of several other groups numbering more than 1 million members each, including the Kanuri, Tiv, and Ibibio. More than 300 smaller ethnic groups account for the remaining 20 percent of the population.
9.	Religion	Christianity, Islam, or indigenous African religions is central to how Nigerians identify themselves. Recent estimates suggest that 46 percent are Christians, 44 percent are Muslims, and 10 per cent adhere to traditional religions
10.	Climate/ vegetation	Climate is tropical with sharp regional variances depending on rainfall. The seasons of the year are hot, dry, and often dust-laden air from the Sahara known locally as the harmattan. Temperatures average from 25° to 28°C (77° to 82°F), and in the Jos Plateau, temperatures average 22°C (72°F). Rainfall varies widely over short distances and from year to year. Parts of the coast along the Niger Delta, where the rainy season is year-round, receive more than 4,000 mm (160 in) of rain each year. Most of the country's middle belt, rainy season starts in April or May and runs through September or October, receives from 1,000 to 1,500 mm (40 to 60 in).
11.	Life expectancy	51.5 (2005 estimate) 99 deaths per 1,000 live births (2005 estimate)
12.	Literacy	70.7 per cent (2005 estimate)

Vision and mission

In pursuance of these mandates and in order to articulate its service charter, FMST has formally defined, and government has accepted the following vision and mission statements of the ministry.

Vision of FMST

The focal vision of the FMST is to make Nigeria, in the nearest future, an acknowledged member of the fast developing scientific and technologically progressive nations of the world and to be Africa's leader in scientific and technological development.

Mission of FMST

The mission of the FMST is to chart a course of scientific and technological development of the nation through planned and guided R&D in all areas of S&T, keeping abreast with latest progress in R&D in science and technology among the industrial and educated communities within the country, sub-region and overseas, assuring that various R&D efforts of the public S&T establishments fit in with overall development plan of the nation, linkages of researchers and their establishments with entrepreneurs for commercialization of successful research results, utilizing the facilities of Technology Incubation (TI) and Science and Technology Parks (STP) to nurture new projects especially in the areas of national priority, promotion of projects that are strategic to the cause of aggregate national development, and advising government on S&T Policy matters.

In order to put each of these operational activities in its correct perspective, the FMST produced four working documents as standard guidelines and operational tools. These were National Policy on S&T, Strategic Plan (2003-2007), a compendium of on-going projects in research establishments under and out of the FMST, and profiles of selected commercializable R&D results.

FMST activities and efforts towards promoting and supporting S&T development

1. Launching of an earth observation Microsatellite Code named Nigeria Sat 1 on 27th Sept, 2003, the first national low orbit remote sensing satellite to be built by a sub-Saharan African country. The launching of Sat 1, at Peseta, Russia, represents the humble beginning of our goal, and the highlights of which include:
   - (*a*) Capacity acquisition for a number of scientists and engineers involved in the designing and building up of all subsystems of Nigerian Sat 1,
   - (*b*) Establishment of a pool of skilled space scientists for Nigerian space capacity for building subsequent generations of satellites by ourselves, *e.g.* Nigeria Sat 2–a communication satellite–the work of which has reached advanced stage for launch in early part of 2007,
   - (*c*) Obtaining membership of an international Disaster Monitoring Constellation (DMC) comprising UK, Nigeria, China, Algeria, and Vietnam. Membership enables Nigeria to receive daily images from DMC dynamic remote sensing activities, and

(*d*) availability of real time data from daily revisit of scenes through DMC partnership to map and monitor environmental problems around the world. Data obtained can be applied to agricultural and environmental studies, National Geographic Information System (NGIS), monitoring of oil and gas pipelines, or mineralization mapping settlements for national census.

2. Launching of mobile Internet unit (MIU)–The unit is locally assembled in 25-seater bus that has been converted into a mobile IT training and cyber centre. It is equipped with Internet access VSAT, printers, photocopier and media facilities. The assemblage was built at a cost 25 per cent of its imported version and over 9,500 youths and adults have benefited from the services of the units. It is an essential tool for computer training/education, community info-base centre, or development of healthcare delivery in the remote areas of our 3rd tier of government (local govt areas LGAs).
3. The re-delegation of the 'ng' top domain for Nigeria to the Director General/ CEO of National Information Technology Development Agency (NITDA) from Mrs. Odusote who invented and registered the 'ngtld' for Nigeria in 1995.

   The centralization of software purchase and application in Nigeria through Software licensing agreement with Microsoft Corporation. The package involved personnel training and supply of PCs.

   Launching of e-government and public service network–Nigeria's ICT policy has been widely acknowledged as laudable by other countries in the sub–region. The policy was approved in 2001, following which NITDA was established. Federal ministries and agencies have been provided standard domain names and websites and establishment of LAN/WAN between them and NITDA.
4. Commercialization of R&D results in cassava-wheat flour composite with 10 per cent cassava flour developed by FIIRO, ceramic porcelain insulators by PRODA for which domestic manufacturers already exist to supply to National Power Outfit PHC, neem processing plant by NARICT with great industrial potentials for pharmaceutical, and bio-insecticides
5. Establishment of National R&D Coordination Council (NRDCC) to be chaired by Mr. President.
6. Reform of National S&T and Innovation System, and related activities.

Status of S&T policy development in Nigeria

Major historical development of S&T in Nigeria

Nigeria has an inherited colonial science structure focused on the exploitation of tropical products and mineral deposits together with sustainable management of agricultural production. Historical record predates the establishment of the first set of research institutes in Nigeria during the colonial rules. The essence then was to boost the output to adequately supply the much-needed raw materials for industries in Europe. Most of the research institutes established at that time became full-fledged national institutes in Nigeria after independence.

Following the international conference on the organisation of research and training in Africa in 1964, Nigeria established the National Council for Scientific and Industrial Research (NCSIR) by decree No. 83 of 1966. The Council was expected to act as an advisory body to the government and as such, lacked executive powers. Its mandate was quite narrow and, therefore, had structural weakness that prevented it from functioning efficiently. The real consciousness of the development of science and technology in Nigeria came into force after the Civil war in 1970.

Federal organs for science and technology (S&T) administration in Nigeria were set up at the end of the civil war (1966-1970) by decree No. 6 of 1970 establishing the Nigerian Council for S&T (NCST), based on the report by UNESCO expert, Dr. N. R. Martin entitled, "NIGERIA: The Science Policy Machinery, July- Sept 1966, Paris, UNESCO", which had four sectoral research councils, *viz.* Agricultural Research Council of Nigeria, (ARCN), Medical Research Council of Nigeria, (MRCN), Natural Sciences Research Council of Nigeria, (NSRCN), and Industrial Research Council of Nigeria, (IRCN).

In 1974, UNESCO organised the first regional ministerial conference on S&T development in Africa called CASTAFRICA I which was held in Dakar, Senegal. The conference identified and recommended among other issues, the need for African countries to formulate their S&T policy to guide national development, especially rural and industrial development.

In 1976, the government of Nigeria, in pursuance of its commitment to the resolution reached at the first regional conference, set up an S&T policy committee which recommended two policy options, namely, the establishment of an executive agency to coordinate S&T policy, and the establishment of Ministry of Science and Technology with all Research Institutes under it.

Selection of the first option led to the establishment of National Science and Technology Agency (NSTDA) by decree No. 5 of 1977 with all research institutes under it. The NSTDA had no Minister and was chaired by Chief of staff Supreme Headquarters, Major General Shehu Musa Yar'Adua.

The search for an ideal S&T structure did not end with NSTDA. In 1980, a full-fledged FMST was created by the Civilian Adminstration of Alhaji Shehu Shagari, who also dissolved the NSTDA. An Economist, Dr Sylvester Ugoh became the first Honourable Minister who was later replaced by Mr. Ademola Thomas, and Dr. M. Bode Olowoporoku during the cabinet reshuffle in 1982 and 1983, respectively.

With the return of the military regime in 1984, the FMST was phased out altogether as an independent ministry, and was placed under the Federal Ministry of Education. It was not until 1986 when it was restored with a pure scientist, a distinguished chemist as Minister–Prof. E. U. Emevon. Under him, the first National Policy on S&T was developed and published and its implementation was the subject of the 1988 Annual Conference of National Committee of Deans of Science and Technology in Nigerian Universities. However, the main financial provision of the policy for national

S&T Fund (NSTF), which was a major aspect of private sector input into S&T development in the country, was never implemented before Prof. G.O. Ezekwe, an engineer became the minister in another cabinet reshuffle in December 1989. By 1992, he introduced an engineering infrastructure (National Agency for Science and Engineering Infrastructures, NASENI), which was identified as non-existent in the FMST policy framework. He also launched the NSTF. The decree establishing the Raw Materials Research and Development Council (RMRDC) was amended in 1992 and provided with a risk fund.

The FMST was re-established in August 1993 after the stepping aside of President Ibrahin Babangida. Prof Bart Nnaji, a famous robotic engineer and technocrat had a short six-month stay under Shonikàn Interim Government. By November he was replaced by Dr Laz Unaogu who set up a committee with code named RAPID Project (Research Adaptation and Production for Indusrial Development) in 1994 to organize the statutory and mandates of the re-established FMST, which was later ministered by General S. Mommah who advocated effective conversion of R&D results into useful goods and services.

After General Momah became Chief Ebitimi Banigo from whom the Current Minster of S&T Prof. Turner T Isuon (popularly known as the Satellite Minster) took over on May 29, 1999. Unlike most other ministers of Science and Technology, Prof T Isuon is a record-breaking minister in terms of stay and achievements. His efforts have seriously elevated the Nigerian science and technology during past seven years. The progress has been evident and is sustainable through the programmes and activities, initiated by him.

A revised national policy on S&T was prepared in 1999 for a 25-year time frame with revision at 5-year intervals. It emphasized Nigeria's commitment to the creation of an independent, integrated and self-sustaining economy, with the national policy itself being the framework for the effort towards the fulfilment of that commitment. Key guidelines included (*i*) S&T should provide the basis for national development, (*ii*) policies should take account of the country's economic situation and cultural milieu, (*iii*) manpower should be developed to provide capability for utilization of natural resources, and (*iv*) socio-political system should reflect the prime position of S&T. Sixteen sectors of the economy were identified as strongly reliant on S&T. The policy extends to the structural issues such as capacity building, popularization, and international collaboration. Specific policies have been drafted for biotechnology, space research, and IT, while others are in preparation.

The 1986 S&T policy

The guidelines of the 1986 S&T policy recognized the fact that scientists and technologists in Nigeria had over the years engaged in their work without well defined national direction (FMST, 1986). Researches were carried out without much effort made towards commercialization, apart from publishing them in journals. Also, entrepreneurs were averse to risk and as such, were not willing to commercialize viable research findings. The policy framework was expected to provide direction and coordination for scientific and technological activities as well as development of manpower, creation of science culture in the society, intensification of basic and applied research, the application of research findings to developmental activities in agriculture, healthcare, industry, defence and security, and to provide employment opportunities for the citizens.

The national objectives in 1986 as captured in the fourth development plan included among others, the achievement of self-reliance through acquiring enough technical know-how, skills and materials to defend the integrity and security of the nation, being technically self-reliant in the production of capital and consumer goods and raw materials, and maintaining a flourishing national export capability of goods and services.

The document addressed the following key sectors of national life

1. Educational system–emphasis on science at all level
2. Mass Movement–stipulates mass movement for S&T
3. Motivation for individuals engaged in S&T projects to include special incentive and/or remunerations
4. Capital Goods–the policy encouraged the production of machinery and spare parts based on local inputs
5. Materials–the policy state the exploitation, processing and utilization of the national material resources to enhance self reliance and export
6. Energy to be developed, protected and optimally harnessed in the overall interest of the nation
7. Technology–advantage to be taken of technology in all aspects of day- to-day activities
8. Other sector included military science, environment, international exchange and cooperation in S&T, national system of S&T (then NCST), financing S&T, research, development, application and marketing.

Within the life span of this policy, achievements were recorded in many aspects of agriculture. However, the policy lacked effective implementation mechanism as well as evaluation and monitoring strategies, in general.

The 2003 S&T policy

The FMST in realizing the changing global policy on S&T and the emerging frontiers of knowledge, embarked on the review of the 1986 S&T policy. The new policy was articulated to:

1. Stimulate initiative and creativity of the S&T community while establishing strong link between their works and requirements of the economy, educational system and society at large.
2. Encourage the mobility of scientists and researchers as a network and explore the best opportunities and alternatives in the country and abroad for use and improvement of their competence.
3. Increase the share of the productive sector in the national effort for scientific and technological development especially in the private productive sector, and

4. Building strong indigenous competence in order to participate as equal in international negotiations in the areas such as protection of intellectual property and rights of access to information, biotechnology and biosafety, environmental control and the establishment of technical standard in international communication networks.

The new policy expanded the frontiers of S&T as contained in the 1986 policy document by identifying three core areas where capabilities should be built. These included information and communication technology (ICT), biotechnology, space technology, and other areas such as linkage between university/research institute and industry, capacity building and technology transfer by multinational companies, energy research, cooperation on implementation and funding of R&D, empowering SMEs through appropriate technology, traditional medicine development, S&T data bank, engineering materials development etc.

Implementation and experiences

Apart from the checkered history of evolution of S&T in Nigeria, which is also applicable to the implementation, a major constraint on the implementation of science-based development has been the funding and lack of continuity or consistency. Nigeria spends far less than the one per cent of GDP advocated by African Union (AU) on R&D. Also, every new government in the country with the ever changing Ministers of S&T have tended to start something new, without recourse to some of the lofty projects of their predecessors.

In the recent past, for any project to be funded by the military or political class in government, it must meet the specific criteria before it can pass through the screening by the Council of Ministers. These include, namely, is the project in national rolling plan? who is to execute it? will it pay for itself? can it wait? what is the cost/benefit profile? who will benefit from it?

It is difficult for R&D projects in science and technology to meet these political and economic criteria without being integrated into the full plan perspective cycle. This lack of integration apparently explains why government funding of R&D projects in S&T has been grossly inadequate to date in the country, though one needs to commend the steps taken already by the present regime of the President Obasanjo to move the S&T out of the seeming vicious cycle.

In a science-based development, the actual operation of S&T transformation requires an operational flow chart that satisfies a different set of criteria. For example, what project must be done, when, where, how, by whom, and with what available resource inventory.

These criteria emphasize scientific and technological capacity building for national self-reliance as well as the relevance and technical feasibility of a project. This underscores the importance of political stability, economic strength, and sustainability for any long–range development plan of the type, which newly industrialized countries have applied for sustainability and to achieve the desired goal of self-reliance. It is also imperative to find a common ground that meets the two sets of criteria outlined above. It is frequently said that policy should be based on sound science. This is more easily said than done. In a system of representational democracy, with strong separation of powers, politics is inextricably connected with policy. This is true for S&T as for any other area of human activity. We have in our society and those of many other nations, a complicated "three-body" problem. The three bodies are science and technology, policy, and politics with forces between them, which we cannot describe in any simple way (Figure 5.1).

Funding S&T

A comparison of S&T funding rates in Nigeria and other countries since the creation of FMST is summarized in the Table 5.2. (Taken from G.O. Ezekwe's Memorandum to the National Constitutional Conference, entitled: A Constitutional Provision for Science, Engineering and Technology Development–February 1994). Nigeria has considerable human potential. It counts 60 universities, 44 polytechnics, and 65 research institutes where researches are carried out. Currently, in the country, funding in Research, Technology Development and Innovation (RTDI) landscape is, principally, from the Federal Government, with institutes normally funded through their supervisory ministry. An exception is the RMRDC, which is funded through 1 per cent import surcharge on capital goods, spare parts and raw materials. Universities are supported through the National Universities Commission. Commercial production arising from S&T is financed by the Nigerian Bank of Industry, venture capital and the Nigerian Bank for Agriculture. Some of the money comes from multinationals. Support from international agencies has come from the IDRC, Canada, DFID, UK and USAID. At present, UNESCO is collaborating in a project aimed at reform of the S&T system.

Research performance is by about 65 research institutes, about 60 universities (including 25 Federal, 24 State and about 10 private universities), and 44 poly techniques recognized as centres for R&D activities.

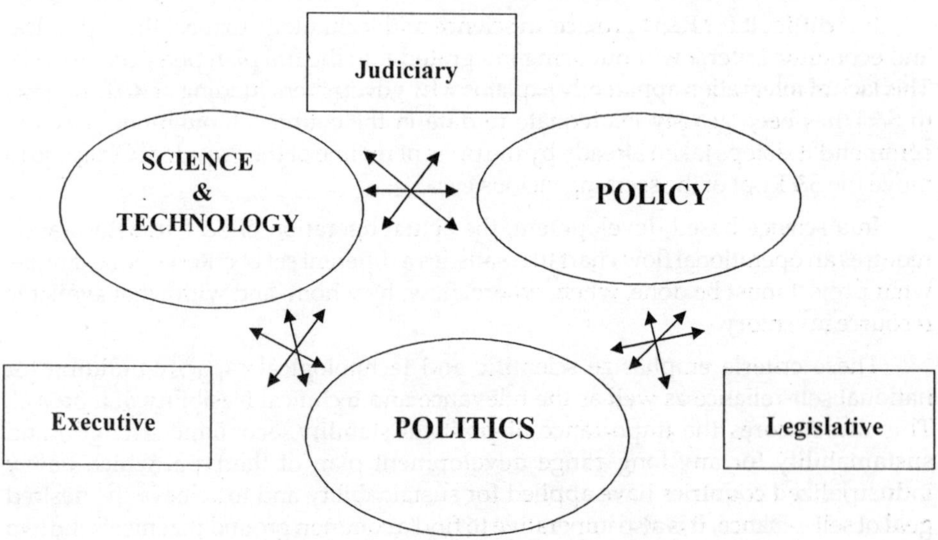

Figure 5.1: Three-body problem of S&T

Some statistics of S&T in Nigeria

Tables 5.2–5.5.

Table 5.2: S&T funding rates in some countries of the World

Country	*Funding rate % GDP/GNP*
Nigeria 1980- 1992	0.1 rose to 1 per cent in 2004
O.A.U (now A.U.) guideline	1.0
Malaysia	0.6–1.0
Tanzania	1.5–3.5
Taiwan	1.35–2.25
India	2.0
South Korea	2.0
U.K	2.3
Germany	2.8
U.S.A	2.8
Japan	2.8
Russia (former USSR)	5.0

Source: Prof A. Animalu's Education, S&T Agenda for the 21st Century.

Table 5.3: S&T orientation in Nigeria

Sl.No.	*Main scientific orientation of S&T in Nigeria*	
1.	Biological sciences	3.7
2	Education and Information sciences	4.3
3.	Environmental sciences	4.7
4.	Earth sciences	15.7
5.	Agricultural sciences (showing greater prospect)	17.4
8.	Exact science and technology	22.0
7.	Medical sciences	32.1

Source: Prof A. Animalu's Education, S&T Agenda for the 21st Century.

Table 5.4: S&T and education

Researchers per 1,000 inhabitants	0.8
Expenditure on R&D as a per cent of GDP	1 per cent in 2004
Pupil/Teacher ratio (primary)	36
Pupil/Teacher ratio (Secondary)	–
Expenditure on education as per cent of GDP	
Student enrolment in major S&T discipline (1998)	155,542

Source: UNESCO Institute of Statistics.

Table 5.5: Culture and communication

Population	127.117 Million	AfDB/OECD 2005 African Economic outlook
Gross national Income per Capita	$471	AfDB/OECD 2005 African Economic outlook, 2004/2005
Literacy rate	69.4 per cent	African ICT Policy Monitor
TVs per 1000 people	68	"
Radios per 1000 people	200	"
Telephone mainlines per 1000 pers.	5	"
Mobile phones per 1000 people	4	"
Personal Computers per 1000 pers.	6.8	"
Internet users (thousands)	115	"
Book production (total)	1314	Country profile, UNESCO Institute of statistics
Daily newspapers: Total average circulation or copies per 1000 inhabitants	24.6	Country profile, UNESCO Institute of statistics

Source: Various references.

Problems and challenges

Apart from low funding of R&D, other major problems and challenges include:

1. Poor state of S&T infrastructures in the country especially the level of deterioration during the period before the President Obasanjo's administration.
2. Problem of linkage between university, research institute and industry leading to poor commercialization of R&D results, while many results lie unutilized in the universities and research institutes,
3. Brain-drain due to poor remuneration/incentives to personnel involved in R&D,
4. Inadequate statistical data for planning,
5. Poor management and inadequate macro-level coordination,
6. Problem of conducive environment (political stability) for private sector participation,
7. Lack of patronage for made-in-Nigeria goods even when they exhibit high quality, leading to irrational importation, and
8. Poverty reduction and sustainable development dominates economic strategy, but there is insufficient understanding of the political leverage and the effect of RTDI on poverty alleviation and eradication and sustainable development.

Strength

The strength of the nation includes the fact that its current growth in the economy is about 10 per cent, Oil and Gas wealth provide about 95 per cent foreign earnings, abundant human resources in R&D and the current efforts of government to tap from Nigerians in Diaspora, and emergence of scientific and technological collaborations with new partners like China, India, Brazil, Japan, Malaysia, and, currently, the NAM S&T Centre, New Delhi.

Weakness

The weakness of the nation includes the pervasive corruption, and inability to use the oil windfall productively to improve social conditions. The strategic role of R&D may have been recognized formally by government but it is not given priority in terms of tangible policy measures and support. The inability to attract the much needed efforts of the private sector in R&D, is another weakness.

Government's new economic initiative and S&T reforms

Globalization

In this day and age, the gap between the technologically wealthy countries and the technologically poor is becoming a more difficult rift to breach than the economic gap. In fact, it has reached to a point where the key to bridging the economic divide is the harnessing of science and modern technology for the economic and social development.

Based on this understanding, the present Government of President Obasanjo has placed great emphasis on the importance of the FMST, and charged it with the responsibility of coordinating various efforts of using S&T for development in a wide array of sectors such as education, medicine, agriculture, biotechnology, and communication. Further to this and realizing the imperfections in our system and the need to build a prosperous and strong nation, the President remarked in his inaugural speech of February, 2005 at the National Political Reform Conference, that "Pluralism is the order of the day and globalization, technology, modernization politics of accommodation dictates that we must reject those ideas and argument that seek to divide us and establish the interest of ethnic entrepreneurship as the sole interest of the people".

Worldwide indicators are devised for signalling or monitoring, for accountability, for legitimization and for awareness on S&T. Some of these indicators currently available are indicated below:

Research and Development (R&D)–measured according to allocated expenditures and human resource input

Human Resource in S&T–measured in terms of science education, and training according to levels and fields

Innovation–indicator of S&T activities in a country in terms of new findings

Publications–measure the technological status in a given socio-economic setting

Patents–measure the impact of S&T activities in a given socio-economic setting

The above indicators are far from being adequate. R&D funding and human resources are receiving growing criticism on the grounds of inadequate proxies for innovative activities at firm level, much of which, is from tacit knowledge. Such arguments have led to S&T indicators review throughout the world. Indicators also have the problem of availability and validity.

Africa in the past 20 years has provided UNESCO with very few and irregular data on R&D. About 80 per cent of the countries have not provided data at all. Less than 4 per cent, which provided data, are regular in a way and are mostly from the Arab States of North Africa (Tchatchoua, 2002 in DPMN Bulletin Vol. X, No.4 of Nov.2003).

African initiative–NEPAD

NEPAD is an acronym for New Partnership for Africa's Development. It is a collective resolve by African leaders to pursue socio-economic development that would be sustainable and bring about peace and stability needed in some countries of Africa. It targets to leap-frog development to the next level in Africa. It is all embracive with S&T contents. For example, NEPAD's research agenda has rekindled the government's interest in biotechnology, ICT, new and advanced materials which are also embedded in the Nigerian S&T Policy. The National Biotechnology Development Agency (NABDA), in Abuja, has created six centres across the six geopolitical zones with research activities domiciled in universities in each of the zones.

NEPAD recognizes the food crises of some African countries and, therefore, also supports biotechnology through the NGO Cooperation and Comprehensive African Agricultural Development Programme (CAADP).

Africa should now devise long-term strategies to build capacity of the Continent's scientific infrastructure, and improve the conditions for scientists by offering them adequate salaries and reasonable career opportunities. The R&D fund situation must need to be changed for the better.

Nigerian initiative–NEEDS

NEEDS is the short for Nigerian home-grown economic development programme tagged "National Economic Empowerment Development Strategy", which is converted to SEEDS and LEEDS in the State and the Local Government, respectively. The NEEDS document is a vision for tomorrow's Nigeria, which states where we want to be, how the vision can be realized, what prevents progress, how NEED will change things and how it will be implemented. NEEDS was launched on 29th May 2004 and builds on progress made between 1999 and 2003. It has made remarkable progress and aims at making the private sector the engine of growth through privatization, deregulation, and liberalization, accompanied by infrastructure development in research institutes, electricity, transport etc. We now have leaders who understand that S&T impacts virtually in all aspects of public policy (*e.g.* health, natural resources and environment, energy, national and internal security, communications, transportation, and other areas of national importance). It is also now understood that demand driven R&D is the process through which nations advance their scientific understanding and develop new technologies. Thus, NEEDS also sees R&D as a matter of S&T policy. The Federal government funds the R&D in the country through Ministries, Research Institutes, Universities and Polytechnic and other agencies of government. The private sector is being encouraged to invest in commercialsable R&D. NEEDS targets 10 per cent growth per annum.

S&T reforms in Nigeria

Prior to the political and economic crises in1985, Nigeria was the most successful country in terms of scientific production in the continent. The crisis had dramatic consequences for S&T establishments, and by 2000, Nigeria had lost its leading position, even though human resources were still available. Future prospects were dim–there were no plans to activate the Nigerian Universities, and no scientific policy could emerge and be committedly pursued to spur technological development. The only means to halt the resulting the brain-drain would be to increase academic salaries.

With the beginning of the Obasanjo administration, it has changed the tide for the better and has put Nigeria on course to reclaim its rightful position in African S&T. The government has brought back the lost hopes for Nigeria to begin to fulfil its enormous economic potentials. Agricultural production has continued to expand, growing at estimated 7 per cent in 2004 compared to 6.1 in 2003. The real GDP growth picked up strongly to about 10.7 per cent in 2003. The mining and quarry sector, long neglected, is also showing tremendous growth. The foreign debt profile of $32.9 billion, at the end of 2003, has reverted to substantial foreign reserve following the debt relief by the Paris Club in support of Nigeria's economic reforms.

At the request of the Federal Government of Nigeria, an international advisory board for the reforms of the country's science, technology and innovation system was established by UNESCO in October 2004. A core activity of the reform programme is a joint review of investment, industry and innovation in Nigeria involving UNESCO, UNCTAD, UNIDO and WIPO. Finance in equal shares by the government of Nigeria and UNESCO and Japan Funds-in–Trust to the tune of US$1million was given to review in part of preparatory work for the donors' conference. Nigeria is planning to call to the fund implementation of a multi–year plan of action on science, technology and innovation. Other international agencies expected to join the reform programme are the United Nations Economic Commission for Africa, the World Bank and the International Association of Universities.

The President also recently announced the establishment of a National Science Foundation using the "windfall" generated by current high oil prices to diversify Nigeria's economic investment in science and technology. He pledged the sum of US$5 billion endowment for the project. The special fund in trust for science will not only benefit Nigeria but also assist other African countries in designing project proposals for the reform of their national science systems and development of managerial capacities.

In June 2006, the Director General of UNESCO launched the Nigerian parliamentarian S&T forum in Abuja. Amongst other, a regional policy forum was recommended for establishment to facilitate an exchange of experiences and know-how in science, technology and innovation policy making, science legislation and ethics, to strengthen partnerships between legislators, scientists, the media, civil society, and public and private sectors in developing national innovation system and to support capacity- building by science parliamentary committees in emerging democracies.

The country is equally investing in the development of Science and Technology Parks. The country through the FMST has also reviewed the National S&T Policy, which is being implementing, currently. The need for the reforms was evident after four decades of military rule marked by state corruption and spiralling foreign debt, following independence in 1960. The reward of the reform could also be immense, for Nigeria which is potentially a wealthy country. Being the world's 13th largest oil producer and the 6th largest in OPEC, Nigeria has enormous gas reserves, which when fully exploited will place her among the world top ten gas producers. The reform comes at an auspicious time. After sluggish growth initially following the end of military rule, GDP rose by nearly 10 per cent in 2003, driven by strong oil receipts and agricultural growth of 7 per cent. Public spending has climbed markedly, from 19 per cent of GDP in 1997 to 50 per cent in 2001 (DFID). One aim of the science reform system will be to use this growth to diversify Nigeria's economy, in order to reduce dependence on fluctuating oil prices, accounting for about 95 per cent of foreign earning compared to 58 per cent in 1970 (UNDAF).

Collaborations

International co-operation is necessary to build our country's human capacity and knowledge base in S&T. The free flow of ideas and information, exchange of high level researchers, and strategic intellectual alliances between individuals and institutions are also necessary to help us find solutions to the developmental challenges both nationally and globally. With this in mind, Nigeria is widely collaborating with national, private sector, international organisations, or NGOs to put her S&T in the desired shape. In the past, bilateral donor agencies such as DFID (UK), DGIS (the Netherlands), USAID, Food and Agricultural Organisation (FAO), WB, Canada, UNESCO, UNIDO, and others have supported substantial S&T programmes in Africa from which Nigeria has also benefited. Today, the list of international collaborations is growing with China, the NAM S&T Centre, New Delhi, and other countries.

Conclusion

Science and Technology are, today, more important than ever for Nigeria, if it is to raise the standards of living of its people, consolidate a modern economy and participate as a significant partner in the global arena. This implies that the economy must be modernized and competitive.

Chapter 6

S&T indicators in India and policy implications

Parveen Arora

Abstract

The present paper highlights the role of S&T system and policies leading to the formulation of S&T Statistical System in India. The generation of various S&T indicators, linkages with other information/statistical systems as well as issues and problems associated with the S&T Statistical System are discussed. The paper also focuses on the policy implications of the existing S&T indicators and the need for the development of new S&T indicators in the context of globalization.

Keywords: India, Indicators, Management information system, Policy, R&D statistics, S&T statistics, Statistical system.

Introduction

The formulation of Science and Technology (S&T) policy, and the development of S&T plans and programmes require up to date, reliable and comprehensive data on a country's scientific and technological potential. S&T indicators serve this purpose, there are a series of data designed to answer questions about the existing state of and/or changes in the science and technology enterprise, its internal structure, its relationships with the outside world, and the degree to which it is meeting the goals set for it by those within or without (Fabian 1979,1984).

Over the past three decades, there has been a growing need in India for an information system and database on science and technology statistics (popularly known as science statistics). UNESCO defines science statistics as field of statistics dealing with the quantitative measurement of the volume and structure of scientific and technological activities of the country. This field includes data pertaining to the human, financial and material resources devoted to scientific and technological activities.

Policy makers particularly those concerned about planning, implementation and management of science felt the need for a comprehensive information not only on the use of input resources which comprise mainly the financial and human resources deployed and infrastructure available to science and technology but also the output of such activities measured in terms of increased productivity, increased economic growth, new products/processes developed and their large scale diffusion and impact on society. Such information is considered useful for undertaking cost benefit analysis and other economic studies as well as for efficient programming, planning and budgeting. It will also help in comparing the national S&T efforts with the other developing/developed countries.

With the growing need of this information, a number of countries including India have started data collection in the field of science and technology. Over a period of time, statistics on S&T has covered not only the financial and human resources deployed but also data on education of S&T personnel, patents, papers, scientific services, foreign collaboration and so on. In the present paper Section 2 deals with the S&T policies, high-level committees and the S&T system while the S&T statistical system in India is discussed in detail in Section 3. S&T indicators as policy tools and the need for development of new indicators are discussed in Section 4.

Section 2: Evolution of science and technology system

Science and technology policies

With the attainment of independence, science became a national pursuit and the government committed itself to fostering science on massive scale by enunciating appropriate policies from time to time. The country adopted the Scientific Policy Resolution (SPR) in 1958. The aims of the scientific policy are to foster, promote and sustain the scientific research to ensure adequate supply of scientists of highest quality and to recognize/reward their work and to secure for the people of the country all benefits from the acquisition and application of scientific knowledge. The Government of India has decided to pursue and accomplish these aims by offering good conditions of service to scientists and according them an honoured position, by associating scientists with the formulation of policies and by taking such other measures as may be deemed necessary from time to time to accomplish the aims of the Scientific Policy Resolution. The Technology Policy Statement was adopted in 1983. The aims of the technology policy are to attain technological self-reliance, to make maximum use of indigenous resources, to provide maximum gainful employment, to develop technologies, which are internationally competitive, and to reduce demands on energy.

The country also revised its educational policy of 1968 and started a New Education Policy in 1986 to make the educational system fulfil the needs of the society in keeping with the technological development. The New Education Policy aims to redesign courses and programmes of higher education to meet the demands of specialization better, to enhance support for research in the universities and ensure its high quality and to develop interdisciplinary research. The Industrial Policy adopted in 1991 gave a major shift to a commercial R&D approach in the wake of liberalization and globalization of economy. The Government has recently come out with a new Science and Technology Policy 2003, recognising the changing context of the scientific enterprise and to meet present national needs in the new era of globalisation. The new policy primarily aims to establish an Intellectual Property Rights regime, high-speed access to information, freedom of functioning for all academic and R&D institutions and encourage research and innovation in the areas of relevance for economy and society with enhanced participation from industry.

High-level committees

A number of high level committees have been appointed by the government from time to time to guide/advise the government on matters related to planning for the growth of science and technology and to monitor the implementation of the aims of S&T policies. The members of the committees represent a wide cross section of S&T fields. Such committees have been created at various points of time since 1948 and their recommendations have led to various mechanisms/programmes in strengthening the S&T capabilities of the country. The first advisory committee was set up in 1948 under the chairmanship of the then Prime Minister, Shri Jawaharlal Nehru.

The National Committee on Science and Technology (NCST) constituted in 1971 is of historic importance. The NCST was instrumental in preparing a comprehensive S&T plan document to fulfil the aims of the SPR and its inclusion into the 5th five-year plan. Also, on the advice of NCST, the Department of Science and Technology was set up in 1971. Presently, there are two high level committees, the Scientific Advisory Committee to the Cabinet (SAC-C) constituted in 1997 and Scientific Advisory Council to PM (SAC-PM) constituted in 2005.

Policy making body for S&T

The Department of Science and Technology (DST) is given the mandate to assist in formulation of policy statements and guidelines for S&T, to implement and co-ordinate activities which are multidisciplinary and involve several agencies/ministries, and to act as Secretariat to apex advisory bodies on science and technology to the Government. Other subject specific policies are coordinated/implemented by respective ministries/departments of the Central Government. The DST was instrumental in bringing out the recent S&T Policy 2003 document.

National S&T plans

The Planning Commission was set up in 1950 which has been responsible for the formulation of S&T plans as well as for the budget allocation of funds for S&T as a part of an overall socio-economic development plan of the country. The Commission has set up various working groups consisting of experts in different fields to formulate the plans for relevant sector in collaboration with Central Ministries/State Governments. In the case of scientific research, the Commission has set up independent committees, working groups and panels of experts and scientists to suggest the plans for various sectors of scientific research. The plan-wise expenditure for S&T has been increasing since independence. It may be noted that the 5th five-year plan witnessed a four-fold increase over the 4th five-year plan as during this period the country's first S&T plan prepared by the NCST was integrated as component of the socio-economic plan. The plan-wise details are given in the Table 6.1 below:

Table 6.1: S&T expenditure in five-year plans

Plan	*Budget allocation (Rs. million)*	*S&T focus*
1st (1951-56)	200	Infrastructure on S&T–chain of labs/research institutes in different parts of country.
2nd (1956-61)	670	Strengthen the existing research facilities, create adequate S&T manpower, reward for S&T manpower, coordinate research between agencies
3rd (1961-66)	1440	Basic research in universities, training of research personnel, research fellowships, development and manufacture of scientific and industrial instruments and investments in pilot plant trials and full scale experiments
4th (1969-74)	3730	Integrating industrial research, improving coordination and evaluation of research programmes periodically at different levels
5th (1974-79)	13810	to achieve self reliance
6th (1980-85)	36680	Growth, modernization and self reliance
7th (1985-90)	82640	Mission oriented projects, identification of thrust/frontier areas in S&T and strengthening and augmentation of existing S&T structures
8th (1992-97)	175290	Basic research in frontline areas, diffusion of appropriate technology and integration of S&T in socio-economic and rural sectors to fulfil the basic needs–water, food, nutrition, etc.
9th (1997-02)	121060*	Development of application oriented research, expansion of existing S&T infrastructure and capabilities, establish linkages between industry academic institutions and research laboratories
10th (2002-07)	252430*	for the benefit of the society

* Plan expenditure/outlay for select scientific departments only.

Science and technology system

The main components of the present S&T system in the country are categorised under the following sectors:

1. Central Government
2. State Governments
3. Higher education sector
4. Public sector industry
5. Private sector industry and
6. Non-profit institutions/scientific and industrial research organisations (SIRO).

At present, there are nearly 3960 institutions under the various sectors engaged in S&T activities in India. These institutions are of varying size in terms of expenditure, manpower and infrastructure, besides variation in the quality of research. The main actors and the various R&D activities performed by these sectors are as discussed below:

Central Government

Major scientific agencies

The component under the Central Government consists of 12 major scientific agencies/councils doing R&D or promotion of S&T as a predominant activity. This is carried out through a chain of laboratories/research institutions under them as well as through research grants/sponsored projects to higher education sector, national laboratories and establishments. These are:

1. Department of Science and Technology (DST)
2. Department of Biotechnology (DBT)
3. Department of Atomic Energy (DAE)
4. Department of Space (DOS)
5. Defence Research and Development Organisation (DRDO)
6. Department of Ocean Development (DOD)
7. Council of Scientific and Industrial Research (CSIR)
8. Indian Council of Agricultural Research (ICAR)
9. Indian Council of Medical Research (ICMR)
10. Ministry of Non-Conventional Energy Sources (MNES)
11. Ministry of Communication and Information Technology (MCIT)
12. Ministry of Environment and Forest (MoEF)

R&D organisations under other central ministries/departments

In addition, there are other Central Government Ministries/Departments, which have a large number of R&D organisation units such as Research Design and Standards Organization (RDSO), Lucknow; Central Power Research Institute (CPRI), Bangalore; Central Water and Power Research station (CWPRS), Pune; under their administrative and financial control. A few public sector industrial in-house R&D units such as Bharat Heavy Electricals Limited (BHEL), Hindustan Machine Tools (HMT) are under the Central Government. A good deal of research in fields like power, irrigation, petroleum, steel, telecommunication, railways, etc., being performed by these institutes/public sector industrial R&D units is controlled by these ministries/departments of the Central Government.

State Governments

The State Governments have their own research institutions, which mainly comprise of agricultural universities, their research stations and state industrial units. Apart from this, there are other research institutions working directly under different departments of the State Governments. State Governments are concentrating mainly on research in agriculture, animal husbandry, public health, irrigation and forestry. The State Governments have also established State Councils on Science and Technology with catalytic support from Department of Science and Technology to promote research and to help in finding solutions to specific problems faced by them.

Higher education sector

Higher Education Sector comprises of the universities, deemed universities, institutes of national importance and colleges, which undertake research activities mostly of basic and applied nature. The majority of funds for research received by them are from the projects sponsored by the Government and industry.

Private sector including SIRO

Private sector industries have established their own in-house R&D units, which are responsible for undertaking research and development activities for the respective industrial organisation. There are non-profit scientific and industrial research organisations and institutions (SIRO), which carry out R&D activities and are supported by either industry or the Government or both.

S&T institutions: Current status

Out of the 3960 institutions carrying out R&D activities in India, there are 822 R&D institutions/laboratories under the Central Government (462 under major scientific agencies and 360 under other Central Government Ministries/Departments which include 115 public sector in-house R&D units), about 834 research institutes/joint sector companies/research stations (big and small) are under the State Governments, 284 universities which include 182 universities and 89 deemed universities including 13 institutes of national importance and about 2020 in-house R&D units of private sector including 248 non-profit scientific industrial research organisations (SIRO).

Section 3: Science and technology statistical system

In order to evaluate S&T policies for proper planning, the country needs a sound S&T statistical system to provide reliable and timely data and analyses both on inputs (expenditure, personnel and infrastructure) and outputs (patents, papers, products/processes etc.) backed by proper documentation.

India has a well-established statistical system for the socio-economic data. The Central Statistical Organisation (CSO) is the major organization responsible for collecting the socio-economic data through various Statistical Bureaus/Directorates at the State level. These statistical organisations do not collect data relating to science and technology except University Grants Commission (UGC) under Department of Education, which collects information on educational statistics. Therefore, a need was felt for the collection of S&T statistics. The progress made in the efforts so far for collection of S&T statistics is given in the ensuing paragraphs.

On the advice of the National Committee on Science and Technology (NCST) the responsibility of undertaking S&T resources studies was entrusted to Department of

Science and Technology (DST) in 1971. Since 1971, National Science and Technology Management Information System (NSTMIS) Division of the Department has acted as the nodal agency in India to collect and disseminate national science statistics at regular intervals.

NSTMIS Division of the Department of Science and Technology undertakes national level surveys on resources devoted to scientific and technological activities biennially using a structured questionnaire. India, being the member country of UNESCO, has been actively participating and contributing in the development and revision of standards/concepts/definitions used for collection of science statistics by UNESCO. Therefore, the definition of various terms used for collection of information to India follows the UNESCO pattern. The data collected are compiled, analysed and published in the form of well documented reports since 1973-74. The latest biennial publication available is 'R&D Statistics 2004-05' (15th in the series) containing statistics and indicators on various dimensions of R&D. It may be worthwhile to notice the growth of R&D institutions, their R&D expenditure and manpower since the inception of national survey undertaken by the NSTMIS. Table 6.2 indicates the phenomenal rise in R&D infrastructure and other R&D input indicators during the last three decades.

Table 6.2: Comparison of input indicators for R&D 1972–73 and 2002–03

Sector	*R&D institutions*		*R&D expenditure (Rs. millions)*		*R&D manpower*	
	1972–73	*2002–03*	*1972–73*	*2002–03*	*1972–73*	*2002–03*
Central+	236	822	1557	120730	25320	49886
State	135	834	108	15284	4319	17993
Private	191	2020	193	36482	4231	25957
Higher education	–	284	–	7505@	–	22100*
Total	562	3960	1858	180002	33870	115936

+: Central Sector includes Public Sector Industries R&D units.

–: Survey for the Higher Education Sector started in 1998-99.

@: Covers 106 universities and 27 PG colleges performing R&D (excludes State Agriculture Universities).

*: Estimated.

Publications

As an outcome of the national survey, various publications are brought out by the Department which enjoy wide circulation both nationally and internationally and act as a source book for S&T policy makers, planners and researchers. These are research and development statistics, research and development in industry, science and technology data book, and directory of research and development Institutions.

Information and analysis

A summary of the information and analysis that are available in the report namely, R&D Statistics is given below:

Primary sources (National survey)

1. Total R&D expenditure and breakdown of this into various sectors
2. Total R&D expenditure as percentage of Gross National Product
3. Total R&D expenditure at current and constant prices
4. R&D expenditure by Major Scientific Departments / Agencies
5. R&D expenditure by Socio-Economic Ministries
6. R&D expenditure by various State Governments
7. Breakdown of S&T expenditure by categories of research-basic applied experimental development and other research related activities
8. Sector wise R&D expenditure by objectives defined by UNESCO
9. Sector wise R&D expenditure by field of research
10. R&D expenditure by public / joint sector companies
11. Sector wise R&D manpower
12. R&D manpower by level of qualifications, income and sex
13. Process and products developed and commercialised
14. Publication data, research papers, books, etc.
15. Consultancy services rendered etc.

Secondary sources

1. Number of higher educational institutions
2. Enrolment by field (disciplines) in higher education
3. Out-turn of science and technology personnel
4. Stock of science and technology manpower
5. Budget allocations for science and technology
6. International comparison of research and development efforts
7. Patent statistics

Issues and problems in the S&T statistics

Some of the issues and problems in collecting and analysing S&T data are indicated below:

1. There are no national laws / mandates / statutory / provisions relating to procurement of S&T statistics from R&D establishments.
2. The major problem in the survey is to receive the completed questionnaire within the stipulated time
3. In higher education sector it is extremely difficult for academic staffs that are performing multifarious activities to apportion expenditure between research, teaching and administration.

4. Many industries have no separate R&D departments and their R&D problems are tackled by production personnel. It is difficult to get information on R&D resources from such industries.
5. The responding units find it difficult to classify clearly the manpower employed for R&D activities and auxiliary (support) activities.
6. A number of responding units find it difficult to provide data on R&D expenditure for the reason that they are not exclusively research institutions and separate accounting for R&D funds are not maintained.
7. At present, the data on social sciences and humanities research are not available for India. However, a beginning has been made in this direction through a pilot study.
8. Actual figures are not available for stock and economically active stock of S&T personnel.
9. The data on classification by age of S&T manpower and R&D expenditure by sector of economy are not collected so far in India.
10. On several occasions, questions are asked whether the output of R&D is commensurate with the expenditure on R&D. It is difficult to establish one to one relationship between input and output of R&D. There is always a scope for improvement of output indicators.

Linkages

The NSTMIS Division maintains a close liaison with international agencies like UNESCO, NSF in the United States and OECD for matters relating to S&T statistics and indicators. Apart form this, the Division continually interacts with various national organizations like Institute of Applied Manpower Research, University Grant Commission (UGC), Confederation of Indian Industry (CII), CSO and other scientific organizations in the country for exchange of information and expertise related to S&T statistics, indicators and science resource studies.

Section 4: Policy implications

S&T indicators as policy tools

S&T indicators are an important policy issue in their own right, and not just an adjunct to other policy initiatives. The use of S&T indicators, and the choice of which indicators to use, are issues that must be included in any major S&T policy initiative. S&T policy analysis, at a bare minimum, requires the development of systematic data to form the foundation for policy deliberation (Holbrook 1992).

S&T indicators serve the policy relevant needs, provide current status of S&T and are useful in an evidence-based assessment of the existing S&T policy for making effective changes. This section shall discuss the policy implication of the existing S&T indicators in the Indian context and the need for the development of new S&T indicators in the present era of globalization.

One of the most commonly used indicators for international comparison of S&T efforts is the proportion of Gross Domestic Product devoted to R&D activities. With the success of USA and some European Countries, it became a common tradition that the level of development of a nation was associated with the percentage of GNP spent on research; 2 per cent to 3.5 per cent being the "ideal", and 1 per cent being a practical goal for developing countries to emulate (UN World Plan of Action). In India, the R&D/GDP ratio has still remained less than 1 per cent, despite the increasing trend in the absolute national R&D expenditure over the years. In terms of the sectoral composition of the national R&D expenditure, the ratio of public to private sector is 80:20 while in case of Republic of Korea it is just the reverse *i.e.* 20:80. In fact in most of the developed nations such as USA, European countries and Japan, the private participation in the national R&D activities is very high in the range of 50-80 per cent. The international comparability of the above S&T indicators has direct policy implications in Indian context. In order to be globally competitive, our new S&T Policy 2003 has envisioned raising the level of S&T investment to at least 2 per cent of GDP with enhanced participation from the Industry. Recently, the Indian Government has announced to achieve the goal of 2 per cent of S&T investment of national GDP by the end of 11th five-year plan (2007-2012). To encourage enhanced participation from industry in science and technology, the government has implemented several fiscal measures to encourage R&D spending through tax incentives, tax holidays, excise and custom duty schemes, special incentives to emerging S&T sectors such as Biotechnology, Pharmaceuticals etc. Further, the private participation has been encouraged through various non-fiscal R&D support measures or mechanisms such as revision of Indian Patent Act 1970 complying with TRIPs, or the Public Private Partnership (PPP) being introduced recently in the post- WTO era.

In terms of S&T manpower, the scientists engaged in active research in India are 110 per million population as compared to 633 for China, 2691 for UK and 4526 for USA. Such a low participation of S&T personnel in the active research has been a great cause of concern and as a result the Government has introduced several programmes/initiatives targeted towards young scientists, performing scientists and various infrastructure development programmes such as setting up of centres of excellence/advanced studies in the universities and academic institutions, creation of new scientific Departments/Organisations and allocation of funds for Improvement of S&T infrastructure in the higher educational institutions. Further, new initiatives have been taken such as inviting distinguished men and women of Indian origin settled abroad for short-term technical assignments to assist in frontier and emerging areas of S&T.

Need for new indicators

In the present era of liberalization, privatization and globalization, the S&T is in a state of dynamic flux then ever before to which India is no exception. In India, this dynamic situation is driven by several factors such as MNCs opening new R&D centres, technology generation and acquisition taking place through FDI route, cross boarder alliances between firms, emergence of public private partnership enhancing industry-university/public research relationships, networking of research organizations, or merger and acquisitions of domestic/overseas firms. These developments are further accelerated by usage of ICT, which have become apparent only in the past few years and are likely to continue in future. As a result, several measurement issues like S&T education and training, mobility of S&T human resources, R&D capacity of the organization, informal links, strategic research partnerships or cooperative R&D and IPR have emerged which need to be mapped in terms of their current trend by indicators.

There has been a paradigm shift from the S&T policy to innovation policy in the global world and in fact the new S&T policy 2003 has also emphasized innovation in all areas of S&T. Innovation is a complex phenomenon and R&D is considered as one of the essential component driving the innovation while non-R&D inputs like acquisition of machinery, external technology, technology licenses from abroad, training and market research constitute a major component. In the case of Community Innovation Surveys of European Union around 60-70 per cent of Innovation expenditure was accounted by the non-R&D inputs. It may be mentioned here that in the Indian context many firms in the small-scale sector carry out innovation without having any R&D unit or perform occasional problem solving R&D. This calls for development of new indicators to measure innovation. In India, recognizing the need for innovation, the NSTMIS has constituted an experts group to evolve a framework to develop the new innovation indicators.

Recognising the constantly changing global S&T scenario, an attempt has been made by the UNESCO calling for a fundamental international review of S&T statistics and indicators for the developing countries to keep pace with the changing techno-globalized era.

Conclusion

S&T statistics collection and dissemination is regularly done in India and the coverage is quite good except for the limitations and problems highlighted. The coverage and reliability have increased over the years. Since India's statistics are fairly timely, there is every possibility that the time gap will be reduced further in future by leveraging the ICT. The NSTMIS Division has acted as a resource centre for countries like China and Trinidad and Tobago in imparting training on science statistics and establishment of S&T statistical system. It recently joined the scientific delegation to Vietnam and Mozambique, which showed keen interest to develop S&T statistics and indicators in their country.

The recognition of the importance of Science Statistics for assisting S&T planning is visible since the users of such statistics are growing. In order to supplement the R&D information received through national survey, pilot projects are being sponsored in the area such as S&T resources, S&T infrastructure, S&T facilities, or R&D output. Recently, the NSTMIS has started several new initiatives to enlarge the coverage and scope of S&T statistics in the area of small-scale industries, higher education sector and social sciences in addition to evolving a framework for initiating innovation surveys.

References

Annual Report (various years). Department of Science and Technology, Government of India.

Giorgio, S., 1998. Conceptualizing and measuring technological innovation. STEP group Idea paper series No. 1 (www. sol.no/step/IDEA)

Hill, S., 1986. From dark to light: seeing development strategies through the eyes of S&T indicators. Science and Public Policy. 13(5): 275-284

Holbrook, J.A.D., 1992a. Why measure science. Science and Public Policy. 19(5): 262-266

Holbrook, J.A.D., 1992b. Basic Indicators of scientific and technological performance. Science and Public Policy, 19(5): 267-273.

Pradosh, N., 2006. Catching-up from behind: lessons for India. DST sponsored project report.

Rahman, A., 1990. Science and Technology in India, Pakistan, Bangaldesh and Sri Lanka. Longman, UK.

Research and Development Statistics (Various Years). Department of Science and Technology, Government of India .

S&T Data Book (Various Years). Department of Science and Technology, Government of India

Smith, K., 1998. Science, technology innovation indicators: A guide for policy makers. STEP group idea paper series No. 5 (www.sol.no/step/IDEA).

UNESCO, 1984. Major S&T indicators derived from a national scientific and technological potential survey. Paris

Website of UNESCO (www.uis.unesco.org).

World Development Report (Various Years). The World Bank .

Yvan, F., 1984. The OECD international S&T indicators system. Science and Public Policy Feb, pp. 4-6.

Chapter 7

S&T indicators as enablers to R&D planning: The case of knowledge intensive CSIR-India

Naresh Kumar

Abstract

The trend of formulation and implementation of the S&T policies is changing throughout the developing world as the new models based on S&T indicators are rapidly replacing the earlier technology centred models. The present paper explains requirement to CSIR, the head S&T policy planning body organization in India, to put the S&T indicators based network approach under consideration which is more focused and also ensure the appropriate utilization of funds.

Keywords: CSIR, India, R&D planning network programme, S&T indicators.

Preamble

Science Indicators, be it the research publications or patents, technologies developed or licensed, R&D expenditure or revenue generation through technology commercialisation. Science and Technology S&T manpower generation or employment, are the direct pointers to the health of any country's scientific and technological activities and its international standing. Several countries across the globe, notably Japan, USA, UK and Germany consider these indicators are of important components in establishing the national S&T priorities as they play, *interalia,* a dominant role while deciding not only funding of various research projects but also for S&T planning. For India, a recent study entitled 'India Science Report' on science education, human resources and public attitude towards science and technology, conducted by the National Council of Applied and Economic Research (NCAER) points out that no systematic and comprehensive empirical assessment of S&T efforts is available in the Indian context, resulting in a relatively chaotic and contradictory picture of national efforts in S&T (Shukla, 2004). The report further indicates that an important factor contributing to such images of S&T efforts in the country is the paucity of reliable data in an accessible and timely manner. Without undermining the efforts being put up by the National Science and Technology Management Information System (NSTMIS), which brings out biennial data sets on Indian science indicators, the NCAER conclusion is quite nearer to the reality. NSTMIS reports give whole data sets but without any value evaluation (http://www.nstmis-dst.org). Thus, in spite of India having a sound S&T base, it is often ranked low in S&T indicators internationally. We therefore need to have a facile and proactive system to record our indicators not only to measure our S&T or R&D capabilities but also as empirical and rational tools for our S&T planning process.

The Council of Scientific and Industrial Research (CSIR) is perhaps the only apex scientific body in India to use its S&T achievement indicators to conceive, plan and deliver its research programmes while formulating successive five year plans.

CSIR: Organisational culture

With its 38 national laboratories, CSIR, one among the world's largest publicly funded organizations, has been occupying the centre stage of most of the scientific and technical endeavours initiated by Indian scientists in various laboratories and in academia. Over the past 60 years or so, it has developed a unique R&D management culture which has passed through several critical and defining periods of time. Whether it were the expectations from it in pre-independent India, or of the socialistic India of post-independence, or an India of open economy, it has shown an organizational adaptation, survival and competence in the face of increasingly discontinuous environmental change. CSIR of today embodies organizational processes that are a synergistic mix of scientific output, as reflected in its quality research publications, a significant portfolio of registered intellectual property–primarily patents, a large number of new products and processes, a number of collaborative, cutting edge research projects with globally known companies, and the creative and innovative capacity of its scientific work force. CSIR of today is a knowledge intensive, multi-locational, multi-sectoral industrial research and development organization.

Indicators to growth: Funding scientific and industrial R&D

Ever since the beginning of planning process in India, CSIR has been a beneficiary of funding for its R&D activities, be it for the projects or for creating new facilities/ institutions, from the Planning Commission of India. From a very modest plan allocation of Rs. 4.67 crore in India's first five year plan (1951-56) the allocation has increased to a plan plus non-plan allocation of about Rs. 6439.00 crore in the tenth plan (2002-2007), which is an increase of about 1400 times (Table 7.1). Such funding, an indicative of the government's faith in creating a S&T base in the country in general and in CSIR in particular has indeed helped the country to build a sound scientific and technological infrastructure. Today, India's research base rests on its more than 350 universities, about 1200 in-house R&D units in industrial undertakings supporting research in their respective industries and almost an equal number of R&D laboratories in the government sector. Such a sound R&D base is due to government's modest yet consistent support to R&D (Figure 7.1). As per the latest NSTMIS data available on the web-site of NSTMIS, the national investment on R&D activities attained a level of Rs. 18000.16 crores in 2002-03, Rs.19726.99 crores in 2003-04 and Rs.21639.58 crores in 2004-05 (http://www.nstmis-dst.org). As a percentage of Gross National Product (GNP) it was about 0.80, 0.78 and 0.77, respectively, showing a decline in relative terms indicates rather stagnation in funding. The per capita national funding of R&D for the years 2002-03, 2003-04 and 2004-05 has been approximately Rs.1722, Rs.1857, and Rs.2005, respectively.

Table 7.1: Five Year Plan allocations of CSIR

Period	*Plan*	*Non-plan*	*Total*
1st Plan (1951-56)	4.67	0.00	4.67
2nd Plan (1956-61)	14.58	6.97	21.55
3rd Plan (1961-66)	33.34	21.59	54.93
4th Plan (1969-74)	37.46	71.41	108.87
5th Plan (1974-79)	54.25	104.96	159.21
6th Plan (1980-85)	216.91	232.24	449.15
7th Plan (1985-90)	359.09	548.01	907.10
8th Plan (1992-97)	703.00	1076.00	1779.00
9th Plan (1997-2002)	1356.00	2583.19	3939.19
10th Plan (2002-07): Anticipated expenditure	3002.00	3437.00	6439.00

Unless specified otherwise, all the financial figures mentioned are in Rs. Crores. One Crore is Rs. 10 million, and One US$=~Rs.47

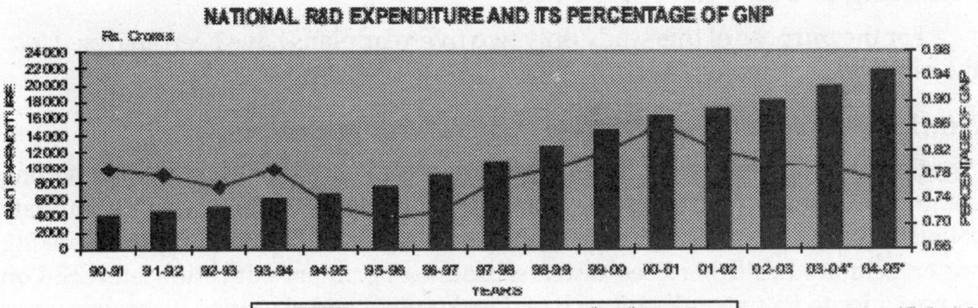

Figure 7.1: R&D expenditure as percentage of GNP

This stagnation in R&D intensity is a major threat even to the European knowledge-based economies where the growth rate of R&D intensity (R&D expenditure as per cent of GDP) has been declining since 2000 and is now close to zero. Europe may miss the objective it has set for itself to boost spending on R&D from 1.93 to 3 per cent by 2010. Europe devotes a much lower share of its wealth to R&D (1.93 per cent of GDP in 2003) as compared to 2.59 per cent in the US and 3.15 per cent in Japan. While China has lower R&D intensity (1.31 per cent of GDP in 2003) it grew at about 10 per cent per year between 1997 and 2002. If these trends in the EU and China continue, China will be spending the same amount of GDP on research as the EU in 2010–about 2.2 per cent, which leads us to conclude that Europe is becoming a less attractive place to carry out research. The same could be true for India too, if the plan funding is not increased substantially. The tentative figures available for India do not reassure us. We are nowhere near the 2 per cent mark of India's GDP being spent on R&D as is enshrined in our S&T Policy 2003. The reason being that research has become highly cost intensive and funding from the private sector too is not to the desired extent. The support to R&D has a positive effect on productivity and economic growth as has been pointed out by a European Commission study. Therefore, if India is to become an integrated research base where cutting edge research can be carried out, and is planning to attract investment from all over the world, there must be a substantial investment in R&D, both in the public and the private sectors. Dwindling investment in research has a direct impact on our science base which in turn impacts our manufacturing and could possibly affect even the IT sector. It is to be recognized that while growth has been a function of capital investments and the employment of human resources, technology is the key to bringing about rapid advancement in the economy of every country in the world.

Eleventh Plan affords us a chance to make amendments and bring in a quantum increase in funding of our indigenous research efforts. This is absolutely necessary to keep the scientific temper alive in our country. Sub-optimal efforts lead us nowhere as otherwise shall lose all our bright students wanting to pursue R&D as a career to institutions abroad. The coming five years plan (2007-12) is, therefore, critical for our R&D, both at national and CSIR levels to close or narrow the gap between developed countries and India. We can not afford to let our R&D institutions stagnate.

Planning process in CSIR India

For the purpose of this study only two five year plans have been analysed, *viz.* the ninth and the tenth plans.

Ninth Five Year Plan (1997-2002): A sectoral approach

For the ninth plan period the Government of India placed emphasis on providing country wide seven identified basic minimum services rather than a single mandate (Ninth Five Year Plan, 1996). These services covered *interalia* providing safe drinking water, primary health services, universalisation of primary education, etc. CSIR on its part adopted a sectoral approach of development and classified its programmes into sixteen broad sectors including aerospace, biology and biotechnology, chemicals, drugs and pharmaceuticals, earth resources and natural hazards mitigation, ecology and environment, electronics and instrumentation, energy, food and food processing, housing and construction, leather, exports of R&D and services.

Under each of the sectoral plans CSIR's ninth plan working group broadly recommended major projects, inputs for modernisation, and deliverables in terms of indicators such as external cash flow, technology development, generation and utilization of new knowledge. The working group also recommended new schemes on (*i*) modernization and up-gradation of R&D equipment and facilities at national laboratories and (ii) intellectual property and technology management.

During this plan period, a major CSIR coordinated programme on discovery, development and commercialization of new bioactive molecules was initiated in the year 2000 which involved networking of 20 CSIR laboratories, 12 universities and 3 indigenous systems of medicine, *i.e.* Ayurveda, Siddha and Unani. This bioactive molecules programme is aimed at bio-resource collection, processing, and evaluation to get leads to newer drugs. In 2000-2001, the Government assigned CSIR the responsibility to manage and operate another programme of immense techno-scientific importance called New Millennium Indian Technology Leadership Initiative (NMITLI). Both these programmes were in a way precursor of the current 'network' approach as being practiced in all the CSIR laboratories.

Tenth plan: The network approach

Towards the end of ninth plan CSIR carried out 'Check' and 'Act' cycles on its ninth plan agenda ('Plan' and 'Do' cycles) through a SWOT analysis for each laboratory and sectorally in biotechnology, chemicals, drugs and pharmaceuticals etc., the sectors where CSIR has large interests and stakes. It was apparent from the SWOT exercises that this sectoral approach did not deliver in ninth plan to the extent that was initially expected. Lack of coordination among stakeholders was perhaps a major reason. The CSIR strategy for the tenth five year plan period was drawn up accordingly on the basis of careful assessment (check and act) of the needs of and the opportunities in the market place in the next five years. Based on the opportunities so derived, CSIR sought to position itself strategically with the Indian industry for innovative research, application and development of technology, commercialization of technology; and technology transfer (Tenth Five Year Plan, 2001).

At this point of time, the Planning Commission of India sought to bring in the 'Network' approach wherein only the new R&D projects would qualify for funding which have a demonstrable network between various CSIR laboratories usually from the same discipline; in a few cases the 'network' partners could be even from other disciplines. This ushered in a new era of doing R&D–from small laboratory centred projects to mega multi-location, multi-objective projects, from hundreds of projects, most of them being resource thin, CSIR had 56 networked projects which are expected to go up to 100 in the coming XI Plan. CSIR proposes to move to XI Plan through a well thought of cycle (Figure 7.2) wherein the output of the individual components of the PDCA (Plan, Do, Check and Act) cycle would be inter and intra-dependent.

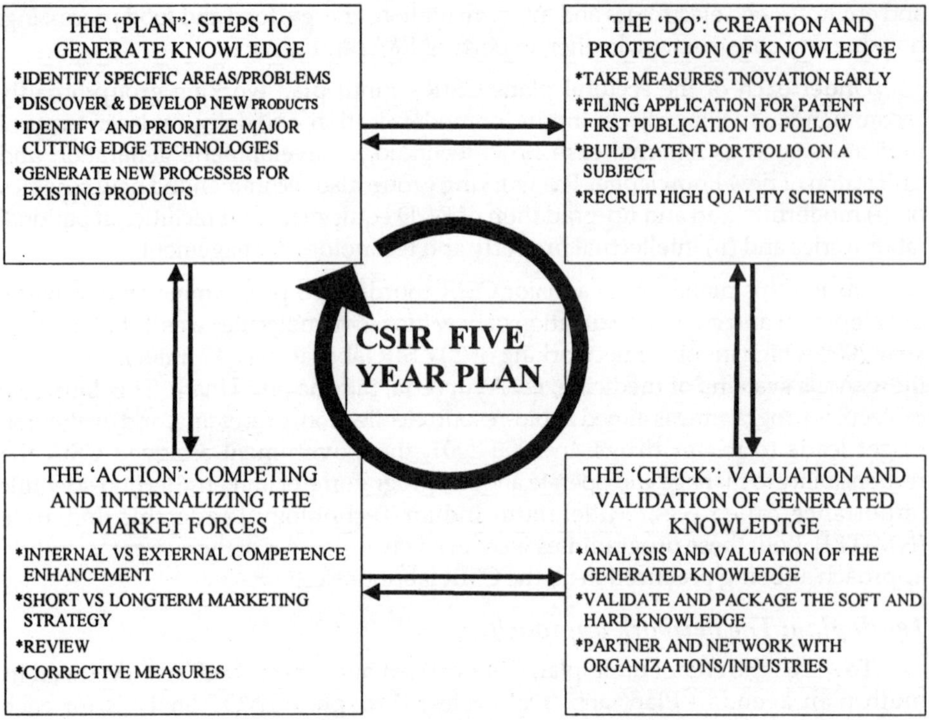

Figure 7.2: CSIR five year flow cycle

Planning and Managing the R&D: Network Style

Thus, it was only towards the end of ninth plan period that CSIR on the directions of planning commission brought in a paradigm shift in the way it was carrying out its R&D programmes. It desired the CSIR to bring in sharp focus, in depth accountability, greater and cohesive networking and an overall relevance to national needs in all its research programmes.

As per the new mandate, CSIR identified 56 network projects. These projects were chosen out of the project list identified by the Planning Commissions' Steering Committee, X Plan Working Group of the CSIR and the Core Committee.

For the selection of these projects several brain-storming sessions were held within the laboratories and at CSIR levels. These sessions not only discussed the science and technology sectors in which these network programmes would be taken up but also their scientific goals, the resources in terms of manpower and finance required. The 'planning' process for these network projects was debated at a meeting of the participating institutions, the 'do' part was to be carried out in the partner laboratories, the 'check' or review was to be taken up in the regular meetings of the partners and any mid course correction or 'action' required was to be carried out in the individual laboratories. A three level system of monitoring has been thus inbuilt into the network project system (Figure 7.3), which has not only sharpened the focus but also has brought in a great measure of accountability.

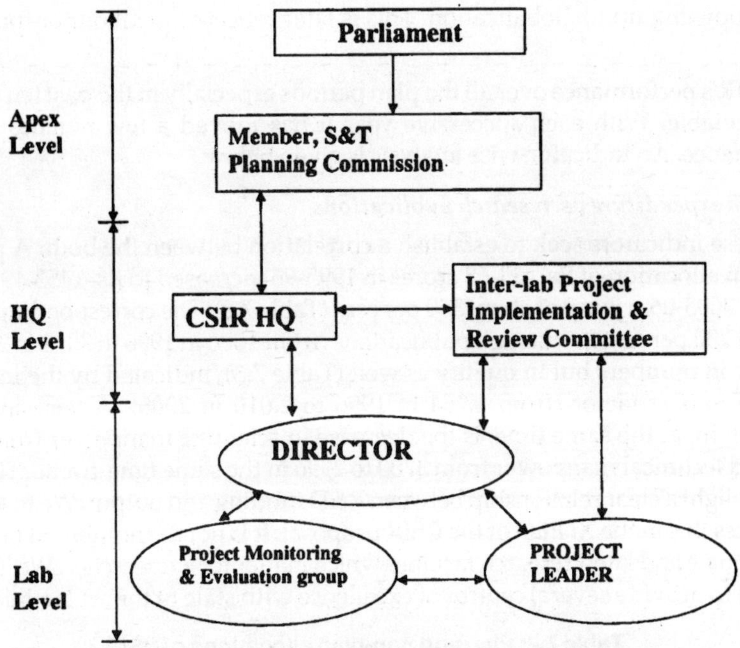

Figure 7.3: Networked R&D project management

This model is now in operation, the results of which when available in 2006-2007 would hopefully indicate the correctness of this network approach.

S&T indicators as a basis for the eleventh plan formulation

Science and technology indicators point out the state of activities in the field of science and technology quantitatively. Since such activities being extremely complex and overarching over a number of science disciplines, their state cannot be represented by a single indicator. A combination of many indicators is necessary, which has been the case in the present study as well.

The indicators regarding S&T can be broadly classified into four groups, namely (i) hard and absolute such as R&D expenditure and R&D personal, (ii) hard and relative such as R&D expenditure per GDP and R&D personal per capita, (iii) soft–technology diffusion, and (iv) soft–S&T human resources. While planning for its five years plans, CSIR has used both the hard and soft criteria to propose very ambitious plans.

As an arm of the government, CSIR follows the five year planning as well as the annual planning cycles. Beginning with the first five year plan in 1951, it has formulated all its perspective plans to secure funding for capacity building and enhancement, and has also planned the new R&D projects in such a way that the funding is secured to reach the targets, both soft and hard. Figure 7.1 indicates funding support, from a mere Rs. 4.67 crores in the first plan to about Rs. 6450 crores in the tenth plan–an increase of 1400 times. Even within this, a noticeable increase in funding has been in the 8th and 9th Five Year plan stages, which points out to the period of

India's opening up to globalization. This is later reflected in all our output indices too.

CSIR's performance over all the plan periods especially in the past ten years has been enviable. With each successive year it has moved a few notches up in its performance. An indicator-wise analysis is given below.

Research expenditure vs. research publications

These indicators seek to establish a correlation between the both. A plan plus non-plan allocation of Rs. 411.68 crores in 1995-96 increased to Rs. 1453.49 crores in the year 2005-06, a jump of about 350 per cent (Table 7.2). The corresponding increase of about 200 percent in research publications (from 1576 in 1996 to 3018 in 2005) was not only in numbers but in quality as well (Table 7.3), indicated by the increase in average impact factor (from 1.264 in 1996 to 2.010 in 2005). A very interesting observation, at the same time, is the decrease in scientific manpower from 5550 to 4635 and technical manpower from 3700 to 2996 in the same time-frame. These data sets highlight a clear relationship between R&D funding and output. We have sought to address this in the XI Plan of the CSIR proposal. It is not the number of manpower but adequate and latest research facilities which matter for a researcher. We, therefore, propose to provide several centres of excellence with state of the art facilities.

Table 7.2: Plan and non-plan allocations of CSIR

Year	Research expenditure (Rs. Crore)	
	Plan	Non-plan
1995-96	160.00	251.68
1996-97	161.00	283.00
1997-98	204.00	360.13
1998-99	204.00	509.32
1999-2000	250.00	548.74
2000-01	303.00	574.88
2001-02	320.80	584.12
2002-03	412.00	597.65
2003-04	430.00	644.54
2004-05	580.77	685.70
2005-06	713.00	740.49

Research expenditure vs. patents

Patent indictors serve to measure the R&D output of CSIR, its productivity, structure and the development of specific technologies (Table 7.4). Patent statistics provide a measure of innovation output, as they reflect the inventive performance of institutions. They are also used to track the level of diffusion of knowledge across technology areas, sectors, client firms and the level of internationalization of innovative activities. Patent indicators have many advantages, mainly, (*i*) they have a close link to inventions, and (*ii*) they cover a broad range of technologies, (*iii*) the contents of patent documents provide a rich source of information, and (*iv*) patents data are readily available from the patent offices. However, patents are also subject to certain drawbacks, *viz.* (*i*) the value distribution of patents is skewed as many patents have no industrial application (and hence are of little value to society) whereas a few are of substantial value, (*ii*) many inventions are not patented because they are not patentable or inventors may protect their inventions using other methods, such as secrecy. Patenting activity in CSIR has shown an appreciable rise during the past 10 years or so. The increase in number of Indian patents filed (from 260 in 1995-96 to 407 in 2005-06) was about 1.7 times, while it is quantum jump in filing of foreign patents, a huge 1100 per cent. This not only is reflective of increased research funding but also of CSIR's coming of age post-WTO. As concluded earlier, it is the funding and provision of latest facilities which have added to the quality, hence, not only a top class publication but also a patent.

Table 7.3: Research publications and average impact factor

Year	*Papers published*	*Average impact factor*
1996	1576	1.264
1997	1563	1.467
1998	1537	1.512
1999	1699	1.538
2000	1667	1.552
2001	1700	1.696
2002	1944	1.632
2003	2188	1.751
2004	2668	1.899
2005	3018	2.010

Table 7.4: Patents filed by CSIR

Year	*Patents filed*	
	India	*Foreign*
1995-96	260	58
1996-97	209	71
1997-98	264	94
1998-99	310	112
1999-2000	377	199
2000-01	408	490
2001-02	413	569
2002-03	421	771
2003-04	406	566
2004-05	415	529
2005-06	407	645

Network projects vs. technologies diffused

Number of technologies generated and licensing deals made have remained the same rather have dipped in some years. This, in one view, points out a steady technology strength built and sustained over the years, in spite of India's globalised economy, per another view, it shows a stationary phase, which is worrisome. However, if viewed against CSIR's earlier work culture of having a large number of small, often unviable, stand-alone projects this was the obvious outcome. And it has been a major consideration, rather compelling force, behind CSIR shifting in its planning process to 'Network' model where synergies are established across several CSIR labs, links forged and outcome driven projects planned and worked with a rigorous monitoring mechanism. This approach is sought to be continued in XI Plan as well. It is sought to be extended by creating intra-lab networks in the 'Supra' mode, where a large group of scientists within a particular laboratory and not necessarily from the same stream would work on a single overarching research problem. This would help create a core competency in each CSIR institution.

Table 7.5: Technologies licensed by CSIR

Year	*Technologies*	
	Licensed first time	*Licence agreement executed*
1995-96	46	190
1996-97	40	160
1997-98	48	180
1998-99	32	150
1999-2000	36	130
2000-01	36	140
2001-02	26	110
2002-03	42	129
2003-04	40	165

External cash flow vs. client base

One important parameter on which CSIR has almost stagnated (except for the year 2005-06) is the external cash flow (Figure 7.4) or the income derived from technology sale, licensing, royalty or sponsored research. This could be either due to reaching our limits of performance or industry (our main clients) itself coming of age or the competition from our international clients. This indicator or the trend we have sought to address through augmenting core competency of each CSIR institution through 'Supra' networks, *i.e.* by involving 25 to 30 per cent scientific work force of each laboratory in a flagship project, which is designed to create a core capability in all CSIR laboratories. In addition, CSIR has plans to reach out to off-shore locations, *i.e.* it would internationalize by physically locating itself through representative offices- in other countries having business potential.

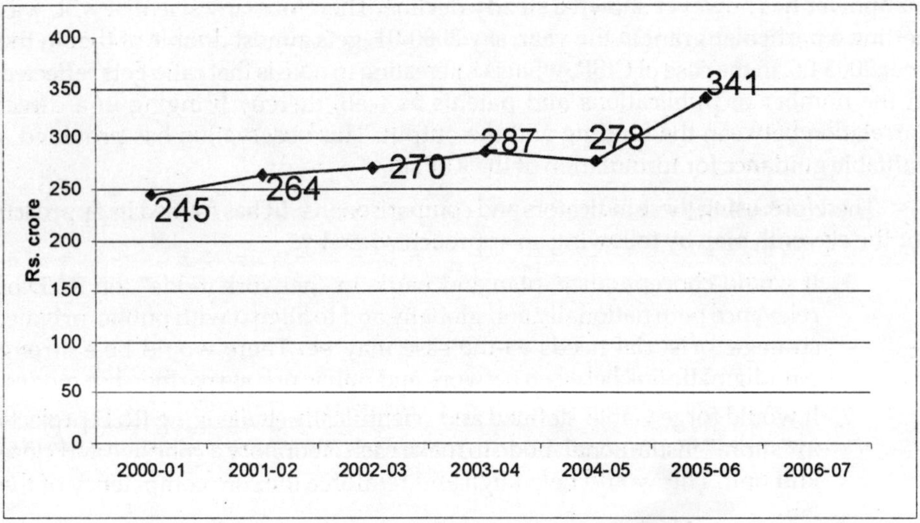

Figure 7.4: External cash flow

Table 7.6: Scientific and technical manpower deployed in CSIR

Year	*Manpower*	
	Scientists	*Technical*
1995-96	5,550	3,700
1996-97	5,860	3,290
1997-98	5,440	3,610
1998-99	5,309	3,612
1999-2000	5,244	3,418
2000-01	5,052	3,241
2001-02	5,112	3,221
2002-03	5,185	3,119
2003-04	4651	2,959
2004-05	4682	2,957
2005-06	4635	2,996

Research expenditure vs. manpower vs. publications/patent

This set of indicators presents yet another evaluation of consequence. Per capita R&D expenditure of CSIR for the years 1999-2000, 2000-01, 2001-02, 2002-03, 2003-04, 2004-05 and 2005-06 has been approximately Rs.3.69, Rs. 4.13, Rs. 4.31, Rs. 4.88, Rs. 5.90, Rs.6.95, and Rs.7.98 lakhs (Rs. 1.00 lakh= Rs.100,000) respectively, which is 10 per cent, 11 per cent, 10 per cent, 11 per cent, 12 per cent, 18 per cent, and 15 per cent higher over the previous year's total allocation, respectively. During these years, manpower has however shown a steady decline. Therefore, any scientist who was getting a particular grant in the year, say 2000-01, gets almost double of that in the year 2005-06. In the case of CISR, what is interesting to note is that ratio gets reflected in the number of publications and patents as well, thereby bringing in a direct correlation between the funding and the output. This observation has provided a valuable guidance for formulation of the XI Plan.

Therefore, using these indicators and comparisons, CSIR has framed its approach for the eleventh plan by following an approach whereby:

1. It would conceptualize, plan and work, in "network mode", on R&D of relevance both nationally and globally and to align it with public, private, strategic or social needs as the case may be. There would be a strong paradigmatic link between network and public private partnership modes.
2. It would forge viable, defined and scientifically challenging R&D projects in "supra" institutional mode to make each laboratory a cohesive and close knit unit. This would help align and reinforce the core competency of the lab.
3. It would plan to build within each lab "centres of sustainable growth", a kind of magnet to attract scientists/technologists of Indian origin, a large number of trainees, industry both national and foreign, and above all, help create a think tank to look at the future with a clear vision and shape blended into its mandate.

This strategy, a three-pronged one, would be used to plan and work out research projects across the whole CSIR system.

Conclusion

Planning and managing R&D processes has changed over the years, moving from the earlier technology centred model to a model based on S&T indicators, interaction and networking style. Thus, the dynamically responsive and knowledge driven research organizations like CSIR, have to renew themselves periodically through strategic system modifications wherein S&T indicators play a crucial and defining role. They set the contours of our total planning focus. Realising that, CSIR has restructured its research programmes from the individual or sectoral approach to 'network' approach. This has not only ensured a proper 'planning', facile 'working/doing', appropriate 'checking/review' and fast 'action' but also to appropriate resource utilization. The programmes initiated in the year 2002 have already started yielding dividend through a more focused approach, appropriate resource utilization and identification of end-users early enough in the whole R&D planning and management cycle.

References

Annual Reports, 2001-2002, 2002-2003, 2003-2004, and 2004-2005. Council of Scientific and Industrial Research, New Delhi, India.

CSIR 2001: Vision and Strategy, published January, 1996. Council of Scientific and Industrial Research, New Delhi, India.

Eleventh Five Year Plan (2007-2012), 2006. Council of Scientific and Industrial Research, New Delhi, India.

http://www.nstmis-dst.org.

India 2004: A reference annual. Publications Division, Ministry of Information and Broadcasting, Government of India.

Naresh Kumar, 2004. Managing R&D in a Knowledge Intensive Research Organisation. The Case of CSIR India". Proceedings 9th Asia-Pacific S&T Management Seminar, Tokyo, Japan.

Ninth Five Year Plan, 1996. Council of Scientific and Industrial Research, New Delhi, India.

Personal Communication with Dr. H.R. Bhojwani, Advisor to Minister of Science and Technology, New Delhi, India.

Personal Communication with Prof. Fujio Niwa. GRIPS, Japan, September, 2006, New Delhi.

Rajagopal, N.R., Qureshi, M.A. and Baldev, S., 1991. The CSIR Saga, Publications and Information Directorate, New Delhi, 1991.

Shukla, Rajesh, 2004. India Science Report. National Council of Applied Economic Research, New Delhi, India.

Tenth Five Year Plan, 2001. Council of Scientific and Industrial Research, New Delhi, India.

Chapter 8

Indonesian S&T policy and development and S&T indicators

Dudi Hidayat

Abstract

Science and Technology (S&T) development has become a serious concern of the Government of Indonesia since late 1970s. This was indicated by rapid development of scientific infrastructure in Indonesia. Science and Technology park in Serpong Tangerang, the satellite city of Jakarta, was built in 1978. The establishment of Agency for Assessment and Application of technology reflected government's commitment to technology utilization for development purposes. The establishment of National Aircraft Industry showed a serious attempt to master high technology. In the area of S&T human resource development, in 1984 the government began a programme to send abroad a significant number of students pursuing degrees in science and engineering. This was the first organized and sustained programme to increase the availability of high quality S&T human resources.

Since early 1980s the Ministry of Research and Technology has implemented various policies and programs to promote S&T development for economic development. However, as in many developing countries, the link between S&T and economic development is still very weak. Public Research Institutions (PRI) in Indonesia, the major R&D performer, are still perceived as not giving much economic benefits to the nation. They are still disconnected from the user of their research, particularly, the industry.

This paper will take a short look at S&T policies in Indonesia. It begins with a view on S&T performance, followed by a description of institutions responsible for policy making. It continues with discussion on R&D performer and funding sources. It further discusses in sufficient detail, the funding mechanism and argues that the mechanism does not provide a correct environment for promoting collaboration between research institutions and private industries. The paper ends with a discussion on the development of S&T indicators in Indonesia, including experiences with R&D and *innovation survey, and suggests an urgent need for a better system for R&D data collection.*

Keywords: Indonesia, Line-items budgeting, R&D funding mechanism, S&T indicators, S&T policy framework.

Indonesian S&T performance

The performance of science and technology sector of a nation can best be indicated by the number of publications in international journals and the citation index. Figure 8.1 shows the development of international journal publications in Indonesia for 10 scientific fields as contained in the database of the Institutes for Scientific Information (ISI) (Hidayat *et al.*, 2005). Two fields stand out above other fields: Clinical Medicine, and Plant and Animal Science.

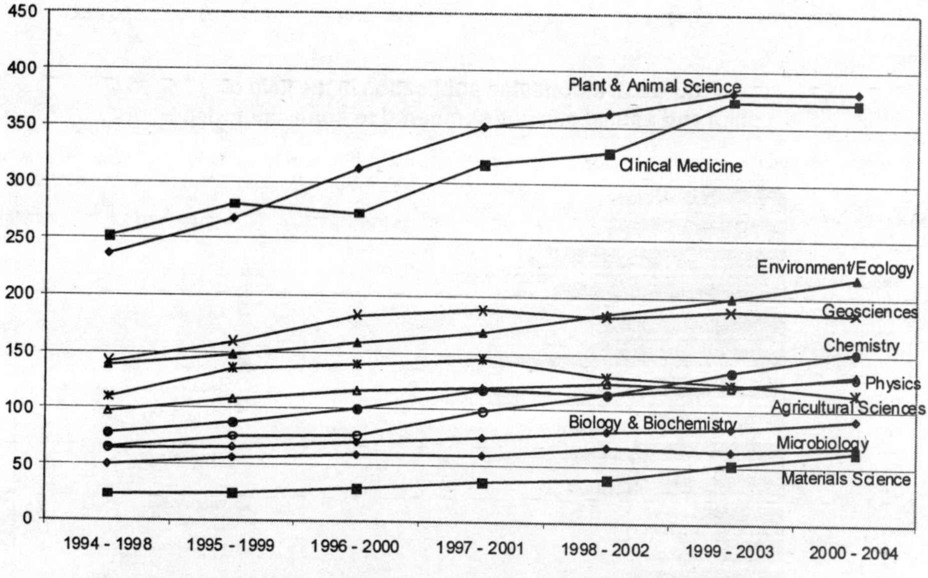

Figure 8.1: Number of S&T publications in Indonesia, 1994-2004

However, compared to neighbouring countries, Indonesian publications in these two fields are still falling behind. Figure 8.2 displays the comparison for the field of Plant and Animal Science (Hidayat *et al.*, 2005).

In terms of the number of citations, Indonesia has two scientific fields with a high citation number per paper. Figure 8.3 shows that in the field of Geosciences and Immunology, among the 6 countries, publication from Indonesia has the highest citation per paper *viz.*, 6.62 for Geosciences, and 14.58 for Immunology.

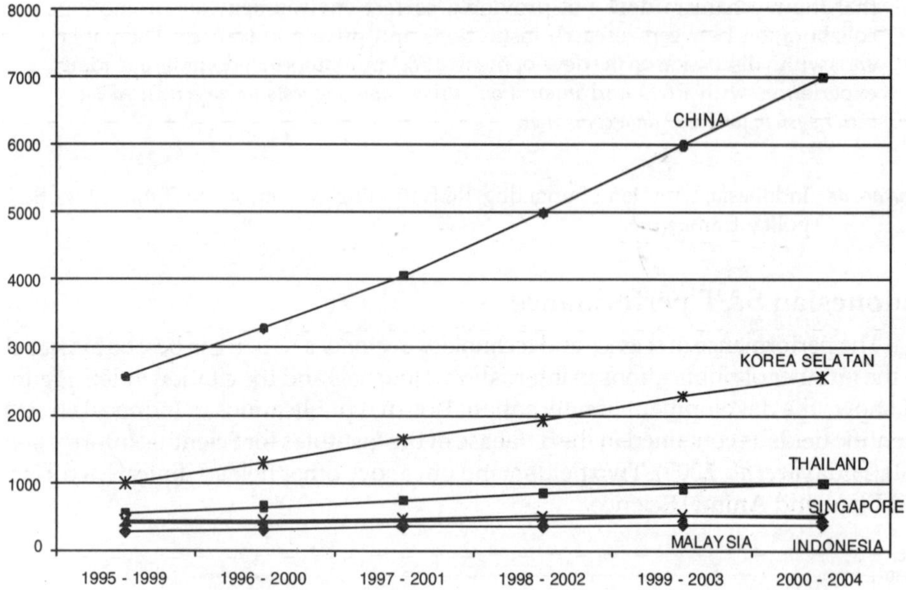

Figure 8.2: Indonesian publication in the field of plant and animal science compared to some countries

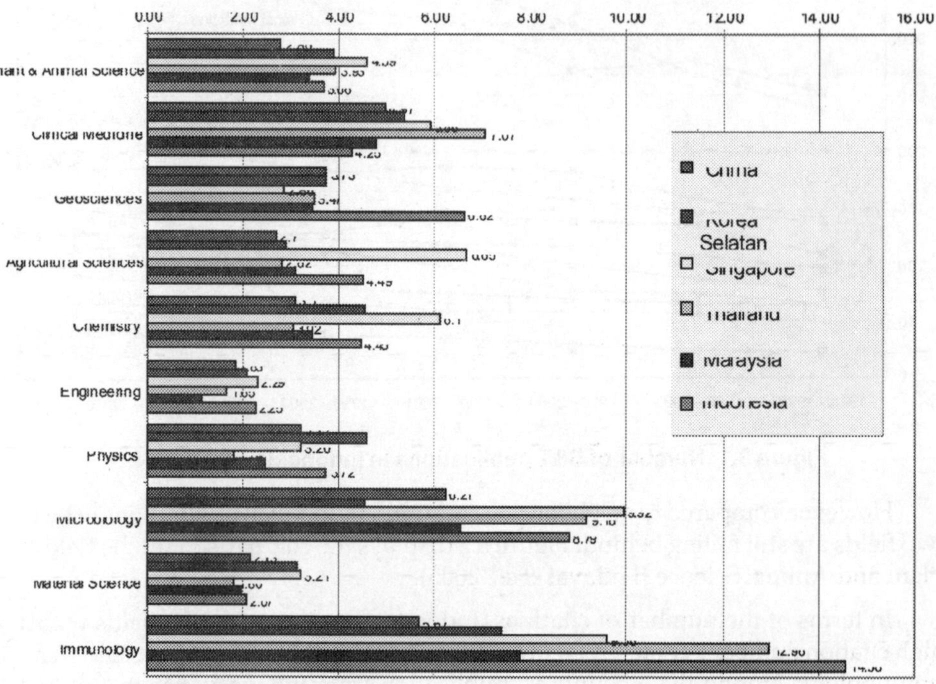

Figure 8.3: Number of citation per paper in a number of fields

The performance of S&T sector is also reflected in the economic performance of the nation. As part of the National Innovation System (NIS), S&T sector should be contributing to the nation's economic performance. The Indonesian NIS is an example of a weak system. Although the major components of the system such as Universities, Government research laboratories and industries already existed for more than four decades since the independence of the country, the linkages among them were still weak. Learning from successful programme implementation in developed countries, the government has made some programmes to stimulate research collaborations between industries and the government research laboratories and universities. Unfortunately, the programs have not been successful in addressing the issue of disconnection between industries and research laboratories and universities. This suggests that policy formulation is country specific. Compared to the situation in developed countries, the intensity of research in industries is very low. With a very loose definition of research, only about 20 per cent of national research expenditure is allocated for the industries. Furthermore, it is most likely that what industries need is not research, but the application of proper technologies in their production line to increase productivity and efficiency. Universities and government research laboratories do not help industries in these technology applications. They prefer to conduct research but, unfortunately, at the end of research programme, it was often found that such results were of no use to the industries.

As a result of the weak system, the system's performance is very low compared to other neighbouring countries such as Singapore, Malaysia, Philippine and Thailand. In 2002, the ratio of the medium and high technology exports to the total manufacturing exports was 31.3 per cent. This is a small amount compared to Singapore, Malaysia, Philippine and Thailand that amounted to 78.9 per cent, 76.2 per cent, 81.8 per cent and 60.3 per cent respectively (UNIDO, Industrial Development Report 2005). Similarly, the number of Indonesian patents registered in the US patent office (UPSTO) since 1976 is only about 200. Singapore has 2,721 patents registered in the UPSTO. For further comparison, Japan has 525,348 patents and Korea has 30,818 patents.

S&T policy making in Indonesia

The major institution responsible for S&T Policy in Indonesia is the Ministry of Research and Technology. It has 5 Deputies covering the following areas of concern:

1. Development of research, science and technology
2. Society dynamics
3. Research, science and technology programme
4. Development of national system of science and technology
5. Utilization and dissemination of science and technology

By law (No. 18, 2002), the Minister is responsible for formulating science and technology policy covering the policy direction, main priorities and policy framework for science and technology development. This policy should be stated formally in a document called "National Science and Technology Strategic Policy" (Jakstra

IPTEKNAS). For the next five years (2005-2009), this policy document has been promulgated on the 08th of August through the Minister's Decree No. 111/M/Kp/VIII/2005.

To assist the Minister in policy formulation, the Minister has issued a decree No 91/M/KP/V/2005, appointing 68 members of the National Research Council. The members come from the various public and private sector organizations. This includes non-departmental and departmental research organizations, government own strategic industries, private industries, public and private universities.

The recent amendment of the Constitution (article 31; verse 5) has stipulated that the government shall promote S&T development to foster the nation's well-being and civilization. This stipulation has substantiated the importance of S&T for the nation, and therefore, the government should set up policy to develop S&T for the purpose of increasing the nation's well being.

Furthermore, to lay a foundation for the improvement of Indonesian National Innovation System (NIS), in 2002 (Law no 18) the parliament had issued a bill called "The National System of Research, Development and Application of Science and Technology". It states the role and the responsibility of four major element of the system, *viz.*, Government, University, Industries and Society in general.

However, the idea of promoting development and interaction between these elements through such a bill is highly contestable. The bill does not mention explicitly how the government can play a role in conjunction with market mechanism. Although the bill states that the private sector shall allocate a portion of its profit for research and innovation activity (article 28), it is clear that the government cannot enforce the private sector to do so. There is no control mechanism whatsoever available to government to enforce the private industry's compliance with this requirement. The decision of firms whether or not to allocate funding for research is mainly determined by the market incentives. Moreover, the experience of developed countries showed that their 'well-established' NSI was not due to the result of a bill. Rather, they were the results of a long historical process driven by market mechanism that is facilitated by the appropriate government choices of various incentive instruments. Thus, rather than a bill, it is the development of various appropriate policy instruments that will influence the firms' decision in regard to research and innovative activites.

The use of data in policy making is not yet a common practice. This is partly due to the strategic planning practice so far in the government institutions and partly due to the difficulty of getting reliable data. However, since the 2004 R&D survey, we begin to see, although not widely, the use of indicators in policy discussion. One indicator is very often heard in policy discussion *i.e.*, GERD as percentage of GDP. It is a very low value *i.e.*, 0.05 per cent in 2001 that has raised the issue of whether the country has invested sufficiently in S&T. Unfortunately, other data such as expenditure distribution across many field of research has not been heard in policy discussion.

Conceptually, all S&T policies and programmes should consider the entire above legal framework. The following picture shows the overall framework for policy making in Indonesia.

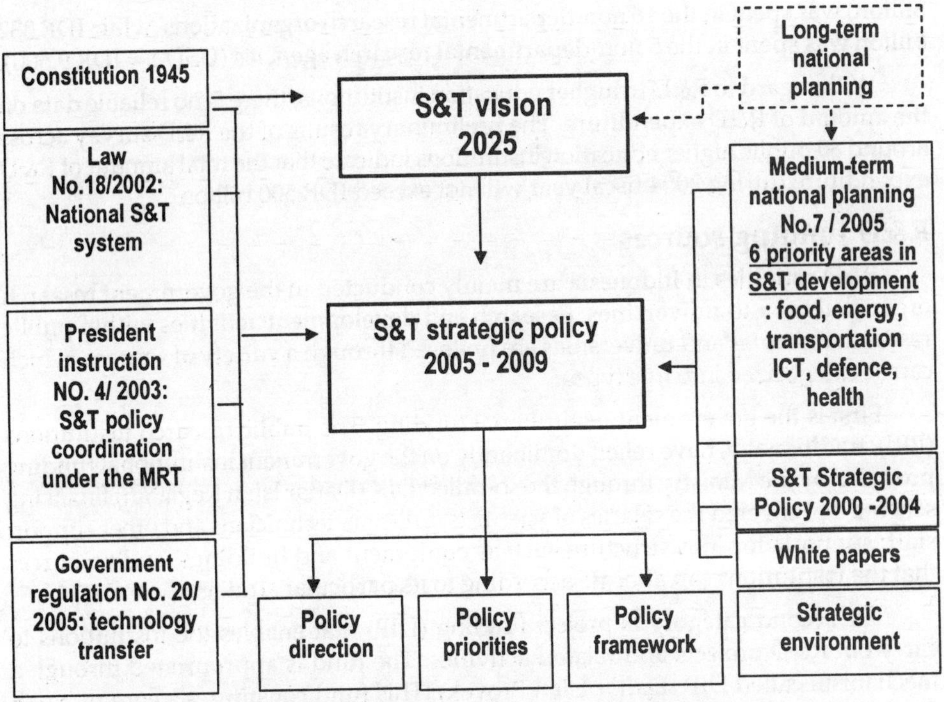

Figure 8.4: Indonesian S&T policy framework

R&D performers

The first thing to note about the Indonesian S&T system is that it is almost entirely funded by the Government and is performed in the public institutions. The very small number of papers and patents from private sources in Indonesia indicates that the R&D expenditure in the private sector is low.

The public R&D system has three components. The first is under the direction of the Minister of Research and Technology non-departmental research organizations, the second under the control of various Ministers departmental research organizations including the Ministers for Industry and Trade, for Minerals and Energy, for Defence, for Forestry and for Agriculture. The third is the universities under the direction of the Ministry for National Education.

The Ministry of Research and Technology directly coordinates 7 major non-departmental research agencies that include Indonesian Institute of Sciences (LIPI), Agency for Assessment and Application of Technology (BPPT), National Aerospace Agency (LAPAN), National Nuclear Power Agency (BATAN), National Standardization Agency (BSN), National Coordinating Agency for Survey and Mapping (BAKOSURTANAL), and National Nuclear Power Supervising Agency (BAPETEN).

R&D survey in 2004 showed that during the 2003 fiscal year, the government sector had spent IDR 1.164 billion on R&D. A great deal out of this amount (IDR 832 billion) was spent in the 16 non-departmental research organizations, while IDR 332 billion was spent in the 5 non-departmental research agencies (USD 1 = IDR 9.500).

With regard to R&D in higher education institutions, there is no reliable data on the amount of R&D expenditure. The preliminary results of the R&D survey across around 80 public higher education institutions indicate that the total amount of R&D expenditure during 2004 fiscal year will not exceed IDR 500 billion.

R&D funding sources

R&D activities in Indonesia are mainly conducted in the government research organizations and universities. Research and development activities within public research institutes and universities are financed through a variety of sources, which can be categorized into five types.

First is the government institutional funding. The public research institutions (PRI), traditionally, have relied dominantly on the government institutional funding provided by the Ministry through the so-called DIK (Daftar Isian Kegiatan) financing scheme that covered the salaries of scientists, full-time technicians and other support staff; supports for infrastructure such as equipment and buildings, and resources that the institutions can allocate according to its particular strategy.

The second category is project funding (DIP) that enables the institutions to carry out R&D projects and related activities. The fund is appropriated through a mechanism called DIP (Daftar Isian Proyek). This fund constitutes a core research funding by which the majority of R&D activities are funded.

The third category of research funding is research or service contracts awarded to institutions by the government departments, agencies and various other organizations, including private sectors. For this category of funding, although the funding does not come from public money, PRI should follow the so-called DIKS procedure that is governed by Law No 20, 1997. The law is well known as PNBP Law that governs the management and utilization of government institutions' external earnings. This law is perceived as a disincentive for getting more external earnings. It is only in May 2005, a government regulation number 20 was issued that exempts the government research institutions from the PNBP law. It remains to be seen whether the new government regulation will give more flexibility and autonomy to the government research institutions.

It has to be noted that since the year 2005, the administration of these three funding sources are combined in one administration called Daftar Isian Pelaksanaan Anggaran (DIPA). This united budget system is expected to prevent inefficiency and to result in a better accountability.

The fourth category consists of various project grants and incentives provided by the Ministry of Education and the Ministry of Research and Technology (MRT) through a number of schemes such as RUT, RUK, RUSNAS, RUTI and various incentive schemes. Recently, beginning this year (2006), all these schemes are transformed into 5 schemes, namely, basic research incentive, applied research incentive, development of S&T capacity of production sector, acceleration of S&T application and diffusion, and the National strategic research programme.

Through these schemes, research organizations are given incentives to conduct research in collaboration with other national or international research institutions as well as with the industries. It is important to note that unlike other funding sources, which are more or less determined on incremental basis, this source of funding is distributed to research institutions on a competitive basis.

Finally, the fifth source of funding is the loans and grants from international agencies or international non-government organizations such as the Ford Foundation and JICA, just to mention a few of them. Loans are appropriated through DIP mechanism as will be described below. Some grants are appropriated through DIP mechanism, some through a mechanism agreed upon by the awarding organization, which is far more flexible than DIP mechanism.

R&D funding mechanism and regulation

This section gives an account on the mechanism through which PRI acquire the above mentioned research funding sources, followed by a short analysis of its implication to the working relationship of PRI with industries.

DIP/DIK

The majority of the funding, that includes loans and grants from international agencies, is allocated annually through DIP and DIK mechanism. The fiscal year starts off at January and is concluded at the end of December. The mechanism to acquire the funding can be shortly described as follows. At the beginning of the second quarter of the fiscal year, every research institution should submit budget proposals for the next fiscal year to the Ministry of Research and Technology (MRT) and the committee under the parliament. Under coordination of the MRT, the budget proposal is discussed and negotiated with the National Agency for Development Planning (BAPPENAS) and the Ministry of Finance (MoF). Final approval should be obtained from the MoF and the parliament (DPR).

The approved budget contains a very rigid and detail description of the kind of eligible expenditures and the amount to be spent for each item. The expenditure items are usually categorized into wages, travelling cost to a particular destination, material, physical equipment, etc. Disbursement is only allowed for those items stated in budget document. One cannot easily shift the use of money from one item to another. Shifting, if allowed, often means long waiting time for approval. This creates inflexibility in the use of money, and is against the very basic nature of research: *uncertainty*. For a research endeavor, one cannot plan with high certainty to which direction research will progress, what material with what volume will be needed, what city to be visited for how long and so on. It is, therefore, preferable for researchers, if they are given flexibility and autonomy to decide on what item the money will be spent. The only requirement is that the item should be eligible for spending and that the spending is technically well grounded, so that at the end of the fiscal year the researchers can justify their spending decisions. Of course, they should do it in a transparent and accountable way by, among others, keeping spending records.

DIKS (in accordance with PNBP law)

Although the new regulation has been issued, DIKS mechanism is still a procedure to be followed by PRI if it is to have collaboration with the private sector. In order to that the new regulation can be operationalized, the article 47 of the regulation states that the Minister of Finance has to issue operational guidelines. Until now the guidelines have not been issued. Like DIP and DIK, the proposed amount of money to be earned from research and service contracts should be submitted one year ahead. This is also against the uncertainty nature of research. Most of the times, revision of the proposed budget is needed to suit to practical problems in the fields. Although revision during fiscal year is possible, the revision proved to be time consuming and therefore impedes the service delivery to the client.

Furthermore, any payment by the client to PRI's account should be transferred to the states treasurers' account, from which the institution can then withdraw the money to finance the activities in the same *bureaucratic* way as DIP or DIK funds. This inflexibility puts constraints on the performance of PRI in delivering its services.

Another important issue regarding this mechanism is that all the remaining external earning at the end of the fiscal year cannot be carried over to the next fiscal year. It should be transferred back to the states treasurers' account, giving PRI no opportunity to utilize the fund for its strategic purposes. Had they been given this opportunity, the capacity of PRI to develop could have been greater.

Apparently, the underlying rationale for justifying this mechanism is that because PRI utilizes public money and facilities, any external earning occurs as a result of utilization of the facilities should be returned to the public, *i.e.* the national treasurers' account. This rationale can be traced back to ICW Law of 1925. Only in this way, it is perceived, that the return to the nation on public investment in research projects and facilities is secured. However, one should not forget a bigger return in the form of improved capability and capacity of research institutions. In fact, this kind of return is a lever that can deliver bigger returns, *i.e.* innovation in industry sector. If this is understood and agreed upon, there is no reason why it is not possible to use the external earnings as an investment in PRI. As such, PRI should be given more autonomy and flexibility to manage external earning as it seems to fit its strategy. It is reasonable and obvious to state that, it is the PRI that know better about the effective utilization of external earning, and not the treasurer of the State.

Given the above discussion, the challenge is how to give more autonomy and flexibility to research institutions without violating the accountability principles. It remains to be seen whether the operational guidelines to be issued will answer the challenge satisfactorily.

Project incentives

With regard to various schemes of project grants and incentives, it could be stated that the schemes are more or less like a block grant mechanism. Every year, the ministry invites research proposals from all the research institutions and carries out a selection procedure to select research proposals, and allocate funding sources to the proposals that are considered having high quality. The selection process should include stakeholders and industry representing potential users of the proposed research.

The agreed amount of funding is disbursed based on the contract that allows the payment of the total amount in three or four payment terms, based on the achievement of agreed milestones. This gives some degree of flexibility to the researchers to manage the financing of their research activities. For this reason, the disbursement based on the contract is preferable to the disbursement through DIP or DIKS procedure.

It should also be noted that unlike other schemes that finance research activities annually, this funding scheme finances research that might last from 1 to 3 years. Given the *uncertain* and *long-term* nature of research, three years budgeting system is preferable to annual budgeting.

It could be concluded that compared to other schemes, this scheme is more competitive, demanding and flexible enough to attract researcher's interest to apply. Unfortunately, only a tiny fraction of the total national research budget is allocated through this scheme. It is recommended that in the future, more research funding should be allocated through a mechanism mentioned above.

Grants

With regard to grants from both national and international agencies, it is sufficed to indicate that the mechanism, by which grants are appropriated, is usually based on the contract arrangement between the awarding agency and the awarded.

To summarize, the Table 8.1 shows the characteristics of various funding mechanisms described above.

Table 8.1: A summary of various funding sources

	Timing	*Basis*	*Disbursement*	*Flexibility*
DIK/DIP	1 year	Incremental	Itemized	Low
DIKS	1 year	Incremental	Itemized	Low
Project incentives	1–3 years	Competitive	Contract	High
Grants	As agreed	Competitive	Contract	High

The foregoing discussion indicates that the government financial management takes the form of line-item budgeting system. There are several features of the traditional line-item budgeting. First, money is allocated to those items considered as input into administrative process like staff, equipment and all incidental items used in running a programme or a department. Secondly, there is a tendency to base the forthcoming budget solely on the record of the previous financial year. This is incremental budgeting where the budget represents an incremental increase on the previous year.

The line-item budgeting has several advantages. As Wilenski points out (Hughes, 1998), the line-item budget is, first, an ideal mechanism to limit expenditure to the amounts and to the items voted in the appropriations. Secondly, it provides a certain degree of flexibility if across-the-board cuts have to be made mid-year for macro-economic purposes. Thirdly, the traditional budget makes budgeting easier than having to justify and more manageable: arguing from a historical base is easier than having to justify each item from scratch; choice is routinised.

However, there are several problems with line items budgeting. First, it stresses input rather than output. There is no necessary relationship between output and input. Although the relationship is stated formally, it is often unclear or even illogical. There is no formal assessment to determine whether the stated output is really can come from the stated input. Secondly, the line-item budgeting is quite short-term, usually only for one year and to be carried out incrementally with inadequate critical appraisal. Thirdly, the budget statement is quite rigid where all expenditure items are specified. The managers have little flexibility in moving resources from one kind of spending to another. Amounts allocated to particular items are often spent invariably, as otherwise the budget for the next year may be reduced to the said items. Finally, as there is only limited data linking the costs to the achievements, there is no satisfactory way of judging whether taxpayers' money serves the desired ends, or is doing so efficiently or effectively.

The inflexible and bureaucratic finance mechanism in dealing with the private sector has created non-conducive working environment for scientists who are willing to collaborate with industries. This situation even discourages PRI to do contract research or 'force' PRI to do contract research without compliance with DIKS procedure. In the latter situation, although it seems to work well, it is not in accordance with the regulations and is not preferable in the long run. In turn, this condition may lead to widening of the disconnection between PRI and industries.

Experience of many successful PRI in a number of countries has shown that PRI should take more active approaches towards the industrial collaboration. The tacit nature and path-dependency of industrial innovation requires the researcher within PRI to work closely with industry if they are to help the industry to create industrial innovation. Therefore, the government policies to promote research collaboration by various schemes should be designed to serve this purpose.

In some countries such as UK and USA, to stimulate commercialization of research result and create a positive attitudes towards commercialization, the government has transferred the rights to exploit research results from the Research Councils to the universities and government laboratories where the research are being conducted. This policy has boosted an increasing number of innovations. Similar evidence can be found in Korea where the country has been very successful in developing technological capability. Since the establishment of Korean PRI in 1970s, autonomy has been given to them to manage research funding from the government.

Furthermore, it is now widely recognized that the PRI that successfully serve the industry, and therefore, has a good linkage with industry, are those PRIs that are managed in a business enterprise style. It is boldly stated by a group of scholars, who conducted a comparative study over nine PRIs in a number of countries, *"Operational reasons for poor Research and Technology Institutions (RTI) performance often stem from a failure to run the RTI as if it were a business"* (Arnold *et al.*, 1998) This is also in accordance with the proposition made by the MRT in the National S&T Strategic Plan, *"Therefore, research institutions should be managed as if it is a business enterprise, it should be led by an executive, not by a Chairman or a Head with bureaucratic mentality and style .... (Jakstra Ipteknas, MRT, page 44).*

With the issuance of New Budgeting Law in 2003 that encourages the performance budgeting system, expectation is high that the financial management in research institutions will be improved. However, as with many of the systems in transition, one cannot expect to see changes in the system very quickly. The challenge remains as usual in the operationalization of the new system.

The development of S&T indicators

Along with S&T development, the development of S&T indicators has been under way. However, the first attempt to systematically develop S&T indicators was only beginning in the early 1990s. As part of the national project on Science and Technology for Industrial Development (STAID), a sub-project was launched to develop the indicators. This project had developed quite comprehensively a wide range of indicators covering three main sectors: Government, Industry and Higher Education sector and a quite comprehensive report of indicators developed during the project lifetime had been produced. Up to present times, the report is still the most comprehensive report ever made about science and technology indicators. It is unfortunate that since the completion of the said project in 1993, no significant and comprehensive survey was conducted. Nevertheless, to a lesser extent, the survey is conducted by the Centre for Science and Technology Development Studies (PAPPIPTEK), which is one of research centres in Indonesian Institutes of Sciences (LIPI).

Until 2004, apart from industry data, all data were collected from secondary sources. In the government sector, R&D expenditure is derived from a budget statement from the ministry of finance. As such, the inclusion of which part of the funding is really R&D expenditure is not always well-defined, and therefore is prone to mistakes. In addition, it had not been possible to identify the field of research and the socio-economic objective of the research. The number of R&D personnel is derived from the national employment office. This data too are not always accurate.

For Higher education sector, the data are derived from the higher education statistics. It is very difficult to get good estimates of R&D expenditure in this sector. Regarding R&D personnel in this sector, it was always assumed that lecturers that hold an S1 degree (Above Bachelor degree, below Master degree) and above are researchers, which may not always be the correct case.

For industry sector, a survey was conducted in 1999 and 2001. It was limited to medium and large companies in the manufacturing sector in Java province.

In 2004, the first census survey in the government sector was conducted by the Ministry of Research and Technology. It covered research centres under the Ministry of Research and Technology and other line-ministries. Almost 90 per cent of government research centres responded to the survey. The design of the questionnaire followed the Frascati Manual. For the first time, it has been able to get more reliable set of data on R&D expenditure and personnel in the government sector. It was also able to identify the field of research and socio-economic objective of the research. The survey data provided information to policy community that how much money is spent for R&D, where the funding comes from and where it was spent at. The survey was repeated in 2005.

In 2005, PAPPIPTEK-LIPI conducted a census survey for the public higher education sector. It covered almost 500 hundreds faculties from around 80 universities. However, the response rate was only around 45 per cent. Part of the problems was a bad record keeping practice in some universities, and the widely spread geographic nature of universities across Indonesian archipelago.

This year, PAPPIPTEK-LIPI commissions the Central Bureau of Statistic (BPS) to conduct a census survey across all medium and large manufacturing industries (number of employees 20-99 is medium, 100 and above is large) across Indonesian archipelago. It will cover around 14,000 out of in total around 20,000 establishments. The preparation for the survey is now on-going. This survey covers not only R&D activities, but also includes some questions related to innovation activities.

Due to the limited budget, enterprises with employees of 20-29 (around 7,000 enterprises) are sampled, but only 700 enterprises will be surveyed.

In the higher education sector, only public universities were surveyed previously. Budget limitation was a prime consideration for this. In addition, it was assumed that private universities barely conduct R&D activities. Only a limited number of them conduct R&D. This year we plan to include the private universities as well.

A need for a better system for S&T data collection

A proper and correct R&D statistics collection requires significant amount of resources. Until recent times, the government did not pay enough attention to this matter. There is no government unit assigned to collect R&D data. It was only in early 2004 that the importance of R&D statistics was widely recognized. There was still no unit officially assigned and therefore responsible for data collection. Surveys conducted in 2004 and 2005 were done by PAPPIPTEK-LIPI (The Centre for S&T Development Studies) and a unit in the Ministry of Research and Technology, which were not specifically designed to do the job.

We think that R&D data collection will be significantly improved if a special unit is set up to carry out the task. Such a unit should be designed properly and given sufficient resources to do its task. The most ideal case is that such a unit be established in the Central Bureau of Statistics. If not, a unit can be established under the Ministry of Research and Technology. The best option is by empowering PAPPIPTEK-LIPI to carry out R&D data collection properly and this option is worth considering due to the long experience of PAPPIPTEK-LIPI in this direction.

References

Arnold, E. *et al.*, 1998. Strategic Planning in Research and Technology Institutes. R&D Management 28 (2), pp. 89-100. U.K.

Hidayat, D. *et al.*, 2005. Study on S&T development and its Determinant. A research report in Indonesian language, Indonesia Institute of Sciences, Jakarta, Indonesia.

Hughes, O.E., 1998. Public Management and Administration, An Introduction. Macmillan Education Australia PTY LTD (4/627Chapel St, South Yarra 3141, Victoria Australia).

The Ministry of Research and Technology, 2000. National S&T Strategic Policies (webmstr@ristek.go.id).

Chapter 9

A glance at the Iranian science and technology (S&T) indicators

Seyed Mohsen Masoumzadeh

Abstract

Iran despite many constraints and limiting factors has done quite well in achieving self-reliance and capabilities in some industrial sectors such as pharmaceutical, biotechnology and new materials but it needs seriously to increase R&D expenditures as well as to improve the other science and technology indicators for better performance. Iran has expanded its universities and the number of faculty members and students has grown manifold.

Keywords: Indicators, Iran, R&D, R&D Expenditure, Science and Technology, S&T

Introduction

The need for science and technology (S&T) development is intensified in the recent years, particularly, with the revolution of the new economic system, which is characterizing by rapid progress of both globalization and the revolution in information and communication technologies (ICT). S&T is becoming not only the key element of economic growth, industrial competitiveness, but also essential for improving the social development, the quality of life and global environment.

Hence, based on this context many countries shift toward such phenomena to obtain better performance. There is need to better link the science and technology infrastructure to the needs of the productive sector.

Iran has relatively developed its own science and technology infrastructure compared to other developing countries in the region. It has also built up a significant industrial base, mainly by licensing technology from abroad. However, the export of high-tech industrial products is narrow. It is, therefore, imperative to diversify the economy and further broaden the industrial and export base. In order to achieve this, Iran needs to improve the performance of its national innovation system. To this end, Iran has to make serious effort to promote its S&T indicators.

Iranian economy

Iran possesses very favourable geological structures and a good potential for mineral development. It has an area of about 1,648,000 square KM (628,000 square miles) with a population of 69.7 million people in 2005 (Economist Intelligence Unit, 2006). During 2003 the population growth was 1.29 per cent annually. According to the last census in 2003, 34 per cent of Iran's population lives in rural areas and about 23.5 per cent of the employed labour forced engaged in agriculture at the time of the 1996 census (Europa World Year Book, 2005).The agricultural sector provides raw materials for such local industries as sugar, textiles and the vegetable oil refineries. In 2005, according to the Economist Intelligence unit, Iran's Gross Domestic Product (GDP) was 177.9 billions of US Dollars.

Crude oil has had a decisive role in the political and economic development of Iran in the last century with petroleum providing a major source of government income. Oil revenues constitute about 70 per cent of all government income. Thus one must accept its fundamental and dominant role in the Iranian economy. Dependence on oil income is now such that its absence causes great difficulties for the economy of the country and, fluctuation in oil income directly affects the performance of the economy. The oil and gas accounts for more than 11 per cent and the agriculture sector accounts for over 11 per cent of GDP. Historically, as national economies mature, the share of agriculture in GDP declines while that of industry rises sharply. The contribution according to GDP of different sectors is given in Table 9.1.

Table 9.1: Contribution of different sectors in GDP (per cent)

Sector	*2000*	*2001*	*2002*	*2003*
Agriculture	14	13	12	11
Industry	33	36	40	41
Services	53	51	48	48

Source: World Development Indicators, 2005.

Industry

Industry (including mining, manufacturing, construction and power) contributed almost 41 per cent of GDP in 2003/04, and engaged 31 per cent of the employed labour force at the 1996 census. Mining (including petroleum refining) contributed 23 per cent of GDP in 2003/04, although the sector engaged only 0.8 per cent of the working population 1n 1996 (Europa World Year Book 2005).

The structure of manufacturing

Iranian industries include handicrafts, small rural and urban workshops, large-scale consumer- goods, and modern heavy industries. Small industrial workshop is those employing 9 individuals or fewer both in rural and urban areas. A majority of rural units are engaged in hand-made carpet weaving and textile, and the rest in food and other activities. Total value-added by this sub-sector to the non-oil GNP is low. Urban small workshops consist of bakeries, blacksmiths shops, carpentry, shoe-making and the like.

Large-scale manufacturing enterprises are defined as those employing 10 or more workers. This sector comprises of 9 major groups:

1. Food, beverage and tobacco
2. Textile, clothing and leather industry
3. Wood and wooden products industry
4. Paper and paper industry
5. Chemicals and chemical
6. Other non-metallic mineral
7. Basic metals
8. Machinery and equipment
9. Miscellaneous industry

Based on the Iran statistical year book 2004 there were 363428 urban industrial units in Iran; no more than 24000 were classified as large formal companies but these accounted for the lion's share of value-added in manufacturing. The manufacturing, value added of GDP in 2003 was only 12.5 percent. This share was much smaller in Iran than in comparable countries (Figure 9.1). It engaged 17.8 per cent of the employed labour force in 1996.

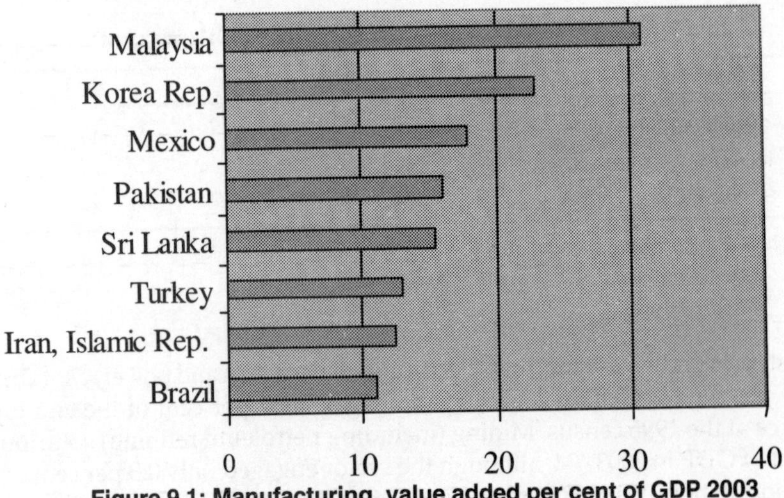

Figure 9.1: Manufacturing, value added per cent of GDP 2003
***Source*: World Development 2005**

Science and technology (S&T) indicators

Science and Technology (S&T) is often defined as a system including all the research institutions/organizations, essential to the education of scientific people, for instance, R&D organization, professional societies link individual scientists to others and to their socio-economic environment.

Research base

In 1920 the first governmental research centre was set up in the field of health care, named "The Pasteur Institute of Iran". The number of research institutes and research units has increased dramatically since the Islamic Revolution. Up to 2001 there were 216 Institutes and research units (excluding university research centres). 112 units are affiliated with the Ministry of Science, Research and Technology. The figures are given in Table 9.2.

Table 9.2: Number of Iranian governmental research institutes

Year	*1920*	*1951*	*1961*	*1971*	*1981*	*1991*	*2001*
Number	1	5	19	57	86	191	216

Source: Science Policy Quarterly 1994+UNDP, 2005.

University education

In 2005-2006, the private sector accounted for 52 per cent of all enrolments in university degree programmes, as compared with 48 per cent for the public sector. 28 per cent of all university enrolments are in engineering and a further 11 per cent in pure sciences. Medical science account for an additional 6 per cent, social science 44 per cent, agricultural 7 per cent and art 4 per cent of students enrolled at the tertiary level.

Iran has expanded her universities and the number of faculty members and students has grown manifold. According to official sources, there were some 751,883 students enrolled in Iran's public universities in 2005-06, in addition to the 1,256,203 students enrolled in private sector, based on statistics of higher education by Institute for Research and Planning in Higher Education (IRPHE) in 2006. The number of faculty members has been increased to 79,617 (40536 full time and 39081 part time members) in comparing with previous years in academic year 2005-2006 according to IRPHE statistics. There has been a significant increase in the number of scientific articles published in international journals, which rose from 281 in 1992 to 3349 in 2004 this shows considerable strengths in scientific research. The comparable data of world outputs during 1981 to 2002 indicated that the world's total S&T papers of 13,278,111, Iran contributed 10,090 or 0.08 per cent. Iran ranked 57 among 178 countries, behind Malaysia (ranked 55), Saudi Arabia (49), Turkey (38), Mexico (35), South Korea (25), Brazil (23). USA is the leading country in S&T knowledge outputs at ranking number 1, contribution 37.64 per cent world science. UK ranks second (9.06 per cent), followed by Japan (8.13 per cent), Germany (8.02 per cent) (MASTIC, 2004).

R&D Expenditure

In the transition to knowledge-based economy, R&D expenditure has been considered as an important indicator for evaluation a country's investment in its knowledge base. The input (resources) indicators are often classified into Financial and Human resources. Financial resource or input indicator includes R&D expenditure. The most widely accepted indicator for evaluating and comparing science and technology efforts in different countries. In addition to financial resources, human resources are central to research and technological change activities.

In last two decades the Iranian R&D expenditures was fluctuated between 0.2 and 0.4 per cent of the GDP. In comparison with the other developed countries is far behind (Table 9.3).

Table 9.3: R&D expenditure (per cent of GDP)

Country	*1999*	*2000*	*2001*	*2002*
Germany	2.43	2.49	2.50	2.52
Iran	0.34	0.31	0.38	0.52
Japan	2.96	2.99	3.07	3.12
Korea, Rep.	2.25	2.39	2.58	2.53
Sri Lanka	-	-	-	-
Turkey	0.63	0.64	0.72	0.66
USA	2.64	2.71	2.73	2.65

Source: World Development 2005+R&D Society of Iranian Industries and Mines 2006.

The major source of Iranian R&D expenditure comes from the governmental sector while the share of private sector is negligible. Human resources in science and technology (such as the number of science and technology and the number of researcher employed in R&D) are one of the key resources for economic growth. Iran with 484 researchers per million in R&D indicates much lower comparing with other countries (Table 9.4). However, this number of researchers may not be adequate to propel the economy to higher level.

Table 9.4: Researchers in R&D (per million people): 1990-2003

Country	*Germany*	*Iran*	*Japan*	*Korea, Rep.*	*Sri-Lanka*	*Turkey*	*USA*
Researchers	3222	*484*	5085	2979	197	345	4526

Source: World Development Indicators, 2005.

Iranian National Innovation System (NIS)

The major elements in the Iranian National Innovation System in terms of education and research institutes, entrepreneurship centres, large enterprises, technology parks, incubators are governmental. The contribution of private sector to innovation and technology development is limited and should be strengthened. All innovation activities are concentrated in research centers and universities. Only a small number of large firms have their own in-house research and development department. The government has started encouraging companies to establish R&D units, but the impact so far has been limited in terms of actual innovation activities at the firm level.

The available resources and various funding schemes are not well understood by the innovators and innovating companies as these need to be well coordinated. There is a need for interaction and coordination among these initiatives of government and other agencies and as it is also required to be strengthened. The process of Innovation activity in Iran is supply-driven that in the sense that government basically set R&D programmes and projects to be undertaken by research institutes/universities. Consequently, decisions on R&D priorities tend to be based on perceived immediate economic or social needs rather research programmes based on consumer demand (UNDP, 2005).

Conclusion

1. Iran with about 0.4 per cent of GDP allocated to R&D is far behind industrialized countries. While increasing the R&D expenditure seems mandatory for developing countries.
2. Contribution of private sector in R&D expenditure is negligible.
3. The number of scientists and Researchers much lower and will have to be increased considerably.
4. Iran has expanded her universities and the number of faculty members and students has grown manifold.
5. The numbers of scientific articles published in international journal has been increased sharply. During 1981to 2002, in comparison with the world's total S&T papers, Iran contributed 0.08 per cent and ranked 57 among 178 countries.
6. The processes of innovation activities in Iran are driven not by demand but by government plans. The research and development and innovative capacity at the firm level are limited while the firm is cantered at the heart of the national innovation system.

References

Economist Intelligence Unit 2006. "Country Report", The Economist.

Europa World Year Book 2005. 64th Ed., Vol.1, Routledge, London and New York.

Institute for Research and Planning in Higher Education 2006 "Statistics of Higher Education in Iran", Ministry of Science and Technology, Islamic Republic of Iran.

Iran Statistical Yearbook 1383, 2004.

Malaysian Science and Technology Information Centre (MASTIC), 2004. Ministry of Science, Technology and Innovation, Malaysia, "Science and Technology Knowledge Productivity in Malaysia Bibliometric Study 2003".

R&D Society of Iranian Industries and Mines, 2006.

Science Policy Quarterly 1994. "National Research Annual Report 1371, 1992" (in Persian), National Research Council of I.R. Iran, pp.8-128.

UNDP, 2005. "Science, Technology and Innovation Policy Review in the Islamic Republic of Iran", UN, New York and Geneva.

World Bank, 2005. "World Development Indicators". CD-ROM, Washington, DC, based on data from the United Nations Education, Scientific and Cultural Organization; aggregates calculated for the Human Development Report Office by the World Bank.

Chapter 10

Statistical development of science and technology indicators in Malaysia

Kamaruhzaman Mat Zin

Abstract

In Malaysia, S&T indicators are mainly used to guide the government policy makers in developing appropriate S&T policies for the country and to benchmark Malaysia's position as opposed to other countries with the main objective of promoting the country's global competitiveness. The paper highlights S&T indicators in Malaysia including input indicators such as financial and human resources devoted to R&D, and output indicators base on publications, and patents. It also delves upon technology innovation indicators and public awareness in S&T. The country is in support of any effort at international or regional level to promote appropriate and systematic usage of statistical indicators in non-aligned and other developing countries.

Keywords: Malaysia, Patents, Publications, R&D, S&T indicators, S&T policy, Technological innovation indicators.

Introduction

Science and technology indicators (S&T indicators) have been widely used by many economies including developing countries such as Malaysia. The primary purpose of these indicators, which are often used in conjunction with other economic indicators, is to evaluate the status, development and progress of S&T system and their contributions to the economic growth of a country. The importance of S&T

indicators in assessing the country's S&T system has long been recognized in Malaysia. The effort to develop appropriate and systematic S&T indicators for the country was first initiated in 1992.

In Malaysia, S&T indicators are mainly used to guide the government policy makers in developing appropriate S&T policies for the country and to benchmark Malaysia's position as opposed to other countries with the main objective of promoting the country's global competitiveness. The industries also find the information very useful to improve their business competitiveness while the general public would use the information on human resources, for example, in deciding their career path in the field of science and technology.

Malaysia had been adopting OECD's guideline (Frascati and Oslo Manuals) as well as practices commonly followed by selected industrialized nations and have since developed the following indicators to measure and map local S&T performance:

1. Research and development (R&D),
2. Human resource for science and technology,
3. Patents,
4. Bibliometric,
5. Public awareness in S&T, and
6. Technological innovation

The above indicators are periodically evaluated and reported. The information for the development of these statistical indicators comes from many sources, namely, government agencies, associations and international databases. Data on R&D, innovation, bibliometric and public awareness is compiled through survey results.

Highlights of S&T indicators in Malaysia

Research and development (R&D)

Among the input indicators, that are currently used, are financial and human resources devoted to R&D activities. Such information is vital in assessing the R&D efforts and whether the level of R&D is keeping pace with the expansion of the economy and infrastructure, in general.

Main sources of data pertaining to overall performance of R&D activities in the country are obtained through the survey results of public and private organizations. MASTIC has carried out and published a series of R&D reports for the year 1992, 1994, 1996, 1998, 2000, 2002 and 2004, respectively. The possible weakness of obtaining statistics through the survey, which is conducted on the basis of voluntary response, is that the possibility of inaccurate or under reporting data may occur, even though many measures have been taken to minimize such cases. Summary of R&D indicators is as follows:

1. Malaysia's GERD has been steadily on the rise since 1996. From RM 549.2 million in 1996, it grew to RM2.50 billion in 2002. In 2004, the GERD stood at RM 2.84 billion, an increase of RM 0.34 billion over 2002.

2. The GERD/GDP ratio, however, decreased from 0.69 in 2002 to 0.63 in 2004.
3. Current expenditure on R&D has also been on an upward trend. From a low of RM 329.1 million in 1996, current expenditure increased to RM 1,375.2 million in 2002 and subsequently, to RM 2, 196.6 million in 2004.
4. In 2004, there was a marked decline in capital expenditure, a decrease of RM478.2 million (or 42.5 per cent) when compared to 2002.
6. Since 1998, the emphasis has been on applied research (55.4 per cent), followed by experimental research (28.5 per cent) and basic research (16.1 per cent). This trend continued in 2004.
6. The three main fields of research (FOR) for 2004 were the same as that of 2000. They were engineering sciences, information, computer and communication technology (ICT), and applied sciences and technologies.
7. In terms of SEO categories, the top 5 in 2004 were manufacturing, information and communication services (ICS), natural sciences, technologies and engineering, plant production and plant primary products and energy resources.
8. The trend in R&D headcount continues to be on the rise. Since 1998, there has been a steady increase in the number of personnel involved in R&D, from 12,127 in 1998, to 23,262 in 2000, to 24,937 in 2002, and 30,983 in 2004. This represents an overall increase of 6,046 (or 24.2 per cent) over 2002.
9. The total number of women R&D personnel increased from 8,009 in 2002 to 10,544 in 2004. Women made up 34 per cent of total R&D personnel, an increase of 1.9 per cent from 2002.
10. In terms of FTE, there has also been a steady increase over the years. Indeed, there is a 66.7 per cent increase in FTE, from 10,731 in 2002, to 17,886.5 in 2004. Researcher FTE increased by 77 per cent, from 7,157.5 in 2002 to 12,669.5 in 2004. In terms of proportion, researchers accounted for 70 per cent of the FTE.

The bulk of the research in 2004 was conducted in Malaysia. Of the total national R&D expenditure, only RM 32.9 million was outsourced, a remarkable decline (by 89.2 per cent) from the RM 306.0 million outsourced in 2002.

Human resources for science and technology

Human resources involved in S&T within the country are also another important input component in the S&T Indicators. This indicator is aimed at measuring the distribution and availability of trained and skilled personnel in various S&T related areas and sectors of the country.

This input indicator also looks at the achievement of training institutions, primarily, Institute of Higher Learning (IHLs) in supplying the demand for the trained, educated and skilled people in the fields of science and technology. Statistics on enrolments and graduations from IHLs are important in predicting the future supply of S&T personnel. Indicators of human resources for S&T are analyzed from primary data provided by the Ministry of Higher Education and they are highlighted below:

There has been a steady increase in student enrolment in public universities over the last three years as shown in Table 10.1.

Table 10.1: Enrolment in first-degree courses at public educational institution, 2002-2005

Academic year	Number of students enrolled	% Increase
2002/2003	184,190	
2003/2004	192,288	4.40
2004/2005	205,314	6.77

In terms of breakdown by field of study, it can be seen from Table 10.2 that for 2002/2003 the enrolment for science-related fields totals 41.41 per cent, for 2003/2004 it is 39.19 per cent, and for 2004/2005 it constitutes 41.23 per cent of the total enrolment. Thus, it can be concluded that in terms of percentage, there has been no increase of enrolment in science-related fields over the last three years.

Table 10.2: Enrolment in first-degree courses at public educational institutions by fields of study, 2002-2005

	Academic Year					
	2002/2003		2003/2004		2004/2005	
	Number of students	Percentage	Number of students	Percentage	Number of students	Percentage
Natural sciences	12,899	7.00	16,040	8.34	13,336	6.50
Engineering	28,591	15.52	31,508	16,39	35,659	17.37
IT and Computer	15,744	8.55	15,373	7.99	21,159	10.31
Medical sciences	7,214	3.92	7,617	3.96	10,439	5.08
Agricultural sciences	11,816	6.42	4,831	2.51	4,054	1.97
Humanities	31,003	16.83	33,274	17.30	31,483	15.33
Other fields	76,923	41.76	83,645	43.50	89,184	43.44
Total all fields	184,190	100	192,288	100	205,314	100

Table 10.3 provides data on the number of students who graduated in first-degree courses from the public universities, for the academic year 2004/2005.

From the Table 10.3, we can see that in terms of fields of study, the percentage of students graduating from the Arts courses still predominate over the science and technical fields, with Arts graduates forming 54.50 per cent of total graduates, 28.75 per cent for science and 16.85 per cent for technical.

Table 10.3: Number of graduate in first-degree courses

Course	Number of graduate			
	Male	Female	Total	
Arts	7,251	16,581	23,832	
Arts and Humanities	1,985	4,815	6,800	
Economics & Business	4,200	9,710	13,910	54.5
Law	471	738	1,209	
Others	595	1,318	1,913	
Science	4,393	8,196	12,589	
Medicine and Dentistry	481	1,015	1,506	
Agricultural sciences	158	307	465	28.75
Natural sciences	1,702	3,458	5,160	
IT and Computer science	1,656	2,672	4,328	
Others	386	744	1,130	
Technical	4,443	2,962	7,405	
Engineering	3,528	2,043	5,571	
Architecture,				16.85
Town Planning and survey	756	769	1,625	
Other	159	150	309	
Total	16,087	27,739	43,826	100

In terms of gender distribution, the number of female graduates exceeds the number of male graduates in both the Arts and Sciences, but not in the technical field. In the Arts, female graduates outnumber male graduates in all the classified disciplines, namely, Arts and Humanities, Economics and Business, Law and other. In the field of science, female graduates again outnumber male graduates in all science fields, *i.e.* Medicine and Dentistry, Agricultural sciences, Natural science, IT and Computer science and others. In the technical field, male graduates exceed female graduates in the field of Engineering, but not in Architecture, and Town planning and survey. As a whole, in terms of gender distribution, the number of female graduates far exceeds the number of male graduates, with the score standing at 27,739 to 16,087.

Patents

Patents have been recognized internationally as one of the important output indicators to measure level, structure and evolution of inventive activities in countries, industries, companies and technologies.

In Malaysia, patents applied and granted both by residents and non-residents are constantly analyzed in order to map technological dependency, diffusion and penetration. The statistics for the purpose of patent indicators analysis is obtainable from the Intellectual Property Corporation of Malaysia (MyIPO), Ministry of Internal Trade and Consumers Affair.

Table 10.4 shows the number of patent applications filed by Malaysians and Non-Malaysians for the period 1996-2000 and 2001-2005. During this period, the total number of patent applications filed has declined by about 8 per cent. The total number of patent applications applied for the period 2001-2005 has dropped compared to the number of patent applications applied for the period 1996-2000. A total of 27,661 applications were received for the period 2001-2005 compared to 30,056 applications for the period 1996-2000. This decline is largely due to the decrease in the number of applications filed by non-Malaysians. The reason for this decline can be attributed to the world economic slow-down following the 1997-1998 currency crises. On the other hand, the number of patent applications filed by Malaysians for the period 2001-2005 has increased by 97.9 per cent compared to the period 1996-2000. A total of 2,013 patent applications were filed by Malaysians for the period 2001-2005 compared to 1,017 applications during the period 1996-2000. The reasons for this increase were, firstly, more people were aware of protecting their inventions due to the Government efforts in promoting intellectual property rights through intellectual property awareness campaigns. Secondly, the tax incentives and rebates schemes for protecting intellectual property were first announced during the 2001/2002 Budget, and thirdly, Government research institutions and public universities became more proactive in protecting their research findings.

Table 10.4: Patent applications filed in Malaysia by Malaysians and non-Malaysians for the period 1996-2000 and 2001-2005

	1996-2000	*2001-2005*	*Variation (Percentage increase or decrease during the period under review)*
Malaysians	1,017	2,013	+ 97.9 per cent
Non-Malaysians	29,039	25,648	–11.7 per cent
Total applications	30,056	27,661	–8 per cent

Table 5: Patents granted to Malaysians and non-Malaysians by the Intellectual Property Corporation of Malaysia (MyIPO) for period 1996-2000 and 2001-2005

	1996-2000	*2001-2005*	*Variation (Percentage increase or decrease during the period under review)*
Malaysians	215	142	- 34 per cent
Non-Malaysians	4,067	9,253	+127 per cent
Total granted	4,282	9,395	+119

Table 10.5 shows the number of patent granted to Malaysians and Non-Malaysians by MyIPO for the period 1996-2000 and 2001-2005. On the whole, the total number of patents granted to Malaysians and Non- Malaysian has increased from 4,282 to 9,935 for the two periods under review. The increase is largely due to the increase in the number of patents approved for non-Malaysians. While the number of patent applications by Malaysians for the period 2001-2005 has increased by

97.9 per cent compared to the 1996-2000 period, there was no corresponding increase in the number of approvals. In fact, the number of approvals for Malaysians declined during the periods under review. The number of patent granted to Malaysian was 215 for the period 1996-2000 compared to 142 for the period 2001-2005.

Table 10.6: Patents granted based on field of technology classified according to the International Patent Classification (IPC) system

Year	*Section*								
	A	*B*	*C*	*D*	*E*	*F*	*G*	*H*	*Total*
1996	285	323	483	31	76	103	178	322	1,801
1997	151	138	196	13	32	45	82	132	789
1998	104	98	141	4	18	31	64	106	566
1999	132	112	191	9	21	49	68	139	721
2000	61	59	110	8	19	42	36	70	405
Sub-total	733	730	1,121	65	166	270	428	769	4,282
2001	155	233	288	18	44	102	231	399	1,470
2002	206	236	334	19	42	104	228	323	1,492
2003	224	242	296	38	38	119	190	341	1,578
2004	325	377	625	25	50	132	321	492	2,347
2005	333	452	600	30	82	164	316	531	2,508
Sub-total	1,243	1,540	2,243	120	256	621	1,286	2,086	9,395
Total	1,976	2,270	3,364	185	422	891	1714	2,855	13,677
% age	14.4	16.6	24.6	1.4	3.1	6.5	12.5	20.9	100.0

Notes: A: Human necessities; B: Performing operations and transporting; C: Chemistry and metallurgy; D: Textiles and paper; E: Fixed constructions; F: Mechanical engineering, Lighting, Heating, weapons and blasting; G: Physics; and H: Electricity.

Table 10.6 shows the number of patents granted by field of technology. The field of chemistry and metallurgy has the highest number (3,364 or 24.6 per cent) of patents granted for the period 1996-2005. Electricity had the second highest number (2,855 or 20.9 per cent) of the patents granted, followed by performing operations and transporting (2,270 or 16.6 per cent). Human necessity had the fourth highest number (1,976 or 14.4 per cent) and the field with the least number of patents granted was textiles and paper (185 or 1.4 per cent). The number of patents granted in the field of chemistry and metallurgy for the period 2001-2005 was 2,243 compared to 1121, for the period 1996-2000. This represents a 100 per cent increase in the number of patents granted in the field of chemistry and metallurgy. In a similar vein, the number of patent granted in the field of electricity for the period 2001-2005 was 2086, compared to 769 for the period 1996-2000. This represents nearly three-fold increase in the number of patents granted in the field of electricity. The reason for the big jump in the number of patent granted in the field of chemistry and metallurgy, and in the field of electricity is largely due to the increase in the number of patent filings for biotechnological inventions, and Information and Communication Technology (ICT) inventions. Biotechnology inventions are classified under the field of chemistry and metallurgy and ICT (including software) inventions are classified under electricity. This trend reflects the importance of biotechnology and ICT in Malaysia and the overlapping of biotechnology and ICT applications in industries. Analysis of the patent documents granted during the last decade, clearly, indicate that ICT is now widely used in biotechnology processes and in biotechnology equipment.

Table 10.7: Number of patent applications applied for and granted in Malaysia for 16 selected countries for the period 2003 to August 2005

Countries	*Applied*	*Granted*
Australia	262	91
Brunei	1	0
China	42	10
France	632	155
Germany	1148	375
Indonesia	8	0
Japan	2470	1654
Republic of Korea	455	157
Singapore	227	26
Switzerland	763	175
Sweden	306	144
Taiwan	367	46
Thailand	26	2
United Kingdom	600	258
United States of America	4588	1732
Vietnam	1	0
Malaysia	1240	80

Table 10.7 shows the number of patent applications applied for and granted in Malaysia for 16 selected countries for the period 2003 to August 2005. For patent applications filed in Malaysia, United States of America had the highest number, followed by Japan and Germany. For the period 2003 to August 2005, the number of patent applications filed in Malaysia by applicants from United States of America, Japan and Germany was 4,588, 2,470 and 1,148, respectively. A similar trend was observed for patents granted in Malaysia. For the period 2003 to August 2005, the number of patents granted to applicants from United States of America, Japan and Germany was 1,732, 1,654 and 375, respectively. United Kingdom ranked sixth after Switzerland and France in terms of number of patent applications filed in Malaysia. However, United Kingdom ranked fourth in terms of the number of patents granted after Germany. The number of patents granted to United Kingdom, Switzerland and France was 258, 175, and 155, respectively. Interestingly, one should take note of Taiwan, a non-Paris Convention and a non-Patent Cooperation Treaty country, where a total of 367 patent applications were filed during the period 2003 to August 2005.

This places Taiwan way ahead of Australia and China who are signatory to the Paris Convention and the Patent Cooperation Treaty. This is impressive especially when compared to China, Taiwan is a small country and has a small population.

Bibliometric

The other widely used output S&T indicator is bibliometric indicator which attempts to measure numbers of scientific publications produced, particularly, the one that was carried out in academic institutions. Scientific articles and publications may be classified according to authors, institutions, fields of research and the country. The impact of such publications is measured by citations indexes.

The focus of this study (Bibliometric) is on S&T knowledge outputs in published international journals as have been recorded in prominent international S&T databases during 1955-2002. This priority of the study is in accordance with Malaysia's Vision 2020 when *"Malaysia must not only be a consumer of technology, but also a contributor to the scientific and technological civilization of the world."*

A total of 13,475 papers were found published in international journals during 1955-2002, covering 59 fields of research. The papers were authored and co-authored by 16,828 scientists and/or researchers from 2,660 institutions in 105 countries including Malaysia.

The above data are analysed in further details as explained below.

Malaysia's papers in international journals, 1955-2002

The country experienced a slow growth of S&T knowledge outputs for the period 1955 to 1980, when less than 100 papers were published in international journals within every two years during the period (Figure 10.1). There was a big surge of outputs from 1981 to 82 with 608 papers, an increase of 419.66 per cent from 117 papers published in the previous two years (1979 to 80). A steady growth occurred after 1989 onwards, with an increase of 21.05 per cent, 20.80 per cent, and 33.17 per cent during 1989-1990, 1991-1992, and 1993-1994, respectively. After 1995, the growth rate declined to 13.59 per cent, 13.63 per cent, 9.22 per cent and 4.46 per cent in 1995-1996, 1997-1998, 1999-2000 and 2001-2002, respectively.

Looking at the outputs under the period of each Malaysia Plan (MP), the eighth Malaysia Plan (1996-2000) appears to be the most productive period, with the highest number of papers, namely 5,688, compared to the outputs of 4,466 papers during seventh Malaysia's plan (MP) (1991-1995), 3183 papers during sixth MP (1991-1995), 1,888 papers under fifth MP (1986-1990), 1,634 papers under fourth MP (1981-1985) and 204 papers under third MP (1976-1980) (Figure 10.2). The first and second MP (1966-1970, and 1971-1975) saw only 14 and 59 papers, respectively while the ten-year span of 1955-1966 had only 11 papers.

Fields of research

Figure 10.3 indicates top ten subject fields with high number of outputs. Physical chemistry has the highest number of papers (1,369 =10.16 per cent), followed by Agriculture (693 = 5.14 per cent), Medical research, General topics (665 = 4.94 per cent), Materials science and engineering (542 = 4.02 per cent), Organic

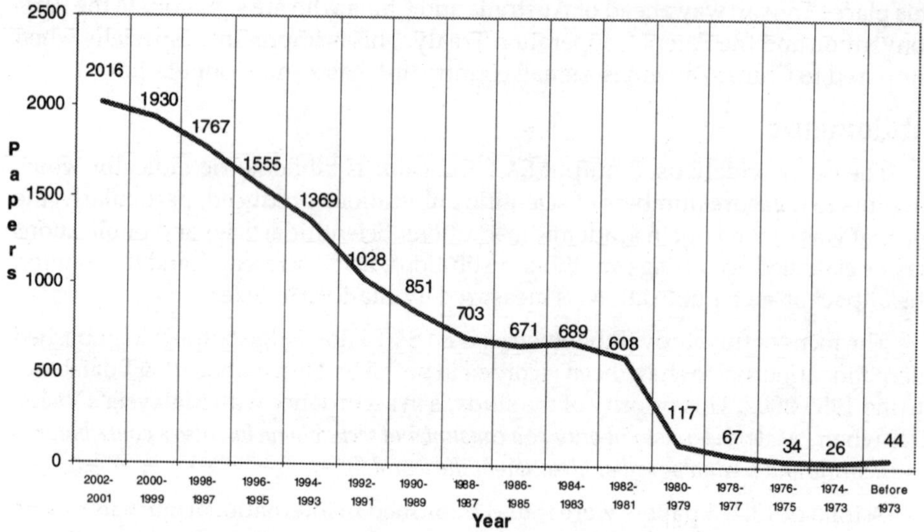

Figure 10.1: Malaysia's papers in international journals by year

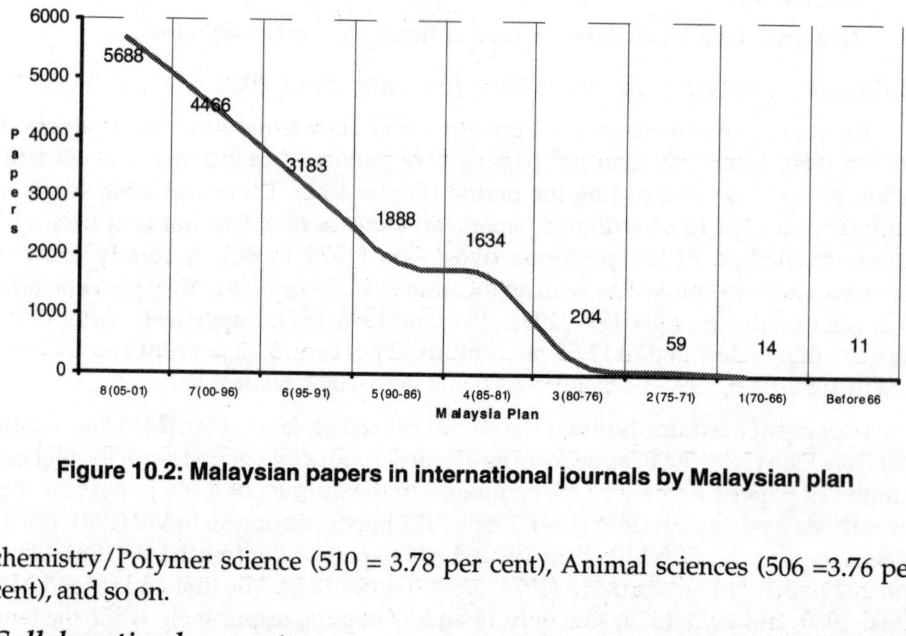

Figure 10.2: Malaysian papers in international journals by Malaysian plan

chemistry/Polymer science (510 = 3.78 per cent), Animal sciences (506 =3.76 per cent), and so on.

Collaboration by country

Scientific research is noted for being a work of collaborative effort among many individuals and institutions. Developing countries, in particular, are learning from the expertise and experiences of renowned scientists/institutions in the developed countries. It is critical to be aware of the status and quality of collaborators indicated as co-authors in scientific papers. Figure 10.4 demonstrates top ten countries where

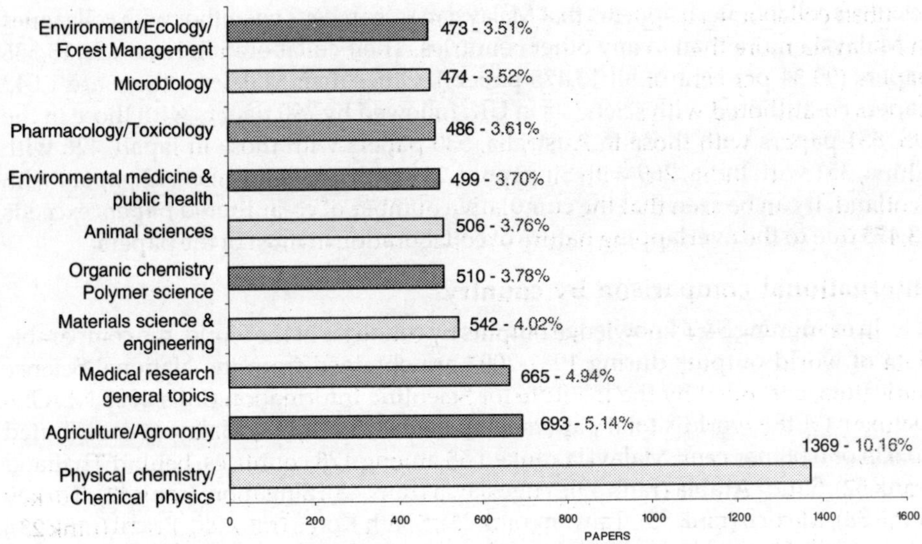

Figure 10.3: Top 10 fields of research ranked by number of papers

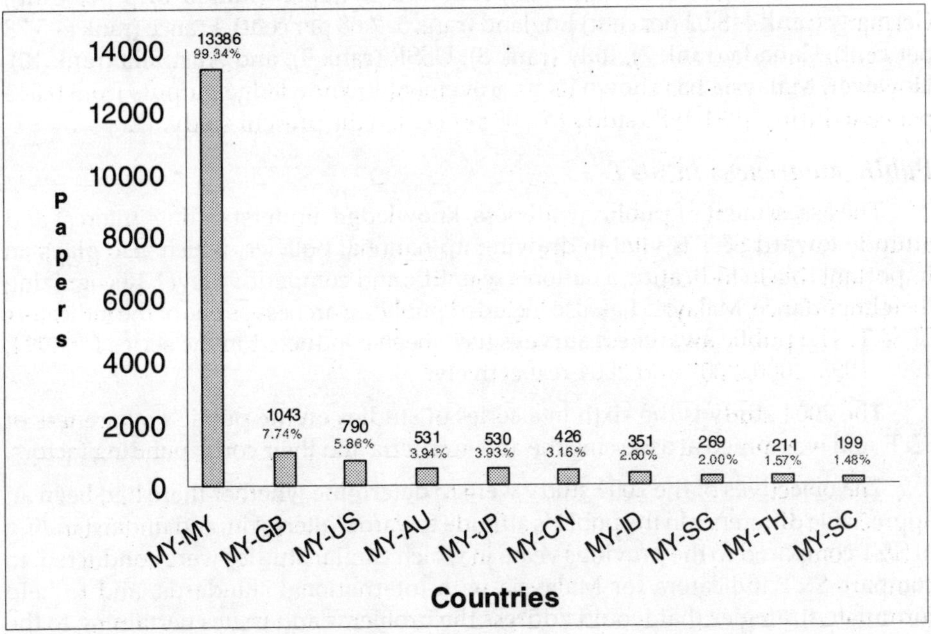

***One of the Rules mentions that, "Counts of co-authorship, and especially international co-authorships, are an indicator of quality, and that scientist who cooperate with their colleagues in other institutions and overseas are more likely to be doing quality research than those that are relatively isolated.**

Figure 10.4: Top 10 country collaborations

scientists collaborate. It appears that Malaysian researcher's co-author with colleagues in Malaysia more than in any other countries. Their collaboration results in 13,386 papers (99.34 per cent of all 13,475 papers). Other than Malaysia, there are 1,043 papers co-authored with scientists in UK, followed by 790 papers with those in the US, 531 papers with those in Australia, 530 papers with those in Japan, 426 with China, 351 with India, 269 with Singapore, 211 with Thailand, and 199 papers with Scotland. It can be seen that the cumulative number of co-authored papers exceeds 13,475 due to the overlapping nature of collaboration in most of the papers.

International comparison by country

In examining S&T knowledge outputs by countries of the world, the comparable data of world outputs during 1981-2002 are obtained from the National Science Indicators, compiled by the Institute for Scientific Information (ISI's 2002 NSIOD-Deluxe). Of the world's total papers in all fields–13,278,111, Malaysia contributed 10,538 or 0.08 per cent. Malaysia ranked 55 among 178 countries, behind Thailand (rank 52), Saudi Arabia (rank 49), Yugoslavia (rank 45), Singapore (rank 42), Turkey (rank 38), Mexico (rank 35), Taiwan (rank 26), South Korea (rank 25), Brazil (rank 23), China (rank 14), and India (rank 11). USA is the leading country in S&T knowledge outputs (rank 1), contributing 37.64 per cent to world science. UK, ranked second, contributed 1,203,514 (9.06 per cent), followed by Japan (rank 3–8.13 per cent), Germany (rank 4–8.02 per cent) England (rank 5- 7.68 per cent), France (rank 6- 5.78 per cent), Canada (rank 7), Italy (rank 8), USSR (rank 9), and Australia (rank 10). However, Malaysia has shown its improvement in knowledge outputs from 0.043 per cent during 1981-1985 study to 0.08 per cent in the present study.

Public awareness in S&T

The assessment of public awareness, knowledge, understanding, interest and attitude toward S&T is vital in drawing up national policies, which also plays an important role in indicating a nation's scientific and competitive level. Recognizing their importance, Malaysia has also included public awareness as one of the indicators of S&T. The public awareness surveys have been conducted in the series for 1994, 1996, 1998, 2000, 2002 and 2004 respectively.

The 2004 study is the sixth in a series of studies on the public's awareness of S&T, and was aimed at assessing the above criteria and their corresponding factors.

The objectives of the 2004 study were to determine whether there had been an appreciable difference in the public's attitude towards, interest in, and understanding of S&T compared to the previous years in which similar studies were conducted, to compare S&T indicators for Malaysia with International standards, and to help formulate strategies that would address the problems and issues pertaining to the public's understanding and awareness, interests, and attitudes towards S&T.

A nationwide survey was conducted on a random sample of 6,869 respondents to assess the Malaysian public's awareness of S&T. The survey was carried out in six zones: The Central zone, Eastern zone, Northern zone, Southern zone, Sabah and Sarawak. The sample was also stratified according to ethnic group, gender and age, based on the 2000 Population Census.

An analysis of the trends from 1998 to 2004 shows that the interests of the Malaysian public towards S&T have remained relatively constant. Malaysians have been between slightly to moderately interested in S&T. This is indicated by a mean of 2.74 for 1998, 2.42 for 2000, 2.41 for 2002 and 2.40 for 2004. The findings suggest that the country has not been taken efforts in this direction which should be a future task.

Malaysians also perceive themselves as a nation having knowledge between a poor and an average in regard to S&T, as indicated by an overall mean of 2.28 for 1998, 2.22 for 2000, 2.32 for 2002 and 2.22 for 2004. This suggests that there is still a lot that needs to be done to increase their knowledge of S&T. Among other things, the country should need to look at the effectiveness of the S&T programs, school curriculum and means of disseminating S&T information.

The attitude of Malaysians towards S&T has improved significantly over the years. For instance, in 2004, 69.1 per cent of Malaysians agreed that scientific research has more positive than negative effects, compared to 62.0 per cent who said so in 2002 and 43.8 per cent in 2000. This trend is encouraging because a positive attitude towards S&T suggests that the public will be receptive to the projects aimed at promoting S&T and to the efforts aimed at increasing their understanding of S&T.

This survey reveals that Malaysians have a relatively good understanding of some basic scientific issues such as "the earth travels around the sun" (70.9 per cent), "oxygen is produced by plants (71.1 per cent)", light travels faster than sound" (62.2 per cent) and "Smoking causes lung cancer" (84.5 per cent).

However, they do not have understanding with regard to issues that reflect a higher level of knowledge of S&T, *viz.* "antibiotics kill not only viruses but also bacteria" (15.5 per cent), "milk that is contaminated by radioactivity will be safe to drink after boiling" (15.1), "all radio activity is man-made" (13.4 per cent); "lasers function by the combination of sound waves" (19.2 per cent), and that "electrons are smaller than atoms" (26.4 per cent).

From 1996 to 2004, the television has been the most popular source of information for Malaysians, which remains relatively constant at about 97.0 per cent. This is followed closely by newspapers, the radio and magazines. Malaysia relied least on the Internet (34.4 per cent in 2004). The findings suggest that we need to capitalize on the use of various media to increase public awareness of S&T. We also need to increase the public's awareness of the Internet's potential as a trustworthy source of information on S&T.

The findings show that the Malaysian, American and European public are interested in S&T but do not feel that they are well informed about S&T issues.

Malaysians also seems to have a lower level of understanding of S&T when compared to Europe, the USA and Japan. This is shown by the mean percentage of correct answers on knowledge of S&T concepts among adults as computed on all the 10 comparable statements, which was recorded 59.9 per cent for USA (2001), 49.8 per cent for Japan (2001), 51.9 per cent for Europe (2002), 34.0 per cent for Malaysia (2002) and 32.5 per cent in the present study (2004).

Malaysians have a generally positive attitude towards S&T and they strongly believe in the promise and benefits of science and technology, as indicated by the index of scientific promise of 71.9 per cent. Indeed, Malaysia seems to have a higher belief in the promise and benefits of S&T compared to the Europeans, whose index of scientific promise is 66.6 per cent but lower when compared to the Americans, whose index of scientific promise is 79.0 per cent.

Generally, Malaysians are not very far behind compared with other countries in understanding scientific concepts, and have a fair understanding on those concepts. They also show positive attitude towards science and technology. However, the perception of knowledge and interest in S&T is rather low.

Technological innovation

Technology innovation indicator measures aspects of the development of technological innovation processes undertaken by industries. It attempts to gather information on resources (costs and personnel) devoted to technological innovation, the sources of ideas, benefit to the companies as well as factors limiting or enhancing innovation. Diffusion of technologies between industries and companies is also analyzed.

Malaysia conducted the third survey (2002-2004) on technology innovation survey. This survey provides an overview of innovation in the Malaysian manufacturing sector. Profiles of both the innovating and non-innovating firms are drawn (Figure 10.5). The profiles include level of innovation, innovation across industries, and characteristics of innovating and non-innovating industries.

Level of innovation

A total of 485 firms responded to the postal survey. Of these firms, 261 (or 54 per cent) indicated that they carried out innovating activities (Figure 10.5). The incidence of innovation is higher for the period of NSI-4 than the incidence recorded in the previous surveys such as 35 per cent in the NSI-3 and 21 per cent in the NSI-2.

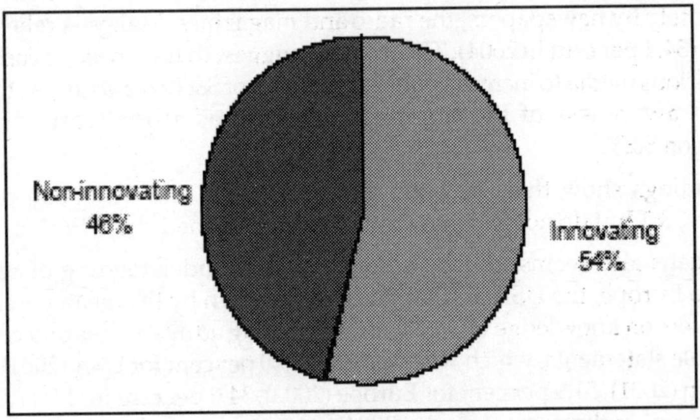

Figure 10.5: Innovating and non-innovating firms in the manufacturing sector, 2002–2004

Innovation across industries

The incidence of innovation varies across the different manufacturing industries. This was observed in the previous surveys as well. The incidence of innovation varies between 25 per cent and 76 per cent.

Industries where the incidence of innovation exceeded 50 per cent included radio, television and communications equipment (76 per cent), textiles (73 per cent), motor vehicles, trailers and semi-trailers (71 per cent), recycling (67 per cent), medical, precision, optical instruments, watches and clocks (67 per cent), Office, accounting and computing machinery (63 per cent), Leather products (62 per cent), rubber and plastic products (62 per cent), basic metals (58 per cent), chemicals and chemical products (57 per cent), electrical machinery (57 per cent), paper and paper products (56 per cent), fabricated metal products (56 per cent), wood products (55 per cent), and other transport equipment (50 per cent).

Generally, these results are consistent with findings from the previous survey (NSI-3) with the exceptions of industries like basic metals, leather products, fabricated metal products and wood products.

Industries with low incidence of innovation of less than 50 per cent included, machinery and equipment (47 per cent), food products and beverages (46 per cent), furniture (43 per cent), publishing and printing (41 per cent), other non-metallic mineral products (32 per cent), wearing apparel (29 per cent), and coke, refined petroleum products and nuclear fuel (25 per cent).

Characteristics of innovating and non-innovating firms

Firm Size

In this study, firm size has been defined in terms of the number of employees. A comparison of the innovating and non-innovating firms indicates that, on the average, innovating firms tend to be larger than non-innovating firms. About 42 per cent of non-innovating firms had 19 or less employees while only 11 per cent of innovating firms were of this size. In contrast, about 38 per cent of the innovating firms had 250 or more employees while only 10 per cent of non-innovating firms fell into this category of firm size.

Local ownership and foreign ownership

The proportion of firms that are locally controlled (100 per cent local ownership or majority ownership) is higher amongst non-innovating firms compared to innovating firms. About 67 per cent of non-innovating firms are 100 per cent locally-owned while the corresponding figure for innovating firms is 58 per cent. The proportion of innovating firms that are 100 per cent foreign-owned is 21 per cent, which is twice the proportion of the non-innovating firms that are 100 per cent foreign-owned (Figure 10.6).

Type of ownership

The most prevalent type of ownership of both innovating and non-innovating firms is the limited company (Figure 10.7). However, the proportion of innovating

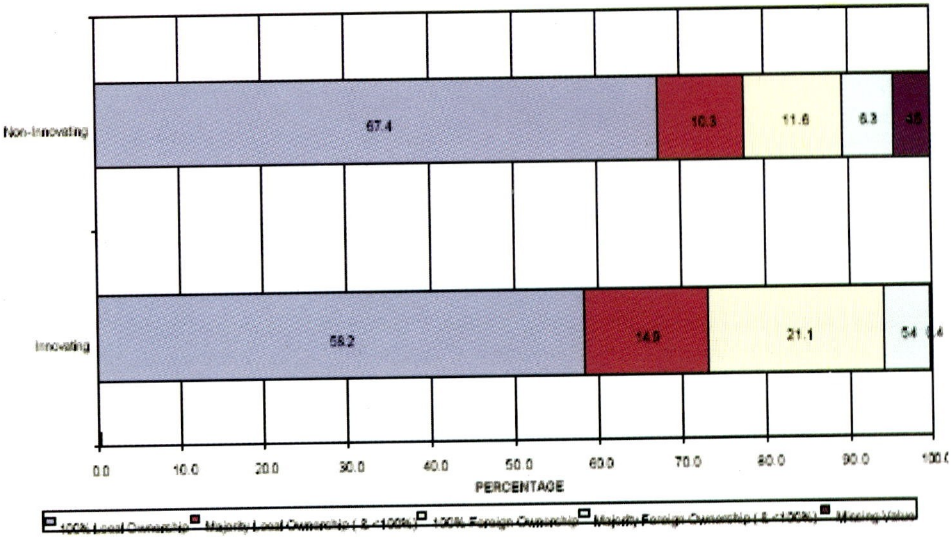

Figure 10.6: Local ownership vs. foreign ownership of innovating and non-innovating firms in the manufacturing sector, 2002-2004

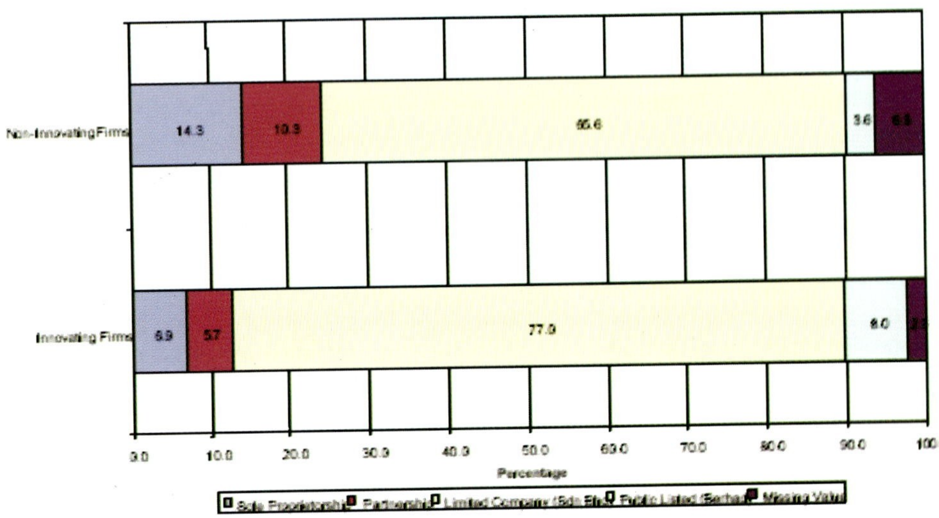

Figure 10.7: Type of ownership of innovating and non-innovating firms in the manufacturing sector, 2002-2004

firms that are limited companies (at 77 per cent) is considerably higher than that observed for non-innovating firms (66 per cent). The proportion of innovating firms that are public listed companies is also higher (at 8 per cent) compared to public listed non-innovating firms (4 per cent). However, more of the non-innovating firms (14 per cent) are of the sole-proprietorship type compared to innovating firms (7 per cent).

Conclusion

It cannot be denied that statistical S&T indicators have been of increasing interest to the government, industry and public as a whole. They serve as one of the important inputs in assessing S&T system of a country so that national competitive advantages could be maintained and promoted.

Malaysia is in support of any effort at international or regional level to promote appropriate and systematic usage of statistical S&T indicators in non-aligned and other developing countries. Issues such as using harmonized statistical indicators, frequency of S&T reports to be produced and other procedural matters should be addressed at appropriate forums so that a more meaningful and accurate comparisons between the countries could be achieved.

Chapter 11

Science and technology in Pakistan: System of governance, status of development and current initiatives

Tariq Bashir

Abstract

Science and Technology Governance System in Pakistan consists of three main components *viz.*, Science and Technology (S&T) legislation, Policy formulation, and Implementation of policies and plans. Since its inception, Pakistan realized the importance of S&T for economic development but the pace of S&T development was slow due to the lack of enough resources. However, despite the constraints, a considerable progress has been made in the field of S&T. In Pakistan, National Commission for Science and Technology is the main institution that provides guidance to scientific and technological efforts at the national level. As a result of the recommendations of the second meeting of the Commission, which are currently serving as the S&T policy guidelines, funding for S&T was increased and a number of programmes and projects, categorized under the broader areas of (i) Human resource development, (ii) Up-gradation of Research and Development (R&D) infrastructure and launching of R&D programmes, (iii) Technology development and industrialization, (iv) Strengthening of policy, coordination and management structure, and (v) Information technology were initiated. Recognizing the fact that Pakistan had missed the first wave of IT revolution in the world, a special attention was paid to Information technology and telecommunications sector. Biotechnology was also declared a priority R&D area, which led to the constitution of the National Commission on Biotechnology and preparation of Biotechnology policy and action plan. In order to enhance productivity in all sectors of economy, initiatives were taken for building technological capabilities that were closely integrated to industrialization.

The present report elucidates the system of S&T governance including mechanism of formulation and implementation of S&T policies in Pakistan, presents S&T statistical indicators depicting the status of S&T development, and explicates the current S&T policy initiatives of the country.

Keywords: Country report on S&T, Pakistan, S&T development, S&T indicators, S&T planning, S&T policy, Status of S&T

Introduction

The importance of science and technology for any country cannot be over emphasized. In today's world the scientific and technological advancements have become a key factor whether it is for strong defence, stable and secure economy, elimination of poverty, social development or even to gain respect from the international community. As scientific and technological developments tend to shape the economic progress of countries, S&T policies are being viewed as an integral part of the overall strategy for socio- economic development.

For developing countries, the key to rapid industrialization, employment generation, poverty reduction and raise the living standards of common people is the formulation of explicit, realistic and target-oriented S&T policies that are closely linked with the socio-economic policies of the country. They need to build a strong S&T infrastructure including properly managed R&D organizations, well-equipped laboratories, adequate funding for research and a stock of highly-motivated qualified manpower. It is also important for them to possess the capability to evaluate, choose and absorb imported technologies and at the same time develop local technological capabilities.

Millennium Development Goals (MDGs), laid down in the United Nations Millennium Declaration of 2000, also stipulate the application of science and technology in achieving these goals. The UNESCO Commission on Science and Technology for Development, in its seventh session in 2004, reaffirmed that the application of S&T is central in facilitating the achievement of most MDGs, and the countries that do not make S&T a top priority on their development agenda are unlikely to achieve the goals. The Commission, *inter alia*, suggested that the developing countries should improve the policy environment for application of S&T, strengthen technology support for institutions and the science advisory mechanisms.

Science and Technology Governance System in Pakistan

Science and Technology Governance System in Pakistan consists of three main components *viz.*, S&T legislation, Policy formulation and Implementation of policies and plans. An overview of S&T governance and the decision making process in Pakistan is presented in Figure 11.1.

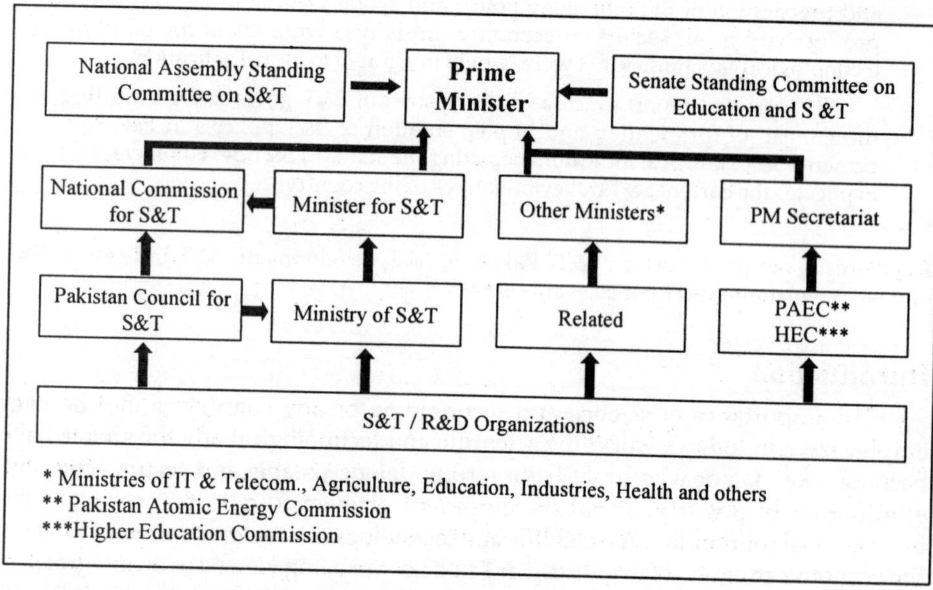

Figure 11.1: Overview of the science and technology governance system in Pakistan (The direction of arrows indicates the flow of information)

S&T legislation

The main role in S&T legislation is played by the Ministers for the concerned Ministries and the Standing Committees of the Senate and the National Assembly on Science and Technology. The Federal Ministries which are mainly concerned with the legislation in respect of S&T matters include the Ministries of Science and Technology; Information Technology and Telecommunications; Food, Agriculture and Livestock; Health; and Industries, Production and Special Initiatives. Ministers and Secretaries of these Ministries have a central position in the S&T legislation as they have the responsibility to supervise the preparation of Bills and/or Acts, concerning these issues and matters relevant to the functions of their particular Ministries. The Bills and/or Acts are finalized, in consultation with the Law Division, and presented in the Parliament by the concerned Ministers. The Ministries also have the responsibility to ensure appropriate legal framework for the organizations working under their administrative control.

Presence of the Standing Committees provides the political oversight over the functioning of the S&T organizations in the country. The Standing Committee of the Senate on Education and Science and Technology comprises 12 members while the Standing Committee of the National Assembly on Science and Technology comprises 18 members. The Minister for Science and Technology represents both the Standing Committees as an ex-officio member. All the bills and other matters relevant to science and technology are referred, by the Speaker of the House of Parliament, to these Committees. The Committees examine the Bills in accordance with the constitution and propose amendments in the bill, if any, within a prescribed time period. The Committees may also examine the expenditure, administration, legislation, public petitions and policies of the Ministries and its associated organizations and make recommendations for policy formulation and improvement of the functions or the work of the Ministries and their associated organizations.

S&T policy formulation

S&T policies are made at two distinct levels, *i.e.* at the national level and at the sectoral level. Policy formulation at the national level means those policies that relate to the overall promotion, development and application of science and technology in the country while the policies that are made at the sectoral or ministerial level are mainly related to the sphere of functions of those ministries.

National Commission for Science and Technology (NCST)

NCST is the main institution that provides guidance for scientific and technological efforts at the national level. The 27-member Commission was created in 1984 as the apex S&T policy-laying body in the country. It has a cross-cutting function of determining policy across many government ministries and agencies but it does not have any administrative function. The Commission is headed by the Prime Minister of Pakistan and includes the Federal Ministers for science and technology, education, agriculture, industries and finance, respectively, along with the Ministers for S&T departments of the Provinces.The Federal Minister for Information Technology and Telecommunications is currently not a member of NCST, as the membership of NCST has not been updated since the establishment of the Ministry in 2002.

The Commission has an Executive Committee (ECNCST) that is responsible to coordinate, oversee and review the S&T policies, R&D development programmes and implementation of the policy decisions taken by the Commission. The Executive Committee is headed by the Federal Minister for science and technology and consists of 29 members with Chairman of the PCST as the secretary of the Committee. ECNCST mainly includes the secretaries of the concerned Ministries and heads of S&T organizations.

Pakistan Council for Science and Technology (PCST)

PCST has the mandate to advise the government on S&T issues, policies and plans and suggest measures for the promotion, development and application of science and technology in the country. PCST conducts scientometric and futuristic studies on various S&T related issues and provides analytical reports, suggestions and recommendations to the government for facilitating its decision making. For seeking expert opinion and advice regarding the development and application of science and technology in different areas, the Council constitutes "Think Tanks" and "Expert Committees" as and when the need arises. These Committees review the existing

[1] Although, Ministers for S&T Departments of Provinces are ex-officio members of NCST, the S&T Departments have not been established in all the four provinces. Only very recently one province, NWFP has established a S&T Department.

status of development in their respective areas and propose short, medium and long-term strategies for research and technological development.

PCST is also the designated secretariat of NCST and in this capacity; it plays a very important role in S&T policy formulation and decision-making process. Chairman, PCST, supported by a small technical workforce, serves as the Secretary of the Commission. PCST prepares agenda and working papers for the meetings of NCST, in consultation with federal ministries, provincial departments, S&T organizations and eminent scientists and technologists. The plans prepared by PCST are reviewed by ECNCST before they are presented to NCST for its consideration.

Ministry of Science and Technology (MoST)

Scientific and Technological Research Division was created in 1964 and placed under the Ministry of Education and Scientific Research. Later on, a full-fledged Ministry of Science and Technology was created in 1972, which now is the national focal point for all-important matters regarding science and technology including developing S&T policies and plans and coordination of S&T at the government level. MoST is also mandated to promote applied research for economic growth and provide guidance to federal and provincial scientific and technological institutes in the national priority areas of research. MoST has five wings *i.e.* Policy and Coordination, Planning and Development, Technology, Electronics and International liaison. The Ministry is pursuing the mission of establishing a sound and sustainable S&T base which would lead to the socio-economic development of the country.

S&T Policy Formulation at the Sectoral Level

In Pakistan, Ministry of Information Technology and Telecommunications, Ministry of Food, Agriculture and Livestock, Ministry of Health, and Ministry of Industries Production and Special Initiatives are the important ministries that are involved in the policy making at the sectoral level along with the Higher Education Commission and the Pakistan Atomic Energy Commission which work directly under the Prime Minister's secretariat.

Information Technology and Telecommunications Division was established in 2000 under the Ministry of Science and Technology. Keeping in view the importance of Information Technology (IT) and acknowledging its central role in the future economic progress, a separate Ministry for IT and Telecommunications was created in the year 2002. This Ministry is now the national focal point for IT and is responsible for formulation of policies for the development and improvement of IT in the country. Preparation of policies and plans regarding agricultural R&D is the responsibility of the Ministry of Food, Agriculture and Livestock. In this regard, the Ministry is assisted by the Pakistan Agricultural Research Council which itself is a prime agricultural R&D organization. The Pakistan Medical Research Council, an autonomous organization under the Ministry of Health, has the mandate to promote, organize and coordinate medical research. The Council plays a leading role in the generation, dissemination and utilization of health knowledge with the ultimate goal of improving health of the people. Ministry of Industries, Production and Special Initiatives is mandated to formulate policies and plans to promote industrial development, enhance value addition and exportability of products at competitive prices through the promotion of applied research for technology up-gradation, training and upgrading skills of industrial personnel and facilitating up-gradation and transfer of technology in industrial sector. The Higher Education Commission (HEC) is an important body where S&T decisions pertaining to higher education, including education and R&D in the basic and applied sciences and engineering and technology, are made. The Chairman, HEC, who has the status of a Federal Minister, works independently of the Minister for Education, and is mainly responsible for the primary and secondary education. The Pakistan Atomic Energy Commission (PAEC) is one of the largest R&D establishments in Pakistan. The Chairman, HEC is a key figure in the defence S&T and R&D policy and the decision-making process in the country. PAEC also has a number of subordinate organizations working in the fields of agricultural and medical research.

Implementation of S&T policies and plans

Organizations that are responsible for implementation of S&T policies and plans in Pakistan can be classified in four main categories, *i.e.* Federal Ministries, Organizations under the PM's Secretariat, Autonomous organizations under Federal Ministries, and Provincial Departments.

Federal Ministries

Federal Ministries are also responsible for implementation of policies within their own spheres of work and administrative control.

Organizations under Prime Minster's Secretariat

Organizations under Prime Minster's Secretariat (*i.e.* HEC and PAEC) are responsible for implementation of policies and plans in their own organizations and their subordinate institutes.

S&T Organizations under Federal Ministries

Most of the S&T organizations working under the Federal Ministries are autonomous bodies. Heads of these organizations, as ex-officio members of NCST, play an important role in the formulation of S&T policies and are responsible for implementation of policies and plans in their organizations. Majority of these organizations are under the Ministry of Science and Technology and the Ministry of Food, Agriculture and Livestock.

Provincial Departments

There are a significant number of R&D organizations working under the administrative control of the provinces, mainly in the areas of agriculture, forestry, livestock and fisheries. Provincial departments make decisions in line with the Federal Government policy guidelines and are responsible for implementation of the policies and plans in the R&D organizations working under their administrative control.

Status of S&T development and S&T statistical indicators

S&T policies and S&T development in historical perspective

At the time of independence on 14th August, 1947, Pakistan did not have enough human and financial resources for science and technology. The present Pakistan (called the West Pakistan at that time), inherited only four laboratories, two universities and one agricultural college. Since its inception, Pakistan realized the importance of S&T for socio-economic development and some important S&T organizations were established in the public sector within the first few years of creation of Pakistan (Table 11.1). A few non-governmental organizations also started functioning for the promotion and popularization of S&T in the country during the same period.

Table 11.1: Public sector and non-government S&T organizations established within the first few years of creation of Pakistan

Name of organization	*Year of establishment*
Public sector S&T organizations	
Central Cotton Committee	1948
Food and Agriculture Council	1949
Department of Scientific and Industrial Research[2]	1949
Pakistan Medical Research Council	1953
Non-government S&T organizations	
Pakistan Association for the Advancement of Science	1947
Pakistan Academy of Sciences	1953
Scientific Society of Pakistan	1955

The first institutionalized effort to formulate S&T policy for the country was made in 1960 when National Scientific Commission of Pakistan was constituted to consider how best scientific research can be promoted and integrated. The Commission was also mandated to consider means and ways for providing adequate facilities to the various research organizations and ensuring the utilization of results of their research for the overall development of the country. The Commission made some useful policy recommendations such as defining priority areas of research, establishment of research councils in important areas, allocation of 2.5 per cent of the national budget for S&T, and making scientific careers attractive. However, due to various reasons, mainly the financial constraints, the constant conflicts with neighbouring India and the political instability in the country, S&T could not become a priority issue of the successive governments. Due to lack of effectiveness of the S&T system in creating effective linkages among education, science, and the production sector, and also in generating the technologies that are appropriate to local needs, the

[2] Department of Scientific and Industrial Research was reconstituted in 1953, as an autonomous organization called Council for Scientific and Industrial Research, which was later on renamed as Pakistan Council of Scientific and Industrial Research (PCSIR).

contributions of scientific research in the economic development, generally, remained insignificant. As a result, the NCST was constituted in 1984 to accelerate scientific and technological capacity building for rapid and sustainable economic growth. However, the NCST met only once for the period of 16 years from 1984 to 2000 although it was supposed to meet twice a year. The fact that only two policy documents[3] related to S&T were approved by the government during the 40 years, since the Scientific Commission's report in 1960, reflects the state of affairs.

However, despite the constraints mentioned above and apparent lack of interest from the government, Pakistan has made considerable progress in the field of science and technology. Presently, there are independent and separate Federal Ministries for S&T and IT and Telecommunications as well as independent Commissions, under the Prime Minister's secretariat, for higher education and atomic energy. There are 85 major S&T organizations with 228 institutes and laboratories under their administrative control. During the last decade, the number of higher education institutions has grown at a tremendous pace (Figure 11.2), and presently, there are 114 universities and other degree-awarding institutions in the country including 11 engineering and technology, 6 agriculture and 7 health sciences universities. In order to increase female participation in higher education, 6 universities have been exclusively dedicated for female education and training (along with opportunities for them to enrol in any other institution of higher education).

S&T manpower

It is essential for any country to have a strong base of highly qualified, trained, motivated and committed S&T manpower for making significant advancement in S&T leading to economic development. Without S&T manpower, it is not possible to achieve good manufacturing capabilities, to take advantage of technological transfers from other countries, to carry out high quality R&D and to make significant innovations

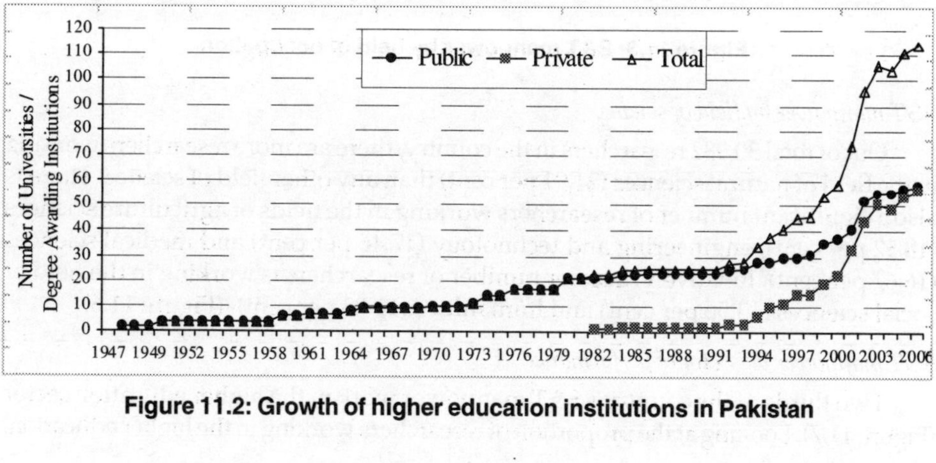

Figure 11.2: Growth of higher education institutions in Pakistan

[3] National Science and Technology Policy, 1984 and National Technology Policy and Technology Development Action Plan, 1993.

contributing to economic activities. Shortage of S&T manpower is, thus, a huge barrier to rapid development. No surprise that the development of human capital has been on the priority agenda of most of the developed countries. Unfortunately, most of the developing countries did not pay due attention to the development of adequate human capital which is a prerequisite for economic development.

S&T Manpower in Pakistan has increased over the years and according to the latest survey of S&T organizations and universities carried out by PCST in 2006, the total number of researchers in the country is 30,982 (Figure 11.3) while the number of manpower with PhD and equivalent qualification is 5,546 (Figure 11.4). The number of PhDs in the country has increased by 70 per cent in the last three years (Figure 11.5). There have been tremendous efforts to increase the number of high level S&T manpower, including PhDs, in the country resulting in the steep increase in the number of PhDs. However, it must be noted that at least a part of this increase may be due to the inclusion of social sciences and humanities[4] in the most recent survey as fields of science.

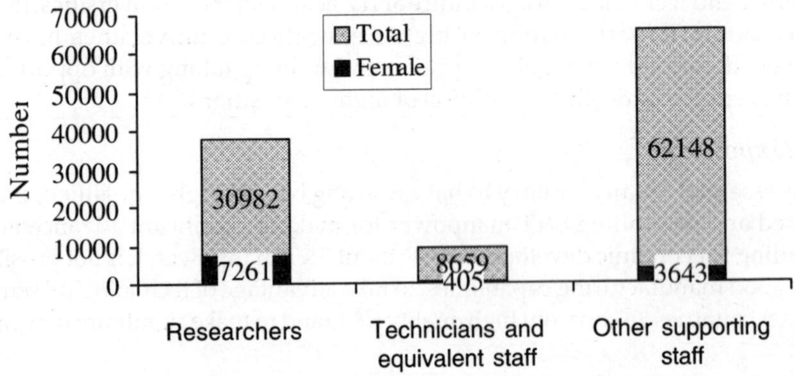

Figure 11.3: S&T manpower by field of occupation

S&T manpower by field of science

Out of total 30,982 researchers in the country, there are more researchers working in the field of natural sciences (21.91 per cent) than any other field of science. There is also a significant number of researchers working in the fields of agriculture sciences (18.52 per cent), engineering and technology (17.46 per cent) and medical sciences (16.67 per cent). Relatively, a lesser number of researchers is working in the field of social sciences (13.56 per cent) and humanities (11.83 per cent). (Figure 11.6)

S&T manpower by sector of performance

Two thirds of the country's S&T manpower work in the higher education sector (Figure 11.7). Looking at the proportion of researchers working in the higher education,

[4] As per the UNESCO's "Questionnaire on Statistics of Science and Technology", the fields of social sciences and humanities are included as the fields of science.

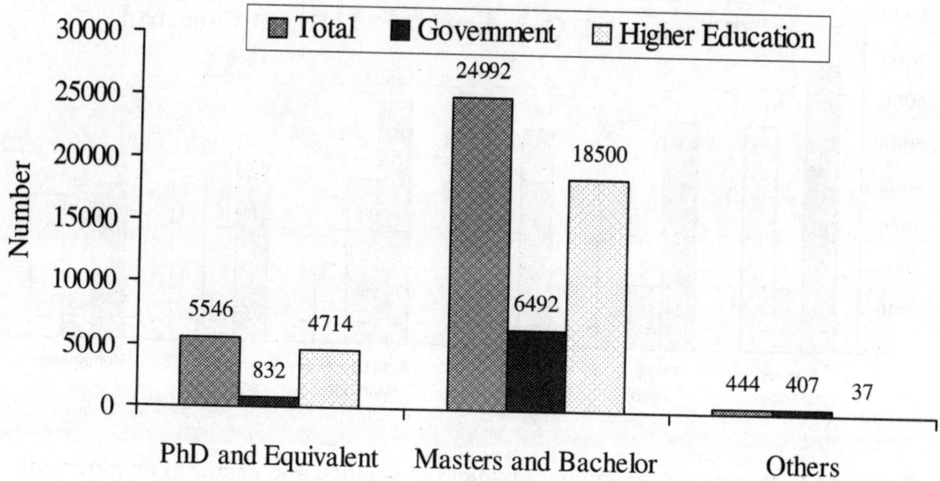

Figure 11.4: Researchers by level of qualification and sector of employment (PhD and Equivalent level manpower includes 4,877 PhDs and 669 researchers with qualifications equivalent to PhD such as FRCS etc.)

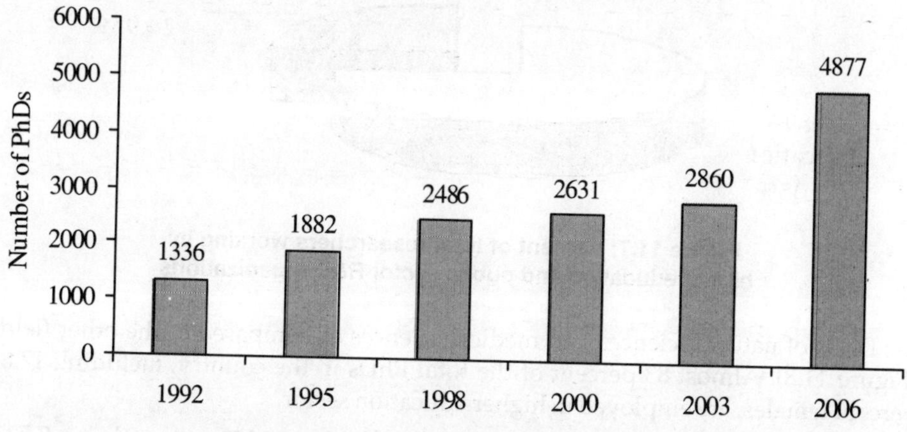

Figure 11.5: Growth of PhDs in Pakistan

it may assume that greater volume of R&D activities is undertaken in higher education sector. However, most of the researchers at the universities are not engaged in R&D activities on full-time basis as their main responsibly is teaching. Higher percentages of researchers in the fields of medical sciences (91.02 per cent); natural sciences (78.67 per cent); and engineering and technology (72.69 per cent), work in the higher education sector while most of the agricultural scientists (67.72 per cent) work in the public sector organizations (Figure 11.6).

Total number of S&T manpower working in higher education sector is 23,251, which includes 27.70 per cent females. There is a greater focus on education and research in

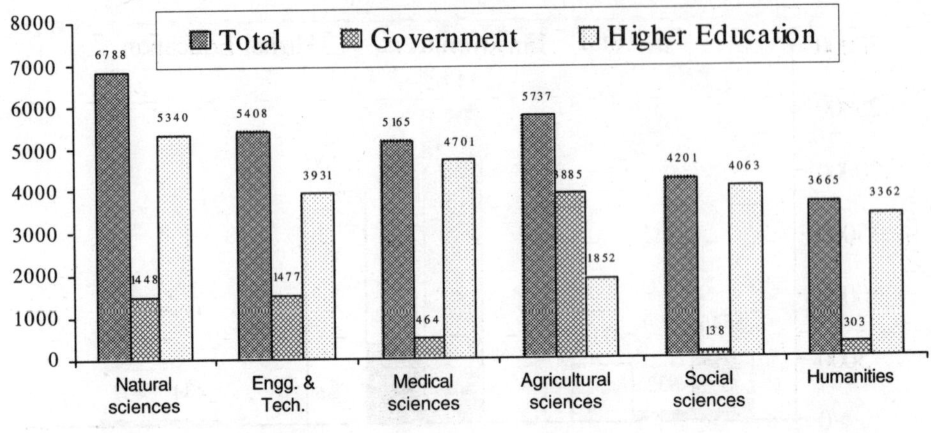

Figure 11.6: Number of researchers by field of science and sector of employment (18 researchers, 16 in Government and 2 in higher education, whose field of science was not provided by the concerned organizations are not shown in the Figure).

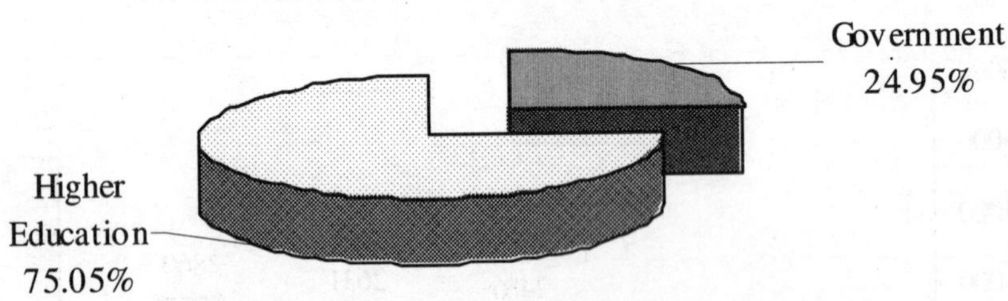

Figure 11.7: Percent of total researchers working in higher education and public sector R&D organizations

the fields of natural sciences and medical sciences as compared to the other fields (Figure 11.8). Almost 83 percent of the total PhDs in the country, including 17.82 percent females, are employed in higher education sector.

Total number of S&T manpower working in public sector S&T organizations is 7,731, which includes 10.61 per cent females. The ratio of females in public sector organizations is less by half of that in the higher education. The main focus of R&D activities in public sector S&T organizations is on agriculture as more than half of the total researchers in the sector work in this field (Figure 11.9).

R&D expenditure

R&D is one of the most important tools for building scientific and technological capacity and a key factor to the economic growth. Government's spending on R&D reflects its commitment to scientific and technological advancement. It also indicates comparative importance a government attaches to R&D in relation to its other economic activities and functions. Private sector's spending on R&D indicates its perceived

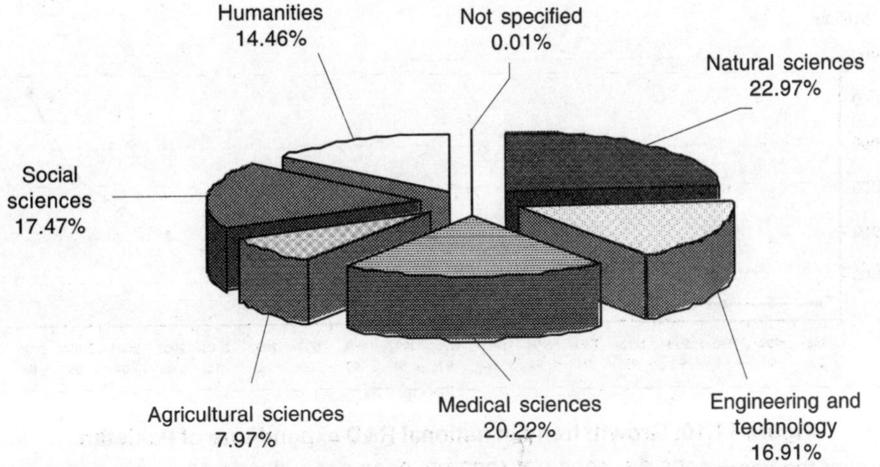

Figure 11.8: Percentage of Researchers Working in Different Fields of Science in Higher Education

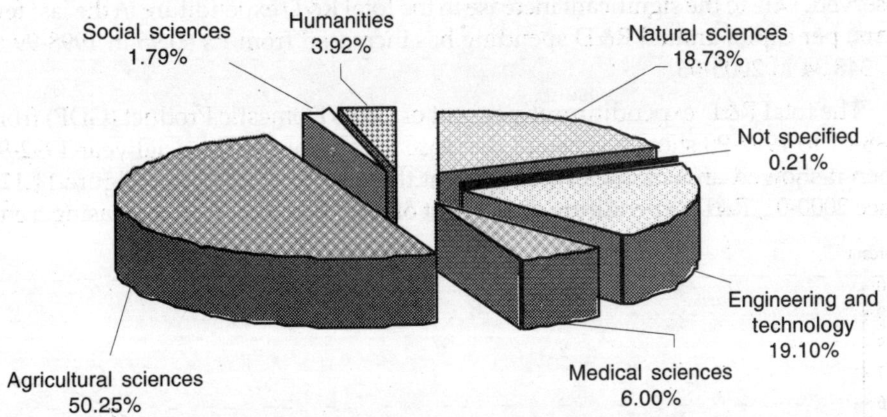

Figure 11.9: Percentage of researchers working in different fields of science in public sector S&T organizations

importance of R&D for wealth generation as the money diverted by companies/firms for R&D could easily be spent on some other business activities. The total R&D spending of a country helps to determine, at least in part, the level and direction of the future technological change in the country.

Total R&D expenditure of Pakistan was Rs 53,631 million in the financial year 2005-06. It has grown at an average annual increase of 17.75 perc ent since 1988-89 (Figure 11.10), however, the increase has been mainly due to the figures of the last six

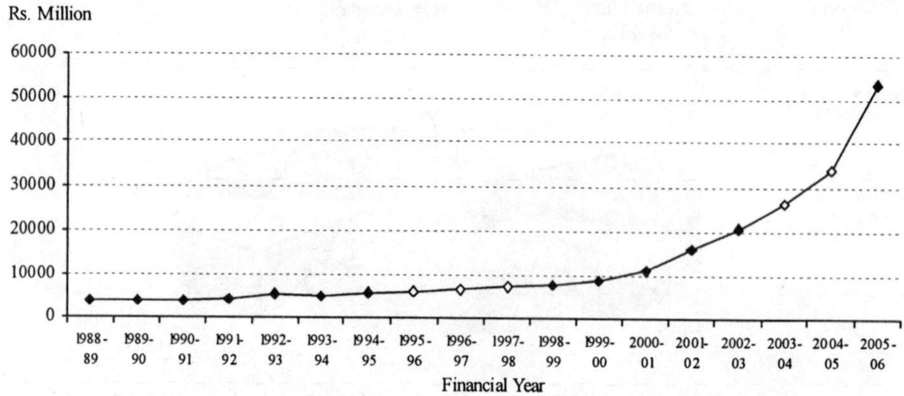

Figure 11.10: Growth trend in national R&D expenditure of Pakistan

(Data for the years 1995-96, 1996-97, 1997-98, 2003-04 and 2004-05 (hollow markers) is estimated on the basis of average annual increase in the previous years)

years *i.e.* from 2000-01 onwards when average annual increase of 35.50 per cent was observed. Due to the significant increase in the total R&D expenditure in the last few years, per capita annual R&D spending has increased from Rs 60.95 in 1998-99 to Rs. 348.34 in 2005-06.

The total R&D expenditure as per cent of Gross Domestic Product (GDP) from 1988-89 to 1997-98 show a constant decrease except for the financial year 1992-93 when it showed an increase of 8.88 percent than the previous year (Figure 11.11). Since 2000-01, R&D expenditure as per cent of GDP has shown an increasing trend

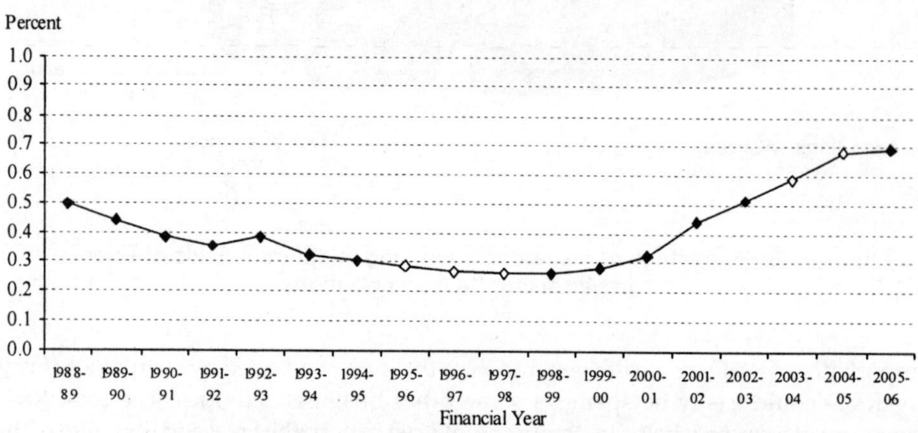

Figure 11.11: Growth trend in national R&D expenditure of Pakistan as per cent of GDP

(Data for the years 1995-96, 1996-97, 1997-98, 2003-04 and 2004-05 (hollow markers) is estimated on the basis of average annual increase in the previous years)

with an average annual growth rate of 16.68 per cent. Due to the substantial increase during the last few years, R&D expenditure has reached to 0.70 per cent of GDP in 2005-06.

R&D expenditure by sector of performance

In Pakistan, organizations that perform R&D activities can be broadly categorized in two sectors *i.e.* public sector S&T organizations and universities. Universities that spend major portion of total R&D budget (Figure 11.12) mainly concentrate on imparting education along with conducting R&D. However, the volume of R&D effort and the amount of budget spent on R&D varies a great deal from one university to another. Public sector organizations are generally involved in R&D and provision of various S&T services. Data is not available in respect of R&D expenditure of private industry and private non-profit organizations. However, it is generally believed that R&D expenditure of private sector in the country is negligible.

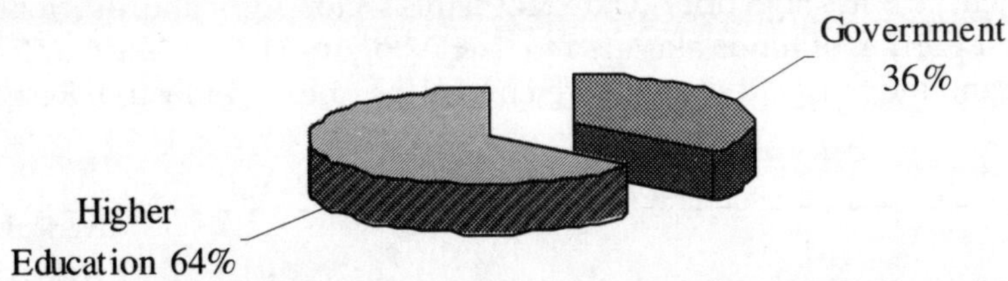

Figure 11.12: R&D expenditure of government organizations and higher education as per cent of total R&D expenditure

R&D expenditure by source of funds

R&D activities in Pakistan are mainly funded by the Government (Figure 11.13), in particular the Federal Government. Contribution of the Provincial Governments towards R&D funding is very little. According to recommendations of the second meeting of NCST held in 2000, the Provincial Governments were required to establish S&T Departments in their respective provinces. This was intended to induce Provincial Governments to make more serious contributions to the S&T effort of the country. However, this has not happened as yet and the share of Provincial Governments in the total S&T funding has not increased. No exact information regarding funds provided by the private industry for R&D is available. However, the information provided by R&D organizations and universities show that the share of industry can be, at the best, 2 per cent of total funding (shown under the title of "Not Distributed" in the Figure 11.13).

Public sector R&D organizations

The total expenditure of public sector R&D organizations was Rs 19,185 million in the financial year 2005-06. There has been a significant increase in the funding during the past few years. This increase is mainly in the development budget of organizations which is utilized for execution of R&D projects.

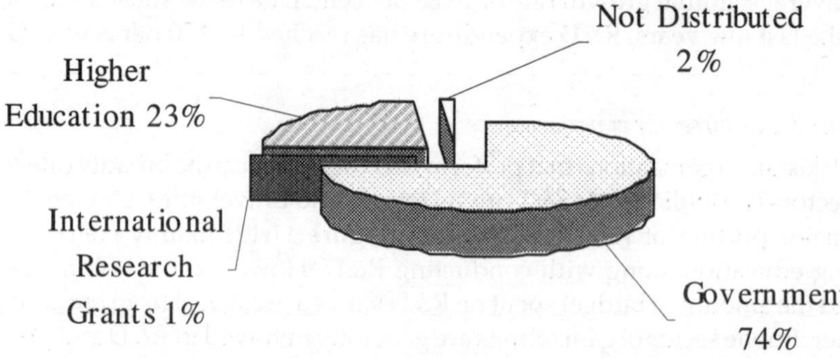

Figure 11.13: Share of different funding sources in total R&D expenditure

Traditionally, Pakistan's economy is based on agriculture. Therefore, the Government attaches high priority to R&D in this sector. Agriculture sector receives the highest portion of funds allocated for R&D (Figure 11.14). Pakistan Council of Scientific and Industrial Research (given under the title of "Industrial Research" in

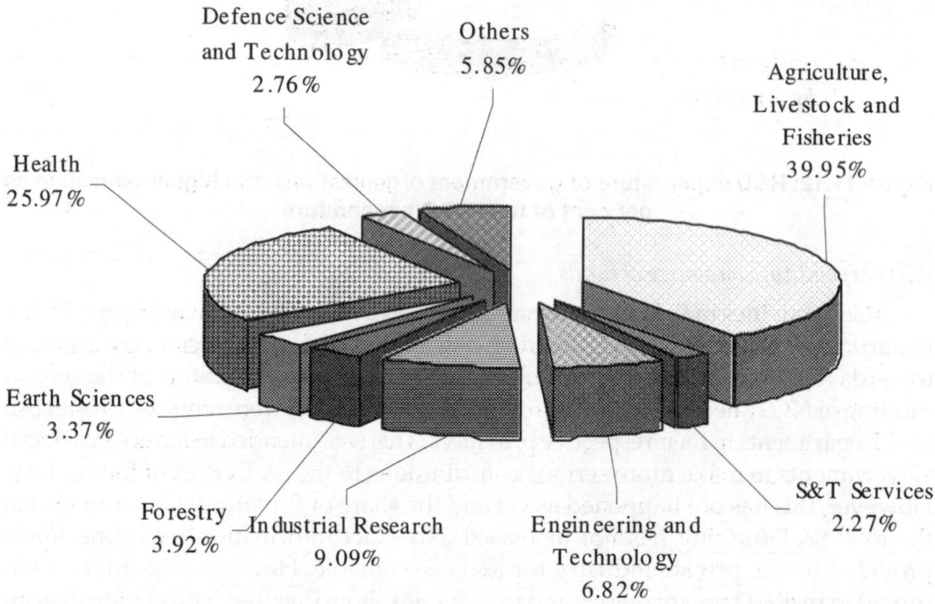

Others:
Housing and Works; Energy; Science Promotion; Wild Life; Irrigation & Water Resources; S&T Policy; Telecommunications; Transport & Communications ; Ocean Resources & Marine Sciences; Metrological Sciences; Environmental Sciences

Figure 11.14: Field-wise expenditure of R&D organizations

(Calculations are based on the data for the financial year 2001-02)

Note: Figure given under the title of Defence Science and Technology relates to the expenditure of non-classified organizations only and does not include the budget of classified organizations the Figure 11.14), which is the premier organization in the country to provide R&D support to the industry, receives 9.09 percent of the total funding of R&D organizations. Engineering and technology organizations receive only 6.82 percent funding which is very low for a country seeking to speed up the process of industrialization.

Higher education

The total expenditure of Higher Education has increased substantially during the last few years and has reached to Rs 34,446 million in the financial year 2005-06. Major portion (60.34 per cent) of funding to Higher Education comes from the Federal Government through the Higher Education Commission. Some R&D funding is also provided by the Ministry of Science and Technology and the Ministry of Information Technology and Telecommunications.

Most of the universities in Pakistan are 'General Universities', which impart education in different disciplines of Science and Humanities and generally these universities, with the exception of a few universities, do not impart education in Engineering subjects. Engineering education is basically imparted at Universities of Engineering and Technology and affiliated colleges. Other mono-discipline universities in Pakistan are in the field of Agriculture and Health sciences. There are some universities that concentrate on a few subjects such as Information Technology and Management Sciences. In this report, all the universities other than mono-discipline universities are categorized under the category of 'General Universities'. Figure 11.15 shows the proportionate expenditure of different types of universities.

Information technology

With many global IT companies having regional hubs in Pakistan and with revenues growing at a tremendous pace (Table 11.2), the IT industry has become the most exciting and dynamic sector in the country. The business community has finally understood the value that automation can add to their commercial concerns, by driving down costs, improving productivity, reducing time to market and providing

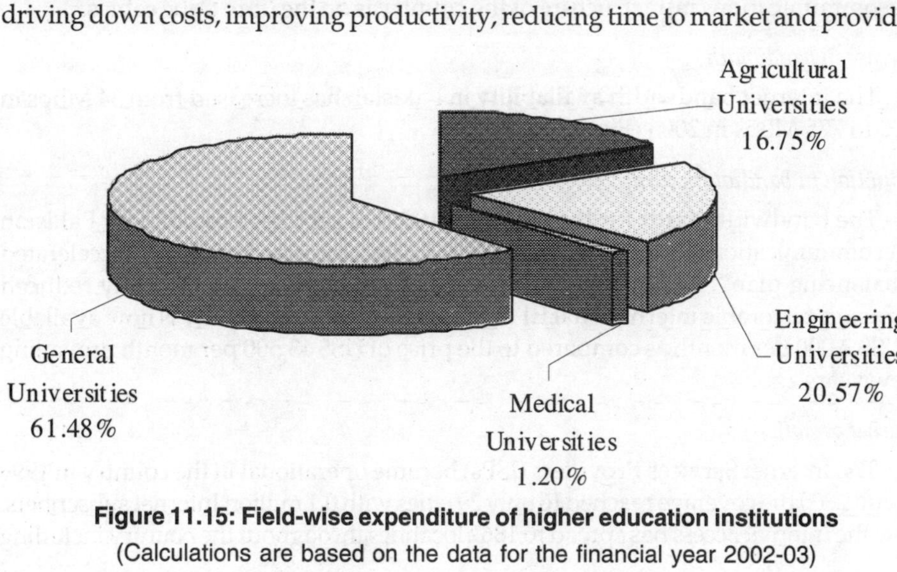

Figure 11.15: Field-wise expenditure of higher education institutions
(Calculations are based on the data for the financial year 2002-03)

quality services to their customers. The textile sector, the largest traditional industry in the country, is investing heavily into specialized software and IT solutions to reduce costs and improve quality. It has ongoing IT projects worth millions of rupees. Some of the international IT companies using Pakistan as a regional hub include

1. NCR: Teradata Global Consulting Centre
2. IBM: Regional support office for Middle East consulting services
3. Oracle: Pre- and post-sales support for the SAARC region, excluding India
4. Hyundai IT: Regional financial services solutions for South Asia and the Middle East
5. Halliburton: Global support on key petroleum software packages

Table 11.2: Revenue growth of some IT companies in Pakistan

Company	*Revenue growth (per cent)*		
	2003	*2004*	*2005*
Oracle	30	40	70
IBM	20	20	25
NCR	25	30	35
Microsoft	20	30	30
Intel	30	30	35
Cisco	20	30	50

Fiber optic connectivity

Considerable progress has been made in the recent past to implement and enhance the fiber optic connectivity in all parts of Pakistan. At present, over 85 per cent of telecommunications infrastructure of the country is on the fibre optic cables.

Increase in bandwidth

The Internet bandwidth availability in Pakistan has increased from 34 Mbps in 2000 to 775 Mbps in 2006 (Figure 11.16).

Reduction in bandwidth tariff

The bandwidth tariff for the entire portfolio of services provided by Pakistan Telecommunications Company Limited (PTCL) has decreased under an "accelerated re-balancing plan". Accordingly, the Internet tariff has been significantly reduced and a non-restorable international E1 (2 Mbps) Internet connectivity is now available at US$ 2,000 per month as compared to the price of US$ 43,500 per month prevailing a few years ago.

Internet growth

The Internet Services Providers (ISPs) became operational in the country in 1996 and till 2000 the coverage reached to only 29 cities with 0.1 million Internet subscribers. Now the Internet access has spread to 1862 localities throughout the country including some far flung remote areas and the Internet usage has spread manifold (Figures 11.17 and 11.18).

Table 11.3: Basic statistics of Pakistan's IT industry

Total number of IT companies working in Pakistan	675 (registered Pakistan Software Export Board (PSEB) members)
Number of substantial IT companies	458 (active PSEB members)
Number of companies International Organization for Standardization (ISO) certified	100
Number of companies CMM assessed	One CMM Level 5 company, one CMM Level 4, two CMMI Level 3, one CMMI Level 2
Total number of foreign IT and Telecom companies working in Pakistan	60
Total industry size–(according to a preliminary study undertaken by PSEB)	US$ 2 Billion (WTO Prescribed formula)
IT and IT enabled services export in year 2005-2006	US$ 72.21 Million
Percent growth in exports over the last 1 year	56 per cent
Export targets for the current fiscal year 2005-2006	US$ 108 Million
Number of IT graduates produced per year	About 20,000
Number of Universities offering IT/Computer Science programs	110
Number of IT professionals engaged in export oriented activities (software development/ call centres etc.)	More than 10,000
Total number of IT professionals employed in the country	90,000
Total IT spending in fiscal year 2005-2006	US$ 1.4 Billion
Total amount of space utilized in STPs	630,000 sq ft

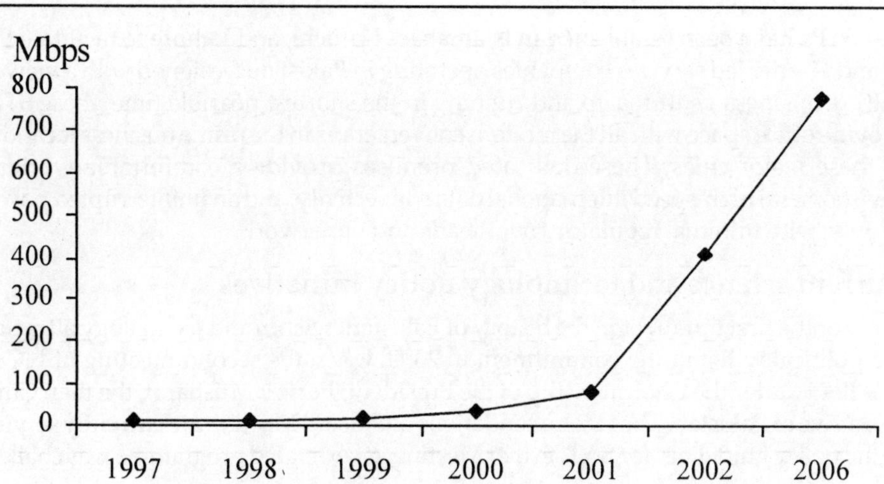

Figure 11.16: Increase in operative bandwidth

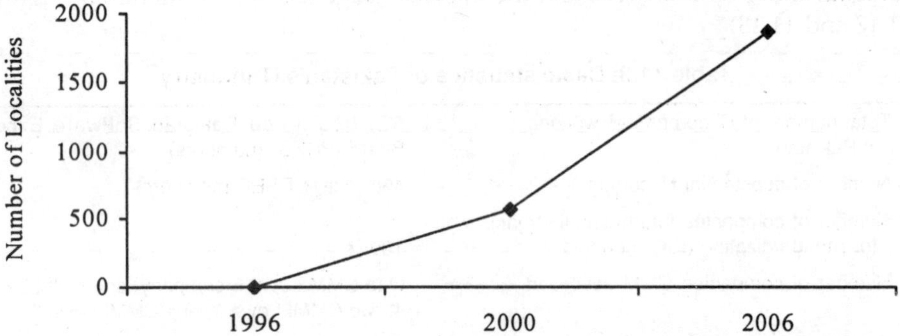

Figure 11.17: Spread of internet services in Pakistan

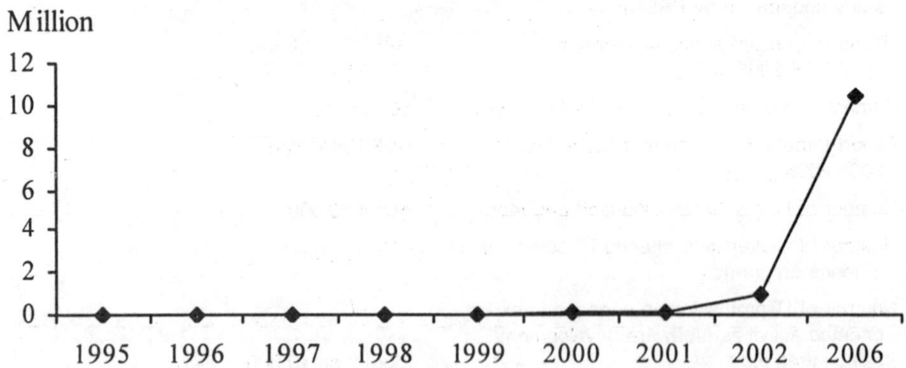

Figure 11.18: Growth of internet users in Pakistan

Software Technology Parks (STPs)

STPs have been established in Islamabad, Karachi, and Lahore to facilitate the IT and IT-enabled services companies operating in Pakistan. Designed with a view to getting business ventures up and running in the shortest possible time, these STPs provide office space with all the modern conveniences in the prime business locations in these major cities. These dedicated premises provide a comfortable working environment, high-speed international data connectivity, and an uninterrupted power supply, with minimal regulatory overheads and paperwork.

Current science and technology policy initiatives

For the first time ever in the history of Pakistan, science and technology attracted the political will and the commitment in 2000, when the second meeting of NCST was held under the Chairmanship of the President, Pervez Musharaf, the then Chief Executive of Pakistan. The recommendations of the said Meeting are currently serving as the policy guidelines for S&T. In that Meeting, an extended programme was chalked out for development of the S&T infrastructure in the country. Consequently, the development budget of the Ministry of Science and Technology was increased from Rs 120 million (in the financial year 1999-2000) to Rs 6.0 billion per annum in the financial year 2001-02–a quantum leap of 5000 percent. Over 300 projects for development of S&T, in general, and for promotion of IT, in particular, were launched. It represented a major step forward towards building an indigenous S&T capability and a knowledge-based economy in Pakistan (Table 11.4). The initiatives were categorized under broad areas of (*i*) Human resource development, (*ii*) Up-gradation of R&D infrastructure and launching of R&D programmes, (*iii*) Technology development and industrialization, (*iv*) Strengthening of policy, coordination and management structure, and (*v*) Information technology (Table 11.5).

Table 11.4: Basic elements of Pakistan's strategy for building knowledge-based economy

Policy Objective	*Strategy*
Knowledge creation	· Strengthening PhD programmes
	· Supporting technical education
	· Improving quality of education through faculty development and revision of curricula
Knowledge cultivation	· Financial support for strengthening research infrastructure of better performing organizations
	· Promoting research in strategic areas
	· Building research infrastructure of universities
	· Restructuring of R&D organizations
Knowledge sharing	· Promotion of collaborative research within the country and internationally
	· Maintaining basic research support data bases
	· Managing basic Science and Technology information
Knowledge diffusion	· Establishment of Venture Capital Companies and Technology Development Fund
	· Establishment of Technology Cluster Councils
	· Promotion of R&D in public and private enterprises
	· Strengthening Metrology, Standards, Testing and Quality System (MSTQ) infrastructure for promotion of quality and standards
	· Establishment of Technology Parks

Human resources development

There is awareness and concern at the highest level regarding the fundamental importance of the human resource development for the socio-economic development of the country. Efforts are being undertaken to devise and implement methods of developing human resources. In order to meet the shortage of qualified PhD level scientific and technological manpower, the government has initiated programmes for PhD level training in science within the country. In fields where facilities are not presently available locally, the brightest teachers or researchers are being sent abroad for training. To keep the knowledge of PhDs, specially the locally produced, up to date a scheme for their exposure and training, at post-doctoral level, abroad has been introduced.

Table 11.5: Allocation of funds for different programme areas for the financial year 2006-07

Sl.No.	Programme area	No of projects	Funds allocated (Rs million)
1.	Human resource development	162	9,603
2.	Upgradation of R&D infrastructure and launching of R&D programmes	114	3,932
3.	Technology development and industrialization	66	2,638
4.	Strengthening of policy, coordination and management structure	11	81
5.	Information technology	83	2,639
	Total	436	18,893

Technical education and vocational training are equally important for supporting the S&T development. Most of the polytechnic institutes and vocational training centres of the country are producing ill-trained technicians, who do not meet required standards. It was, therefore, decided that such sub-standard institutes and training centres should be strengthened and new ones, on the pattern of the Pak-Swiss Training Centre, Karachi,[5] should be established.

Indigenous PhD programme

In order to enhance the national output of PhDs, universities are being supported to upgrade their research infrastructure. PhD supervisors, accredited by HEC, are provided a small honorarium and a grant of Rs 500,000 per student per year for research materials.

Merit scholarship scheme for PhD in S&T

A complimentary scheme to the Indigenous PhD Programme is the Merit scholarship scheme, which provides research and training opportunities for bright young students as well as teachers and researchers leading to award of PhD degree at national universities.

Split PhD programme

Aims to train high-level manpower in strategic areas where local universities lack the requisite technical expertise. This involves one year training at local universities and 2-3 years at universities abroad.

[5] Pak-Swiss Training Centre, Karachi, established with the cooperation of Switzerland, is believed to be producing high quality technicians who are readily employed on higher salaries by the private sector.

Teachers and researchers overseas scholarship scheme

In order to ensure an adequate supply of skilled teachers, scientists and engineers to meet the growing needs of educational institutions and research organizations, this program has been initiated to train teachers and researchers at universities and research institutions abroad.

Post-doctoral fellowship programme

This project provides opportunities to PhD teachers and researchers working in universities and R&D organizations to update their knowledge at top institutions in advanced countries.

Consultancy programme for strengthening faculty in universities and R&D organizations

The shortage of highly skilled manpower in R&D organizations and universities is also being addressed through a scheme of contractual appointments of 6 months to 3 years at salary scales matching top private sector organizations.

Career development of young scientists and technologists

To ensure that young PhDs trained abroad are gainfully employed when they return to the country, "starter grants" provide them immediate access to research funds. They are allowed to use the laboratories of public sector institutions, provided a grant of Rs 200,000 and an honorarium of Rs 15,000 per month (for the unemployed).

Availing services of expatriate Pakistani scientists and engineers and foreign academicians

To upgrade the quality of research and teaching at national institutions, renowned expatriate Pakistani and foreign scientists are being invited to work at Pakistani institutions.

Curriculum development at graduate and post-graduate level

Outdated curricula of the universities are being updated in various disciplines of science.

Strengthening of polytechnics (introduction of new technologies/trades in selected institutions)

In order to train technical graduates in the need-related fields such as agriculture, textiles etc., and in the production related techniques, various new technologies are being introduced in the technical colleges.

Training of laboratory technicians

In order to train manpower for the maintenance of equipment, the technicians working at various laboratories of the universities and R&D organizations are being trained locally as well as overseas.

Participation of scientists and technologists in international conferences, seminars and workshops

Conferences and seminars provide excellent opportunities for knowledge sharing and generation of new ideas. This project provides financial assistance to scientists, technologists, doctors and engineers to present their research findings and interact with their colleagues at national, regional and international conferences, seminars and workshops.

Special S&T allowance

Scientists and engineers working in public sector institutions were paid PhD allowance of Rs 1,500 per month since 1986. On the recommendation of NCST, the Government has increased PhD Allowance from Rs 1,500 to Rs 5,000 per month. This is meant to encourage students to opt for research careers.

Research productivity allowance

Research productivity allowance has been introduced to reward those scientists, who despite the meagre resources, have been producing quality research.

Upgradation of R&D infrastructure and launching of R&D programmes

Peer review of R&D organizations

Presently, there are 85 major R&D organizations with 228 institutes, and laboratories in the country. In order to rationalize their functioning, a peer review programme was initiated to critically examine each organization. The recommendations of the peer review committees included strengthening of 147, reorganization of 6, merger of 33, streamlining of 21 and closure of 6 organizations. No recommendations were given in respect of 15 organizations. Seventeen organizations/S&T departments were selected, on the basis of their past performance, for upgradation as Institutes of Technology or Institutes for Pure and Applied Sciences.

Restructuring of R&D organizations

The restructuring involves rationalization of legal framework, administration/ structural reforms and financial reforms. A summary submitted to the Cabinet on restructuring of R&D organizations has been approved. A draft ordinance has been prepared and is currently under review.

Increase in the recurring budget of R&D organizations

On the recommendation of NCST, the Ministry of Science and Technology has requested the Finance Division for a substantial increase in the recurring budget of R&D organizations under the Ministry. The government has granted exemption from payment of duties and taxes to public sector universities and S&T institutes on scientific equipment, consumables and books/journals; thereby saving almost 50 per cent of the costs of import of these items.

Repair and maintenance of laboratory equipment

Most university departments and R&D organizations required support for maintenance of laboratory equipment. This problem is being addressed through various projects providing financial assistance for the repairs of out of order equipment and training of laboratory technicians.

National science libraries

Six major libraries in national institutions in the fields of engineering, physical, agricultural, earth, chemical and biological sciences are being upgraded to become designated "National Science Libraries". These libraries are being equipped to access databases around the world so that they can share the library resources and provide services to other libraries in various departments/institutions in these fields in the country.

Establishment of endowment funds for engineering universities

Establishment of endowment funds for engineering universities has been recommended for promotion of postgraduate research at engineering and technology universities.

Support for scientific societies

The work of Scientific Societies is being promoted through financial assistance for holding national and international conferences, seminars and workshops and for publication of scientific journals.

Establishment of COMSTECH research and training centre

Establishment of COMSTECH research and training centre in Islamabad will provide training to executives and policy makers from less developed OIC member states in the areas of science and technology policy, talent management, research management and IT related fields. The full fledged centre is in the planning stage, however, the first training course has already been held during the last year.

R&D programmes in priority areas

Many of the programmes of NCST have been focused towards human resource development and strengthening of the research infrastructure. However, these efforts can only be fruitful if these contribute to the country's production sector and, in turn, to economy benefits. Creation of knowledge markets in selected areas where Pakistan has the potential to achieve a competitive edge is the next step forward. The priority areas that have been identified are information technology, biotechnology, engineering, industrial electronics, mineral development, pharmaceuticals and renewable energy. Several projects for developing the requisite manpower and R&D infrastructure in these fields have been initiated. The Expert Committees constituted in each of these fields have identified several projects of strategic importance.

Technology development and industrialization

Pakistan Technology Board (PTB)[6]

PTB has been established with the Minister for Science and Technology as its head. The Board members are drawn from relevant Ministries and the private sector.

---

[6] PTB was established as an autonomous body under the administrative control of the Ministry of Science and Technology in 2000. However, its regular manpower could not be recruited and the Board was not made fully functional. Now (since April, 2006), PTB has been placed under the administrative control of PCST. Process for recruitment of first batch of regular staff of PTB has just been initiated (in October 2006).

The main objective of the Board is to identify, classify and assess technology needs and capabilities, advice on technology transfer and foster public and private partnership in commercializing locally developed technologies.

Technology development fund

For providing financial assistance to the R&D activities related to improvement in the quality of existing products and processes, and development and commercialization of new products and processes. It is envisaged that this would enhance productivity and competitiveness of industrial enterprises in the domestic and international market.

Venture capital fund

Venture capital companies provide a risk capital to entrepreneurs on equity participation or on loan basis and are considered essential for commercialization of indigenous technologies. A seven year exemption of income tax to the investors has been approved.

Technology cluster centres

Technology cluster centres owned and managed by the respective industry associations are being promoted. These Centres act as a bridge for public and private partnerships and provide services to small and medium enterprises for training, product development, consultancy and quality improvement.

Pakistan Standards and Quality Control Authority (PSQCA)

PSQCA has been established to strengthen MSTQ system in the country. PSQCA is working closely with the industry to introduce quality standards for industrial products and processes to meet the demands of high standards and to increase competitiveness of the global economy, especially with enhanced role of World Trade Organization (WTO).

Establishment of Pakistan National Accreditation Council (PNAC)

PNAC accredits agencies providing certification of ISO-9000 and ISO-14000 standards, laboratories for testing and calibration, and registration of auditors and training courses in quality control.

Crash awareness raising and training programme in quality management

A national campaign is being undertaken to raise awareness among exporters, industrialists and suppliers of services about the World Trade Organization's policy and regulations. PNAC has planned to organize 400 seminars/workshops and 400 training courses in all major cities in the fields of quality, environment, product certification and testing and calibration.

Industrial analytical centres

Industrial analytical centres are being established so that the industry could be provided with the analytical facilities for their products and processes. A programme for strengthening 8 laboratories of R&D organizations located in major industrial cities has been initiated.

Science and Technology for Economic Development (STED)

Under this programme joint projects are being initiated between research institutions and industries for technology-based production of high value-added goods. This is a public and private partnership programme where contribution of the public sector is 80 per cent and the private sector 20 per cent. The implementation of the STED programmes is likely to set up the trend for production of commercially viable products and processes.

Strengthening of policy, coordination and management structure

Strengthening of MoST and PCST

MoST being the national focal point on all S&T matters and PCST as the secretariat of NCST, both are being strengthened through induction of manpower.

Provincial coordination

A proper coordination system is required to link provincial S&T effort at the national level. In this connection Provincial governments have been approached, interim responses have been received and are being reviewed.

International cooperation

Scientific collaboration between countries is considered an important tool for building capacity by sharing and transferring knowledge. Pakistan has signed agreements with 21 countries. It also has joint ministerial commissions with 4 countries, executive protocol with one country and memorandums of understanding with 6 countries. However, active collaboration between scientists and R&D organizations has only been achieved with USA, China and some European countries. A joint revolving fund for S&T cooperation has been set up with China and efforts are being made to establish similar funds with other countries.

Information technology

Realizing the over-riding and pervasive nature of IT and Telecom in today's rapidly changing economic, technological, social and political environment, the Government of Pakistan has taken a large number of measures for the promotion, development and application of IT in the country. The government has approved a dynamic IT Policy (Table 11.6) which lays emphasis on the main areas of human resource development, enabling measures, e-government, industry support and policy initiatives to encourage foreign direct investment.

Human resource development

1. Seven new IT universities are being established in the public sector to meet the growing demand of IT professionals in the country.
2. A virtual university has been established on the concept of distance learning utilizing the latest web technologies and digital television. The most prominent feature of this program is that it provides the top-class faculty from across the country to impart IT education. Broadcast of the lectures

over television is free for viewing, but the students enrolled for a degree have to register, attend classes and undergo examinations. The classrooms are setup all over the country in a public-private partnership.

Table 11.6: Main objectives of Pakistan's information technology policy

Policy objective	*Strategy*
Human resource development	· Produce 50,000 high quality IT graduates annually within 7 years to meet the multi billion dollar export target
	· Impart vocational training to over 100,000 workers for IT enabled services within 2 years
	· Develop world class institutions of higher learning in IT
	· Strengthen faculty and infrastructure of existing universities
Infrastructure development	· Develop modern countrywide Telecom infrastructure
	· Encourage active participation of private sector through accelerated deregulation
	· Develop Pakistan into a regional hub for Telecommunication and Information flow
	· Launch a geo satellite to extend telecommunication-broadcasting facilities to remote areas within Pakistan and the neighbouring countries
	· Establish a chain of modern self-contained software technology parks throughout the country
	· Increase Internet accessibility through the universal Internet access programme and lowering of bandwidth prices
Efficiency and transparency in Government	· Re-engineer archaic governmental processes to improve efficiency
	· Introduce IT to speed up flow of information and track processing
	· Convert data from paper to electronic form so that it is available on-line for improved transparency and accountability
	· Measure the functional performance of Government machinery in quantifiable terms
Improve services to citizens	· Make Government information and forms available to citizens on-line
	· Bring significant improvement in services like bill payment, agency specific applications, and complaint handling
	· Eliminate time spent by citizens waiting in long queues
Stimulate the domestic economy	· Create jobs and economic growth by strengthening IT industry
	· Apply IT to traditional industry to increase its productivity
	· Create environment to facilitate e-commerce
	· Facilitate automation of banks and other sectors to improve the general business climate
Increase exports	· Generate a 50-fold increase in direct export of IT and IT-enabled services within 5 years
	· Leverage IT to help traditional export industry become more competitive in the international market
	· Widen the export base with the help of an e-commerce infrastructure

3. Thirty public sector degree-awarding institutes have been provided funds to start IT programs and to increase the intake of Bachelors and Masters candidates in IT. As a result of this program each institute will contribute to the national intake of BS/MS students by additional 100-200 students per year.
4. Faculty of international calibre is invited to join local universities. The selected faculty member is matched with the most suitable institution based on the requirements of the institution and expertise of the IT professional. These faculty members are expected to introduce R&D culture as well as the advanced learning techniques in local universities. Salaries of between Rs 100,000 per month to Rs 300,000 per month are offered to competent foreign professionals.
5. A large number of scholarships are being awarded for the Bachelors and the Masters courses in IT and computer science.
6. For promoting IT education at school and college level, IT laboratories are being set up and teachers are being trained. Twelve thousand school teachers will be trained in the first phase in a joint programme with INTEL.
7. For producing workforce in the IT-enabled services sector, training of shorter duration is being provided to government employees (in basic IT education), doctors and medical graduates (in medical transcription), java developers, inter-networking engineers (Cisco) and legal transcriptionists.
8. Projects for training high level IT professionals (PhDs) in the best national universities and abroad are under process. The objective is to create a pool of high level IT manpower to meet the growing needs of the universities and industry.
9. National testing service is being set up to provide an affordable, indigenous service similar to Graduate Record Examination (GRE) and Scholastic Aptitude Test (SAT) provided by the educational testing service, USA. The testing service can be used as entrance criteria by educational institutions, for the award of scholarships, an aid to employers in recruitment, or for any other situation where a widely accepted standard test is required.

Enabling measures

1. IT law has been drafted to facilitate e-commerce and other on-line transactions requiring digital signature and authentication. This law opens the way for electronic transactions by providing it a legal cover.
2. IT accreditation programme has been initiated to protect students against enrolling in substandard institutions as the proliferation of IT training institutes all over the country is suspected to introduce a varying degree of quality standards. The program will be supported by an Ordinance that has been drafted and is currently being reviewed by the concerned Ministries.

E-Government

1. The web portal for the Government of Pakistan, web sites of 34 federal ministries/divisions and 3 special purpose websites have been developed. The websites focus on providing information that is useful to the general public.
2. Office automation of government ministries is being undertaken to streamline the functioning of the government and increase its efficiency.
3. Automation of Salary withdrawal of federal employees will allow federal government employees to withdraw their salaries from Automatic Teller Machines (ATMs) situated in 13 locations within the capital.
4. The automation of the district governments has been initiated by selecting 2 model districts in each of the 4 provinces of Pakistan, and 2 in Azad Jammu and Kashmir.
5. Internal e-mail and mail tracking system is being developed for improving the accountability of action handling and speeding up intra-ministerial communication. The system will initially be set up at the Ministry of Science and Technology and later on will be provided to other ministries.

Industry support

1. Quality standards: The Government shall subsidize IT companies in preparing for certifications for ISO 9001 and CMM, the two quality standards applicable to the IT industry.
2. PSEB has launched several programs to assist the local IT companies in international marketing.
3. PSEB is opening its branch offices in major cities to assist IT companies in execution of business plans for domestic marketing, remove bottlenecks, and organize local IT events and exhibitions.
4. An information centre for IT industry with an up-to-date library will be setup in major cities.
5. PSEB will set up a pavilion in the software technology park, Islamabad where all participating companies will set up a coordinated stand, which will be flown to international exhibitions abroad.
6. Industrial automation programme is being initiated which includes automation of business processes supporting the industry. The cost of hardware will be borne by the industry while the Government will provide funds of up to Rs 500,000 to local companies for developing the software.
7. Measures are being taken to increase investments in the IT industry, especially in the area of IT infrastructure, IT education, medical and legal transcription, IT call centres and wireless communications.

Funding of R&D in IT

1. PTCL's R&D fund, managed by a 17-member committee, annually receives

one per cent of the gross revenue of all the telecom service providers in Pakistan, to be utilized for R&D.

2. Typical grants range from US $200,000 to US $600,000. Proposals that are submitted jointly by a university and an IT/telecom company are preferred.

Policy incentives to encourage foreign direct investment

1. A reduction in bandwidth rates from US$ 43,500 per month per Mbps to US$ 2,000 per month per Mbps, thus, making bandwidth costs in Pakistan one of the lowest among developing countries.
2. IT companies have been allowed a 15-year income tax exemption.
3. Income tax rebate of 50 per cent is offered on income of IT professionals.
4. Software exporting companies are allowed to retain 35 per cent of their earnings in foreign currency accounts.
5. A reduction in royalties charged on payphone companies from 4 per cent to 1.5 per cent and ISPs from 1.5 per cent to 0.66 per cent.
6. ISPs intending to start business in the province of Balochistan get a reduction of 50 per cent in their license fee for 5 years.
7. Zero per cent import duty on computers and computer parts.
8. Full repatriation allowed for capital gains, dividends and profits.
9. Processing period for license applications in the deregulated sector has been reduced from 12 months to 7 days only.
10. Tax relief for first year allowance @ 90 per cent of machinery cost (50 per cent on equipment for IT services).
11. Grant of seven-year income tax holiday to venture-capital funds has been approved. Banks are also allowed to invest in the form of equity in dedicated ventures.
12. Lower bandwidth rates for universities, educational institutions, software exporters and ISPs.
13. Deregulation of several financially attractive IT and telecom value-added services.

Biotechnology

Pakistan has declared biotechnology as a priority area and has funded projects of more than one billion rupees in this area. There are 28 organizations/university departments that are engaged in R&D in biotechnology and genetic engineering with a combined manpower of 504 researchers including 238 PhDs. To harness this potential for development of biotechnology industry in the country, the National Commission on Biotechnology was established in 2002. The Commission has prepared a draft National Biotechnology and Genetic Engineering Policy and Action Plan which has been submitted to the Government for approval. The thrust areas of the policy and their objectives are presented in Table 11.7.

Table 11.7: Thrust areas of national biotechnology and genetic engineering policy

Thrust Area	*Strategy*
Human resource development	· Introduce biotechnology curricula at secondary, undergraduate and post graduate level
	· Collaborate with biotechnology industry for training of eligible individuals
	· Identify and acquire possible funding resources
Research and development	· Identify possible focus areas
	· Establishment of network of research facilities
	· Coordinate communication between all the centres
	· Promote research in economically viable areas
	· Identify and acquire possible funding resources
	· Development of commercial projects for global market
Marine biotechnology	· Improvement of marine stock
	· Improvement of productivity of fishing industry through new and improved varieties
Industrial biotechnology	· Encourage manufacturing of affordable and cost effective drugs and devices for the treatment of common prevalent diseases
	· Encourage vaccine production
	· Encourage establishment of State sponsored and private manufacturing facilities for bio-pharmaceuticals and raw materials
	· Create conducive environment for investment in establishment of manufacturing units
Health	· Encourage R&D of generic bio-pharmaceuticals
	· Encourage R&D of bio-diagnostics based on national needs
	· Promote research and utility of medicinal plants of Pakistan
	· Encourage development and manufacturing of vaccines for tuberculosis, hepatitis, malaria and cholera
	· Secure emerging technologies
Agriculture	· Encourage development of new and improved useful varieties of animal and plants
	· Augment feed and fodder availability and processing
	· Improve overall nutritional security
Environment	· Develop programmes for conversion of waste to energy
	· Develop programmes for bio-remediation of heavy metals
	· Develop programmes for leaching of ores
Bioinformatics	· Establishment and promotion in educational institutes
Public awareness	· Organize public awareness programmes through popular media in local languages
Intellectual Property Rights (IPR) and ethical considerations	· Manage IPR of local R&D
	· Develop a national code of ethics

References

SCP, 1960. Report of the Scientific Commission of Pakistan, Government of Pakistan

Qurashi, M. M., and Kazi, A.Q., 1997. Fifty Years of Research and Development in Pakistan. Pakistan Council for Science and Technology, Islamabad, Pakistan, pages. 134

PCST., 2000. Agenda and Working Paper for the Second Meeting of National Commission for Science and Technology, Pakistan Council for Science and Technology, Islamabad, Pakistan

MITT., 2000. IT Policy of Pakistan, Ministry of Information Technology and Telecommunications, Islamabad, Pakistan

PCST., 2001. Agenda and Working Paper for the Third Meeting of National Commission for Science and Technology, Pakistan Council for Science and Technology, Islamabad, Pakistan

Naim, S.T.K., Nasir, A., Bashir, T., 2002. Science and Technology: The Way Forward, Pakistan Council for Science and Technology, Islamabad, Pakistan

NCB., 2003. Draft National Biotechnology and Genetic Engineering Policy and Plan, National Commission on Biotechnology, Islamabad, Pakistan

PCST., 2005. Science and Technology Indicators of Pakistan, Pakistan Council for Science and Technology, Islamabad, Pakistan, pages. 70

Howard, A., 2006. Pakistan Biotechnology: Agricultural Biotechnology Report, GAIN Repot PK6009, USDA Foreign Agricultural Services, USA

Website of Higher Education Commission of Pakistan http://www.hec.gov.pk (accessed on 14 Oct. 2006)

Website of Pakistan Software Export Board http://www.pseb.org.pk/page.php?page_id=78 (accessed on 19 Oct. 2006)

Website of Pakistan Software Export Board http://www.pseb.org.pk/page.php?page_id=79&PHPSESSID=2e9042929027e7a0b18ce83b3698a117 (accessed on 19 Oct. 2006)

Website of Pakistan Software Export Board http://www.pseb.org.pk/page.php?page_id=77&PHPSESSID=2e9042929027e7a0b18ce83b3698a117 (accessed on 19 Oct. 2006)

Website of Pakistan Software Export Board http://www.pseb.org.pk/page.php?page_id=102&PHPSESSID=2e9042929027e7a0b18ce83b3698a117 (accessed on 19 Oct. 2006)

Website of Asia Internet Usage Stats and Population Statistics http://www.internetworldstats.com/asia.htm#pk (accessed on 19 Oct. 2006)

Website of IT World http://www.itworld.com/Net/2614/050705pak/(accessed on 19 Oct. 2006)

Chapter 12

The status of science and technology in Myanmar

Thidar

Abstract

The government of the Union of Myanmar is led by the State Peace and Development Council with its objectives to enable the nation to become a "Peaceful, and Modern Developed Nation". The government acknowledges the role of highly qualified intellectuals. The intelligentsia is indispensable and very important in building a modern and developed nation. Realizing the importance of the human resources development as key to the progress of a nation, the Ministry of Science and Technology was established on 2nd October 1996 as the 30th Ministry of the Union of Myanmar, to directly support the development of the country. Further, the State has been implementing national education promotion programmes for human resources development. As a top priority the government has been opening new technological universities, Colleges, new computer university, colleges and Myanmar Aerospace Engineering University throughout the country to fulfil the increasing demand of the country, and to conduct courses in new academic fields to enable qualified students to pursue advanced technologies. This paper presents the current status of S&T in Myanmar and initiatives taken to uplift the R&D through science and technology towards the socio-economic development of the country.

Keywords: Higher education, Human resource development, Myanmar, Research and Development, Rural growth, Science and technology.

Introduction

Myanmar or by the formal and official name, 'the Union of Myanmar' is known to visitors as the "Golden Land" not only due to its golden pagodas but also for its rich natural resources including its vast fertile lands, forests and precious stones to name a few. Myanmar is also rich in cultural heritage and blessed by the nature with many attractions. Its glittering pagodas, historical monuments, snow capped mountains, picturesque lakes, unspoiled beaches and its traditional arts and crafts combined with the charm and the hospitality of its people make Myanmar one of the most fascinating destinations in Asia. The Union of Myanmar is lead by the State Peace and Development Council. The Union of Myanmar has seven states and seven divisions. The capital is Yangon. Myanmar is a member of the United Nations since 1948, and also, became a member of Association of South-East Asia Nations (ASEAN) in 1997. Accordingly, Myanmar has friendly relations with neighbouring countries as well as other nations of the world.

Myanmar is located in the Indochina Peninsula of Southeast Asia region. It has an area of 2, 61,228 square miles. Its neighbours are People's Republic of China on the north, Laos on the east, Thailand on the south east, Bangladesh on the west and India on the north west. It is also bordered by the Andaman Sea on the south and the Bay of Bengal on the south west. The country has a population of over 54 million. Its natural resources are teak, oil, natural gas, minerals, gems and marine resources. Myanmar has a tropical monsoon climate with three seasons. They are the rainy season (May to October), winter (November to February) and summer (March to April). The average mean temperature is between 70° F to 82° F and annual rainfall is 100 inches.

Myanmar is a union of many nationalities with as many as 135 national races. The term Myanmar embraces all nationalities. The main national races are Kachin, Kayah, Kayin, Chin, Mon, Bamar, Rakhine and Shan. The Bamar is the largest national race constituting 70 per cent of the whole population and the yearly population growth rate of the country is 1.88 percent. Over 80 per cent of the Myanmar people are Buddhists though there are also Christians, Muslims, Hindus and Animist. However, there is full freedom of worship for the followers of all religions.

Objectives of science and technology in Myanmar

With the aim of enhancing successful implementation of development programmes in science and technology, the State Law and Order Restoration Council established the Ministry of Science and Technology (MOST) according to the declaration No.30/96 on 2nd October, 1996. The objectives of the Ministry of Science and Technology are to carry out research and development programmes for the national economic development, to strengthen the national economy utilizing the national resources, to enhance production in agricultural and industrial sector through technology transfer and distribution of knowledge gained from research and development work, to produce and nurture human resources for advanced science and technology, to carry out tasks such as laboratory analysis, quality control and standardization of industrial raw materials and finished products, and to conduct applied research on the utilization of atomic energy.

Organization structure of MOST

The Ministry of Science and Technology (MOST) consists of various departments *viz.*, Scientific and Technological Research Department (MSTRD), Department of

Technical and Vocational Education (DTVE), Department of Advanced Science and Technology (DAST), Department of Atomic Energy (DAE), Department of Technology Promotion and Coordination (DTPC), and Materials Science and Materials Engineering Research Department (MSMERD).

Implementations

The implementation programs of the MOST are divided into two sectors, namely, (*i*) developing human resources sector, and (*ii*) research and development sector.

Developing human resources sector

In order to develop human resources, the MOST has taken several measures with momentum. These include extending and upgrading Technological Universities and Colleges, Universities of Computer Studies and Government Computer Colleges, and Government Technical Institutes, that offer special courses and new specializations, providing advanced teaching aids and laboratory equipment, providing up-to-date reference materials, and providing up-to-date machines and other facilities to be used in the practical workshops which help support in applying theoretical knowledge.

Before the MOST was set up, the universities, institutes and technical high schools in the field of science and technology included Yangon Technological University (YTU), (formerly known as Yangon Institute of Technology–YIT), Mandalay Technological University (MTU), (formerly known as Mandalay Institute of Technology–MIT), University of Computer Studies, Yangon (UCSY), 12 Government Technical Institutes (GTI), and 17 Technical High Schools (THS). According to the human resources planning laid down by the MOST, after being set up, the new universities, colleges, and institutes that have been established, with the aim of recruiting qualified teaching staff, include Pyay Technological University (PTU), Yangon Technological University (YTU/Hline Tharyar), Myanmar Aerospace Engineering Universities (MAEU), University of Computer Studies, Mandalay (UCSM), 26 Government Technological Colleges (GTCs), 24 Government Computer Colleges (GCCs), and 9 Government Technical Institutes (GTIs).

Conducting new course

Apart from the existing courses, other postgraduate courses, undergraduate courses, diploma courses, and special courses including PhD programmes, which were not available in the past, have been initiated at the Technological Universities in order to support the nation-building tasks. New courses in biotechnology, nuclear technology, and mechatronics, which directly support the state development, are also offered at these institutions.

Extended courses

Besides the existing courses, the extended courses have been established, which include PhD courses, special master courses, special postgraduate diploma courses, special undergraduate courses, special undergraduate diploma courses, special technical teachers training courses, and B-Tech Degree- conferring courses.

Purchasing modern laboratory equipment

To upgrade Myanmar Technological Universities, GTCs, Universities of Computer Studies, GCCs, and GTIs at par with other advanced universities, colleges, and technical institutes in the Asian countries, modern advanced researches and teaching materials have been purchased to fulfil the needs. An amount around US$ 14.048 million has been used in purchasing teaching aids, modern laboratory equipment and machinery to fulfil the needs of the universities, colleges, and institutes under the MOST.

Library sector

It is planned that Technological Universities, GTCs, Universities of Computer Studies, and GCCs must have libraries with modern books. According to the plan, subject-wise journals and books have been purchased from the USA, England, Japan, and India.

Purchasing machinery for practical work

In the sector of nurturing technological human resources, it is of paramount importance to link the acquired technologies with practical work. To relate theory and practice, modern machinery such as Computer Numerical Control (CNC) machines, including Lathe, CNC Milling Machine etc., and CNC Machine Centre have been purchased and installed at the CNC Machine Centre. Technologies are being taught theoretically as well as practically. In addition, the required machinery is also being produced.

Moreover, subject-wise references for Doctoral and Master' programmes are facilitated by the use of Internet. In order to increase the use of the Internet, more Internet facility has been established at YTU, MTU, UCSY, UCSM, and the Faculty of Biotechnology.

For subject wise references, the establishment of the Digital Libraries is also in progress. The Digital Library and the Internet will soon be used for collection of data and references that are necessary for doing research work.

Research and development sector

Research works that are being done by the MOST are presented in four sectors. These are research works on rural development; research works on industrial development; research works for Master's and doctoral programs, and regular research works relevant to the research departments.

Research works on rural development

To develop rural areas, the following research works are being carried out under close supervision of the MOST.

1. Research on natural fertilizer, fish fertilizer, and biofertilizer production
2. Research on power supply to household in the country using biogas
3. Project on gasifiers
4. Research on production of multi- functional 22 Hydro-power engine

Research works on industrial development

As research works on industrial development, the research on constructing models of mini Hydro, CNC, computer Numerical Control machine, the research on constructing a model robot and a robot, the research on constructing a model medium-size engine and medium-size engine, the research on constructing models of electronics accessories and electronic accessories used in industrial services, and the research on constructing a model induction furnace and an induction furnace for steel smelting are being carried out.

Master and doctoral research programs

To apply Master's and PhD research programs practically such tasks as research on production of farming tools, research on production of machine tools, research on production of military equipment, and research on social affairs are being carried out.

Regular research works of research departments

Regular research works are being carried out at the Myanmar Scientific and Technological Research Department and the Metallurgical Research and Development Centre. Subject-wise research works are being carried out at the Applied Chemistry Research Department, the Pharmaceutical Research Department, the Food Technology Research Department, the Polymer Research Department, the Metallurgy Research Department, the Ceramics Research Departments, and the Physics and Engineering Research Department under MSTRD.

Objectives and functions of the department

Myanmar Scientific and Technological Research Department (MSTRD)

Brief history

The 'Myanmar Scientific and Technological Research Department' was first established in 1947 and was reorganized as the "Union of Burma Applied Research Institute" (UBARI) in 1954. It was later also known as the "Central Research Organization"(CRO). It was re-named as the "Myanmar Scientific and Technological Research Department" (MSTRD) was adopted in May 1994.

Objectives of MSTRD

The objectives of MSTRD are to carry out research works in order to utilize the national resources more abundantly for the development of the national economy, to carry out research works in order to established more industries based on advanced scientific and engineering techniques, to conduct applied research for the domestic production after acquiring the modern scientific and technological know-how from foreign countries, and to conduct scientific and technological research activities for the benefit of the nation.

Responsibilities and undertaking of MSTRD

Nine research and development departments and six service departments not only assist the technical departments of MSTRD but also offer technical services to other government departments according to their requests. Moreover, these departments are participating in HRD programmes of MOST, by supporting and supervising the thesis of post-graduates and PhD candidates.

Myanmar Scientific and Technological Research Department is organized with the following departments:

The nine Technical Research and Development Departments are Applied Chemistry Research Department, Ceramics Research Department, Food Technology Research Department, Metallurgy Research Department, Pharmaceutical Research Department, Physics and Engineering Research Department, Pilot Plant Research Department, Polymer Research Department and Pilot Plant Complex.

The six Technical Services Departments are Analysis Department, Fine Mechanics Department, Standards Department, Technical Information Department, Mechanical Workshop and Administrative Department.

Department of Advanced Science and Technology (DAST)

Brief history

The Department of Advanced Science and Technology was established on 2nd July 1997.

Objectives of DAST

The Department of Advanced Science and Technology was set-up to work out sufficient production of engineers, architects and computer technologists for a modern development nation, and to encourage and supervise advanced technologists and technicians to do research, thereby inventing mechanical appliances according to the plans for the development of science and technology so as to build up an agriculture-based industrialized nation.

The colleges and universities that are under the guidance of the Department of Advanced Science and Technology, include Yangon Technological University (YTU) (*Gyo Gone*), Yangon Technological University (YTU) (*Hline Tharyar*), Mandalay Technological University (MTU), Pyay Technological University (PTU), Myanmar Aerospace Engineering University (MAEU), University of Computer Studies, Yangon (UCSY), University of Computer Studies, Mandalay (UCSM), and 24 Government Computer Colleges (GCCs).

Responsibilities and undertaking

The Department of Advanced Science and Technology (main office) consists of three departments undertaking assigned destinations, *viz.* Administration and Finance Department, Department of Human Resources Development for Advanced Science and Technology, and Department of Planning for Development of Advanced Science and Technology.

Yangon Technological University (YTU)

The present Yangon Technological University (YTU), first set up in 1924 as the Department of Yangon University, was upgraded to the level of the Faculty of Engineering in 1946. In 1961, it was moved to the present location, *GyoGone, Insein* Township. Yangon became a separate university in 1964. Then, the Yangon

Technological University, the Mandalay Technological University, and the Computer University, which were earlier under the Ministry of Education, were transferred to the Ministry of Science and Technology with effect from January 1, 1997.

The main objective is to produce engineers and architects, applied scientists, technologists, and those proficient in English for carrying out researches for the benefit of the country.

The following courses (the new system) are offered at the Yangon Technological University.

For B.E./B.Arch. degrees the duration is 5-years and 6-months whereas for M.E./ M.Arch. degree it takes 2 years.

As for the Ph.D. Courses it takes 3 to 5-years after graduation whereas Postgraduate diploma courses last for 1 year.

The Bachelor and Master courses in Engineering and Architecture that are offered include Civil engineering, Mechanical engineering, electrical power engineering, Electrical engineering, Information technology, Chemical engineering, Textile engineering, Mining engineering, Nuclear technology, Petroleum engineering, Metallurgical engineering and materials science, Aerospace engineering, Architecture and Mechatronic.

The Ph.D. courses offered, which normally take 3 to 5 years, include the following areas, *viz.*, Civil- Structural, Civil-Water resources; Mechanical; Electrical power; Electronics; Chemical; Architecture; Metallurgical; Textile; Mining; Petroleum Engineering; Nuclear Technology; Information Technology; and Mechatronic.

The Postgraduate diploma, Master, and PhD courses in applied science available are Engineering mathematics (Master and PhD), Engineering chemistry (Master and PhD), Engineering physics (Electronics) (Master and PhD), Nuclear technology (Master and PhD), Biotechnology (Postgraduate diploma, Master and PhD), Engineering geology (Postgraduate diploma, Master and PhD).

Besides, the 1-year duration postgraduate diploma courses and the 2-year duration Master degree courses also offered *viz.*, Business and technology management (Postgraduate diploma and Master), Environmental engineering (Postgraduate diploma and Master), Environmental planning and management (Postgraduate Diploma and Master), Food technology (Postgraduate diploma and Master), Settle planning and design (Postgraduate diploma and Master), Aqua-technology (Postgraduate diploma and Master), and English (Postgraduate diploma and Master).

Since September 1997, two-year part-time diploma courses offered include Diploma in Technology (Civil), Diploma in Technology (Mechanical), Diploma in Technology (Electrical Power), Diploma in Technology (Electronic), Diploma in Technology (Chemical), Diploma in Technology (Textile), Diploma in Information Technology, Diploma in Technology (Mining), Diploma in Technology (Petroleum), Diploma in Technology (Metallurgical), Diploma in Technology (Aeronautical), Diploma in Technology (Architectural design and planning), and Diploma in Technology (Mineral).

Mandalay Technological University (MTU)

State Peace and Development Council has given instructions to establish universities in upper Myanmar parallel with the ones in lower Myanmar. Mandalay Institute of Technology was established on 1st October, 1991. Then, it was renamed Mandalay Technology University on 1st July, 1998.

The main objective is to produce qualified engineers and architects as well as applied scientists with proficient knowledge in English.

It now offers eight Bachelor degree programmes in engineering and architecture. The duration of the courses is 5-years and 6-months.

The degrees offered at MTU are the Bachelor of Engineering in the areas of Civil engineering, Mechanical engineering, Electrical power engineering, Electronics, Information technology, Mechatronic Engineering, Chemical Engineering, and the Bachelor of Architecture.

In the case of the engineering field, those who complete the bachelor's degree are given the opportunity to follow the two-year postgraduate course towards the master degree.

The MSc courses offered at present are Civil engineering, Mechanical engineering, Electrical power engineering, Electronic engineering, and Chemical engineering.

Moreover, the Master, and the PhD Courses in applied science are also available in Engineering mathematics (Master and PhD), Engineering Chemistry (Master and PhD), and Engineering physics (Electronics) (Master and PhD).

In Science, since academic year 1999-2000, the special courses offered at MTU with the support of Ministry of Science and Technology include, First Year BE and B.Arch Sanpya engineering course, Second year B.E and B.Arch Sanpya engineering course, Third year BE and B.Arch Sanpya engineering course, Fourth year BE and B.Arch Sanpya engineering course, Fifth year BE and B.Arch Sanpya engineering course, Special technical teachers training courses (B. Tech.) 1st Batch, Special technical teachers training courses (B. Tech.) 2nd Batch, and Postgraduate diploma in English for Specific Purposes (Dip. ESP).

In addition, since academic year 1999-2000, Diploma in Technology, part-time, courses have been offered. These are Diploma in Technology (Civil), Diploma in Technology (Mechanical), Diploma in Technology (Electrical Power), and Diploma in Technology (Electronics).

Pyay Technological University (PTU)

To provide chances for those high-school graduates, living in the western part of the country, higher engineering and architectural subjects, PTU came into existence on 18th December 1997.

The main objective is to offer Bachelor and Master of Engineering degrees in engineering and architecture, like YTU and MTU.

PTU started on 27th December 1999. Students from states and division in lower Myanmar, apart from Yangon Division are now attending PTU, and studying their engineering and architectural courses.

Myanmar Aerospace Engineering University (MAEU)

Myanmar Aerospace Engineering University was established on 15th February 2002.

The main objectives of MAEU are to contribute to the development of the country through aerospace engineering, to produce highly qualified aerospace engineers and technicians in the country, and to plan and strive for the continuous development of the knowledge in aerospace engineering in the country.

The motto of MAEU is "To Produce Highly Qualified Engineers through Quality Education". Its entrance requirements are the highest scores in matriculation examination and one must also pass the entrance examination.

The Degree and Diploma courses are offered at MAEU and the durations are 5-years and 1-year respectively.

University of Computer Studies, Yangon (UCSY)

The Universities' Computer Centre was established in April 1971. Courses leading to Master of Computer Science degree and Computational Diploma were introduced in 1973. The Institute of Computer Science and Technology was founded on 29th March, 1988. It was renamed as the University of Computer Studies, Yangon on 1st July, 1998.

The Degree courses in computer studies are offered at University of Computer Studies, Yangon and the durations are 3-years for Ph.D, 2-years for Master, 1-year for postgraduate diploma, 3-years for bachelor and 1-year and 4 months for diploma.

University of Computer Studies, Mandalay (UCSY)

Experts of Computer Science and technology have been steadily produced after the establishment of the University Computer Centre in April 1971, the Department of Computer Science of Yangon University in 1986-87 academic year, and the Institute of Computer Science and Technology in 1988, and the University of Computer Studies, Mandalay, Upper Myanmar in September 1997. The Degree courses in computer are offered at University of Computer Studies, Yangon and the durations are 2-years for Master, 1-year for postgraduate diploma, 3-years for bachelor and 1-year and 4 months for diploma.

Department of Technical and Vocational Education (DTVE)

Brief history

The Department of Technical and Vocational Education was first established in 1954 and was later reorganized as the "Directorate of Technical Education" and then, in 1972 as Department of Technical, Agricultural and Vocational Education. It was renamed as the Department of Technical and Vocational Education on 1st January 1997.

Objectives of DTVE

The main objective is to nurture and train middle-rung technicians and performance-base well trained engineering technicians and engineers who are well educated and efficient in practical work as well.

Before the DTVE was set up, the existing Institutes and schools were Technical Teachers Training Institute (TTTI), 12 Government Technical Institutes (GTIs), 17 Technical High School (THSs), 11 Handicraft School (HSs), and 2 Machinery Repair and Maintenance School (MRMSs).

Annual intakes of about 4,000 were admitted to the Institutes and schools offering A.G.T.I Diploma.

Extending and upgrading of colleges and institutes

According to the human resources planning laid down by the MOST 26- Government Technological Colleges (GTCs) and 11- Government Technological Institutes (GTIs) were established with the aim of recruiting qualified teaching staff. The Engineering Departments/Courses that were established for teaching and doing practical jobs at GTCs and GTIs included civil, electronics , electrical, Mechanical (MP, MT), Chemical (PR, FC), architecture, mechatronic, textile, metallurgy, petroleum, aeronautical, mining, nuclear technology, information technology, biotechnology, food technology, aqua technology, environmental, applied sciences, and industrial management.

Presently available courses at GTCs and GTIs are civil, electronics, electrical power, mechanical power, machine tools and design, mining, plastic and rubber, mechatronic, food and chemical, information technology.

Certificates, diploma and degree offered at GTCs and GTIs

The Certificates, Diploma and Degree courses in Engineering are also offered at GTCs and GTIs and duration include BSTC (Module 1&2) for 1-year, Associateship Government Technical Institutes (A.G.T.I) for 2-years, Basic Technicians (B.Tech) for 2-years, and Bachelor of Engineering (BE) for 1-year.

Department of Technology Promotion and Coordination (DTPC)

Brief history

In order to become an industrially developed nation, Union of Myanmar has to build many industries. This process of industrialization requires importing machineries from abroad or construction of machineries locally. Establishment of the Department of Technology Promotion and Coordination was approved by the Government Cabinet Meeting (26/98) on 29th July, 1998.

Objectives of DTPC

Main objectives of the DTPC are (i) to fulfil, utilize and upgrade the local technological potentials in the construction of machineries, and (ii) to collaborate with international organizations for promotion of technology.

Responsibilities and undertaking of DTPC

The Department of Technology Promotion and Coordination is organized with the following departments, namely, mechanical technology promotion and inspection department, chemical, metallurgy and textile technology promotion department, electrical technology promotion and inspection department, design and construction department, and administration and finance department.

Mineral Research and Development Centre (*Ela*), which was previously under Myanmar Scientific and Technological Research Department, was transferred to the Department of Technology Promotion and Coordination on March 1, 2000. The Mineral Research and Development Centre is organized into administration department, basic research department, applied research department, and technical services and consultation department.

Biotechnology Department was transferred to the Department of Technology Promotion and Coordination on 16th August, 2001. The Biotechnology Department is organized with the areas such as environmental biotechnology, biochemistry, microbiology, plant biotechnology, plant tissue culture, industrial biotechnology, aqua technology, and animal biotechnology.

Activities of DTPC are to carry out research on natural fertilizer, fish fertilizer, and bio fertilizer production; research on power supply to household in the country using biogas; project on gasifies; and research on production of multi-function 22 H.P engine.

Department of Atomic Energy (DAE)

Brief history

The Department of Atomic Energy has its roots in the Union of Burma Atomic Energy Centre (UBAEC) established in 1955 as a division under the Central Research Organization (CRO) which again got the new name of Myanmar Scientific and Technological Research Department (MSTRD). DAE was established as a new separate Government Department on 2nd July 1997.

Objectives of DAE

Main objectives of the DAE are (*i*) carrying out research work for the development of nuclear technology in the country, (*ii*) carrying out research, development and training in the fields of atomic energy to keep abreast with developments at international level, (*iii*) implementation nuclear radiation protection, (*iv*) to represent as an organization on behalf of the State in International relations in atomic energy matters, and (*v*) coordinating with the Government Departments and the private sector for their nuclear applications and promotion.

Conclusions

MOST is implementing the tasks of nurturing required technicians and scientists by means of short-term and long-term plans outstrips the target goals, and successfully, carrying out in drawing plan and doing state–level research works for rural development, defence, public health, social affairs, industrial development, and technological development.

Acknowledgement

Our special thanks are due to *His Excellency Minister U Thaung*, Ministry of Science and Technology. I would like to acknowledge U Zaw Win, Head of Office, Ministry of Science and Technology, Nay Pyi Taw, Dr. Ko Ko Oo, Director General, Ministry of Science and Technology and Dr. Aye Myint, Director General, Myanmar Science and Technological Research Department for their permission to attend this conference.

I also would like to express my deepest appreciation to Daw Mya Kyi, Director, Department of Advanced Science and Technology, Dr. Sayne Lei Shwe, Assistant Director, Ministry of Science and Technology for their guidance, valuable advices and their help.

Finally, I would like to thank Centre for Science and Technology of the Non-aligned and Other Developing Countries (NAM S&T Centre), New Delhi, India, National Science Foundation, Colombo, Sri Lanka and National Science and Technology Commission, Colombo, Sri Lanka for invitation and sponsoring me.

Chapter 13

Science and technology in Nepal

Dilli Raj Joshi

Abstract

Science and technology plays a vital role in economic and social development of any nation. Nepal has grappled for years to come up with its own science and technology policy. This paper presents the history of S&T in the country since 1950s, which has mostly been based on its indigenous traditional technologies. However, the inclusion of S&T in the development plans of the country occurred only after 1980s, but with a slow progress in implementation. The country entered towards a new era after the establishment of the Ministry of Science and Technology followed by the institutionalization process of S&T. The paper describes the mandate of the S&T Ministry and other key institutions such as the Nepal Academy of Science and Technology (NAST), National Science and Technology Council (NSTC). The newly prepared S&T Policy in the country in 1990 has identified many key areas in the field of S&T as key to development. The major programs undertaken in the plans included development and promotion of appropriate and indigenous technology, human resource development, establishment of science and technology information service system, promotion of professional societies related to science and technology, mobilization of private sector, promotion of information technology, promotion of alternative technology, and institutional strengthening (institutions, policies). However, the non-availability of information, statistical data and relevant indicators in science and technology have been identified as major gaps in this direction.

Keywords: Nepal, Science, S&T indicators, S&T institutions, S&T policies, Technology and development.

Background

Nepal, situated between latitudes of 26°22' and 30°27'N and longitudes 80°40' and 88°12' E on the globe, is a small landlocked country with area of 147, 181 sq km with total population of 24.2 million. The average length from East to West is 885 km and North-South varies between 130-245 km. It borders with two technologically giant countries, *viz.* India in the east, west and south and China in the north. The altitudes range from 60 meters above the sea level in the south to the highest peak of the world 8848 meters high Mount Everest in the north. The country has great variety of topography which is reflected in diversity of weather and climate. Physiographically, Nepal can be divided into three regions: mountain, hill and tarai. The mountain region, ranging from 3000 meter and above, covers 35 per cent of the total area of the country and 7.5 per cent of the population live in this region. The hill region which comprises a belt of moderate sized mountains and low lying valleys covers 42.0 per cent of the area and 44.5 per cent of the people live in the region. Similarly, the low land tarai region is situated in the southern part of the country and covers 23 per cent of the total area and 48 per cent of the total population live in this region.

Administratively, Nepal is divided into five development regions, *viz.* Eastern, Central, Western, Mid-western and Far-western Development region. They are further divided into 14 zones and 75 districts. The districts comprise of Village Development Committees (VDCs) and Municipalities which are the lower administrative units in each district. Each VDC comprises of nine wards and these wards in a municipality range from 9 to 35 depending upon the size of population living in a particular ward. At present, there are 3195 VDCs and 58 municipalities including one metropolitan (Kathmandu) and four sub metropolitan cities in Nepal.

Nepalese economy is a developing economy, gradually, transferring from rural to urban and traditional to modern market oriented. By the beginning of 1990s, the Government had adopted the policy of economic liberalization, and hence, the concept of open market economy came into practice. Nepal is predominantly an agricultural country, about 80 per cent population deriving their livelihood from this sector. It is the main source of gross domestic production (about 40 per cent) income and employment generation. Though, there is a changing trend in the methods and technology used in the agriculture, traditional methods of cultivation are most common in the rural mass. Arable land is 15.8 per cent of the total area of Nepal; irrigated land is about 30 per cent. Thus, dependency on monsoon is one of the main problems that the agriculture sector is facing.

Science and technology system in Nepal

Development of science and technology in Nepal

The history of science and technology in Nepal began with the end of self isolation from the rest of the world in the early 1950s. Until then, the country relied almost entirely upon its indigenous traditional technology for its development. Nepal is rich in many traditional technologies. Some of the traditional technologies in Nepal are at least two thousand years old. With the modernization of the country and the increasing adoption of imported technologies, some of the traditional technologies appeared to have lost their importance, especially in urban areas of the country. Some noteworthy traditional technologies are listed below:

1. Nepali craftsman had a sound knowledge of the basic principles of metallurgy which is proven by the excellent artifacts, mostly images and statues of bronze, copper, and gold, found in the temples of Kathmandu valley,
2. Clay works *i.e.* pottery. Various clay articles are produced such as bricks, roofing tiles, paving tiles, or different types of pots,
3. Traditional technology was also remarkable in the field of architecture and civil engineering. The application of these technologies was often blended with art and culture of Nepal society mostly in the form of temples, pagodas and stupas,
4. Traditional methods of textile manufacture using spinning and weaving devices such as handlooms,
5. Traditional method of paper manufacture, and
6. Food technology including the technique of brewing beers, wine and hard liquor, preserving meats, fruits and vegetables, preparation of dry foods from milk, vegetables, fruits and grain.

Science and technology and development plans

Although the country initiated its planned development plan in 1956, it was only after the sixth five year development plan (1980-1985) that science and technology was recognised and the planning documents contain a separate chapter on science and technology. Since then the development plans have constantly focused on the development of science and technology. The major programs undertaken in the plans included development and promotion of appropriate and indigenous technology, human resource development, establishment of science and technology information service system, promotion of professional societies related to science and technology, mobilization of private sector, promotion of information technology, promotion of alternative technology, and institutional strengthening (institutions, policies).

Science and technology institutions

Though establishment of National Institute on Science and Technology to coordinate science and technology activities in the country and to formulate a national science policy was recommended in 1966, it was only in 1976 that such a body was formed. The National Council for Science and Technology (NCST) was formed under the chairmanship of the Vice-chairman of the National Planning Commission in January 1976 as an advisory body to the government. In December 1982, the then Royal Nepal Academy of Science and Technology (RONAST), and recently, renamed as Nepal Academy of Science and Technology (NAST) was established to coordinate science and technology in the country. After the restoration of democracy, it was strongly felt that a Ministry of Science and Technology be formed. Accordingly, the Ministry of Science and Technology was set up in 1996.

Ministry of environment, science and technology

Ministry of Science and Technology (MOST) was set up on May 15, 1996. After the dissolution of the Ministry of Population and Environment in March 2005 and with the merger of Environment, the Ministry has been named as Ministry of Environment, Science and Technology (MOEST).

Objectives

The major objectives of the MOEST are to promote environmentally sustainable economic development of the country, preserve natural and cultural environment and protect life supporting system (*i.e.* air, water and soil), identify new technologies that contribute to achieving the national objectives regarding poverty alleviation, developing appropriate and new technology through the research, develop and promote traditional and indigenous technologies, and encourage the intellectuals working in the field of science and technology by creating appropriate opportunities.

Functions

Broadly, the functions of the Ministry are related with following matters:

1. Formulation and implementation of policies, plans and programmes pertaining to environment, science and technology,
2. Liaison and coordination with national and international organizations pertaining to environment, science and technology,
3. Pollution control, environmental conservation and balance,
4. Exploration and research on the progress made in the field of science and technology,
5. Promotion of alternative energy,
6. Study, research and forecasting on hydrology and meteorology, and
7. Development and expansion of information technology, planetarium, observatory and science museum.

The role of the Ministry, in particular, is to be a promoter, facilitator and regulator.

Nepal Academy of Science and Technology (NAST)

The Act provides NAST with wide range of mandate for the development of S&T in the country. Since its inception, the Academy has been involved in research and promotion. The main functions of the Academy include to undertake and promote studies and research in priority areas, advise the government on formulation of technology transfer policy and its implementation, implement S&T programmes in collaboration with national and international organizations, collect and disseminate S&T information through a central S&T information system, organise seminars and conferences on S&T topics and help facilitate S&T publications. It also aims to establish and strengthen linkages with regional and international institutions in order to promote mutual cooperation, and to mobilize internal and external resources, both financial and technical, for S&T development as well as to carry out the activities of the Academy. The Academy gives recognition to outstanding individual and institutional contribution for development, promotion and application of S&T. The functions of the Academy also include to increase public awareness of the importance and usefulness of S&T as well as to create conditions conducive of S&T application, and undertake studies and submit reports on the programs and achievements of S&T related national institutions. The Academy advises the government on S&T development programmes, on the establishment of new institutions or laboratories for S&T related research and development. It provides financial and technical assistance for the establishment and development of new S&T institutions/ laboratories, sets up exhibit centers for S&T related inventions and innovative models, and undertakes all necessary measures to achieve the goals and objectives of the Academy.

Present activities of the academy

Research activities of the Academy include the areas of biotechnology, natural products, environment, alternate energy, high altitude science and technology, and science and technology policy. Its S&T Promotional activities include science popularisation programmes such as science exhibitions to create awareness on science and technology among the general mass, and weekly science radio programme over Radio Nepal which has nation wide coverage. The radio programme encompasses topical commentaries, question-answers, project reports and interviews with scientists and technologists. This programme has recently been aired through FM Radio (Kantipur) as well. The other activities are a weekly TV programme on various S&T related topics, on-the spot quiz contest for high school students in different parts of the country, Meet your Scientists Interaction Programme – a forum to discuss the contemporary issues with scientists, media personnel and other concerned people, and school science programmes, which are designed for school level students like science quiz, interaction with teachers, local leaders and professionals, and science video shows. This program is community based and is conducted with participatory approach encouraging the local communities and the clubs to conduct them. The TV programmes telecasted range from documentary on indigenous technologies to various advanced technologies. Local, national and regional level seminars and workshop are held from time to time. For example, a seminar on 20 years Perspective Plan for Science and Technology, series of seminars on technology transfer at national and regional level, seminars on intellectual property right were held. The Academy has organized four national conferences, which were participated by national and international scientists and technologists. The other activities include organisation of regular consultative meetings and talk programmes, providing regular research grants and scholarships, awards, regular support to professional societies, and S&T publications. It brings out Vigyan Lekhmala, (the popular science magazine), books and booklets, and the research journal: Nepal Journal of Science and Technology.

National Science and Technology Council

The National Science and Technology Council has been formed under the chairmanship of the Minister of science and technology. The Council comprises the representatives from different relevant organizations, universities, professional organizations and private sectors. Its functions are to review and revise the policy relating to the development and extension in the field of science and technology, to review the progress of the work carried out in the field of science and technology, to find out measures for resolving the problems as per the requirements, and coordinate and monitor all the actions done in this field.

However, this Council is not yet operationally constituted. Besides these national institutions there are four universities which produce high level manpower including Ph.Ds, in natural sciences, medicine, agriculture, engineering, forestry, and biotechnology. Though Nepal has been a late starter in the generation of science and technology manpower and infrastructures, during the last 50 years, the country has created basic infrastructure of science and technology and produced significant number of related manpower. It is estimated that there are about 30 thousand science and technology related manpower in Nepal, which is about one per thousand of population.

Science and technology policy

Nepalese Science and Technology Policy for the first time was adopted in 1989 with a view to improving the country's science and technology capabilities for overall development and scientific innovations. In the Constitution of Nepal (1990), the Directive Principles and Policies of the State envisage an explicit policy on science and technology, which states that the state shall adopt policies giving the priority for the development of science and technology and the local technologies as well.

Vision

To build the country as a developed, dynamic and prosperous state by raising the living standards through the appropriate development and use of science and technology.

Objectives

The objectives of the policy are to enhance the national capacity through the appropriate development and use of knowledge, skill and efficiency in the field of science and technology, to assist in the poverty reduction activities by utilizing natural means and resources in a sustainable manner through the use of science and technology and by promoting social and economic status of the people and protecting and preserving environment, and to elevate the country to a competitive position through the optimum development of science and technology.

Policy

Different policies shall be initiated for the accomplishment of the aforesaid objectives. Some of these policy initiatives include to link up planned development of science and technology with the overall development process of the country, to assist in poverty reduction programmes through development of science and technology, and to use science and technology as a powerful mean to increase production and productivity of the country. The aim is to create an environment for maximum utilization of knowledge and skill pertaining to science and technology available in regional and international arena by promoting mutual cooperation and coordination with bilateral, multilateral, regional and international organizations in the field of science and technology. The policies aim to encourage participation of private sector in the development of science and technology and facilitates contribution to the development of science and technology through organizations associated with science and technology, and people's participation. The policies also include the development and mobilisation of skilled human resources in the field of science and technology, extension of the development of technology to rural levels by giving priority to the activities relating to development, extension, mobilization and utilization of indigenous technology, means and resources, as well as to perform a role of a facilitator, promoter and coordinator to the government for developing and extending all sectors to be involved for the development and use of science and technology. The different policies aim to create a conductive environment in order to maintain high morale of scientist and technologists, to attract internationally renowned Nepalese scientists living abroad and to minimize Brain Drain, to link up tertiary education pertaining to science and technology with research, and to provide quality education pertaining to science and technology from the primary level.

In order to implement these policies, the strategies and action policies include infrastructure development, human resource development, research and development, provision of other sectoral policy guidelines in applying science and technology, and coordination.

Gaps in science and technology

The issues that need to be addressed to make a more visible impact of science and technology in the national development are given below:

1. Poor public understanding of science and technology: Interactions between political leaders, planners, decision makers with the scientific community to gain mutual confidence and support,
2. Lack of coordination and cooperation between science and technology organisations: National Science and Technology Council needs to be activated,
3. Less priority for R&D: research culture is yet to take its root in Nepal,
4. Lack of integrated long term master plan for the development of science and technology,
5. Gaps between policies and practices and between planning and performances need to be avoided to achieve desired result,
6. Technology development and transfer need to be better institutionalized,
7. State funding very low: dilemma of the developing country (or least developed country like Nepal) in investing for S&T,
8. Lack of economic development: no funding for R&D from the industrial and private sectors,
9. Lack of minimum research facilities, and Brain Drain.

Conclusion

It is well accepted that science and technology plays a vital role in economic and social development of any nation. Nepal has grappled for years to come up with its own science and technology policy. It is, thus, important to have a strategically designed science and technology policy at the national level, which fits into the circumstance of the economy and industrial needs in the globalizing society. However, in many developing countries, S&T policy is rather focused on basic, scientific and sometimes academic researches, which normally are not appropriate to local economic conditions. Moreover, most developing countries do not have proper human resources to design their own national S&T policies, and thus, the experience in S&T policy implementation.

S&T policy of any country today cannot be sustained in isolation. The country has to be aware on what is happening around. Then, the situation could be analysed as to what kind of S&T policy would lead to economic growth and enable development of our own talents and skills. One should also try to find out how far the S&T policy drawn would suit local conditions.

Experience of NAM S&T member country's S&T policy and its implementation are not well shared among the member countries. S&T policy experience of each country could be shared with other member countries with a view to learn lessons in science and technology policy during its industrialization process, to share S&T policy experience and national innovation system of each NAM S&T member country, and to draw recommendations for the direction of S&T policy for the member country in terms of science and technology policy.

Nepal is yet to achieve major science and technology accomplishments. As a late starter and with new political developments in the country, it is expected that the decision makers would not only give priority to science and technology but also implement the newly prepared Science and Technology Policy. To achieve the objectives of the science and technology policy, information on various aspects of S&T needs to be collected. Such information may include S&T statistics and indicators related to expenditures and human resources devoted to R&D, education of R&D personnel, measurement of technological innovation, and S&T outputs like publications, patents, technologies developed and commercialized, or technological balance of payments, in which Nepal is still lagging behind.

References

Bajracharya, D., 2001. Science and Technology in Nepal, Royal Nepal Academy of Science and Technology, Kathmandu, Nepal

MoST, 2005. Science and Technology Policy, Ministry of Science and Technology, His Majesty's Government of Nepal, Kathmandu, Nepal

Part II

Leveraging S&T Policy for Innovation

Chapter 14

Measuring science, technology and innovation in developing countries: The UIS experience

Ernesto Fernández Polcuch

Abstract

After developing a comprehensive strategy, based on a worldwide consultation, the UNESCO Institute of Statistics (UIS) has implemented its renewed science and technology (S&T) statistics programme in 2004. The programme includes a series of activities comprising not only data collection, but also capacity building and methodological development. The UIS S&T statistics programme seeks to take into account the developing countries' specificities while aiming at improving quality and cross-national comparability.

In this framework, three different approaches were taken to develop statistics on careers of Doctorate holders (CDH), innovation, and research and experimental development (R&D). These approaches are classified as "development ex novo", "adaptation", and "adoption", respectively. The characteristics and implications of these different approaches provide strategic guidelines for future action in the development of methodologies for the measurement of science and technology in developing countries.

Keywords: Developing countries, Measuring science, technology and innovation, R&D statistics, S&T statistics, UNESCO.

Introduction

The UNESCO Institute for Statistics (UIS) has re-launched its science and technology statistics programme in 2004, resuming a programme that had published the first S&T statistics in 1968, but had been in a consultation-mode for more than a decade (Fernández Polcuch, 2006). After a thorough consultation process involving more than 200 experts from all over the World, *viz.* UNESCO International Review of S&T Statistics and Indicators, a strategy paper, namely, Immediate, Medium and Longer-Term Strategy in Science and Technology Statistics (UNESCO, downloaded from web) was developed to guide the Institute's work on this issue for the short, medium and long term. The following priority areas were set for the UIS work:

1. Human and financial Resources (Immediate term)
   - o R&D statistics (R&D personnel and expenditure)
   - o Human resources in S&T, including International mobility and gender issues
   - o Science education
2. Innovation (Medium term)
3. Output and Impact (Longer term)

The various priority areas had to be approached with three different and concurrent lines of action, *viz.*

1. Data collection and "guardianship", analysis and publications
2. Training and capacity building
3. Establishment of appropriate methodological guidelines

However, the application of the lines of action to the various priority areas does not admit a one-size-fits-all strategy. Every priority area merits a different approach, as discussed in this paper.

While UIS is the lead agency for S&T Statistics in the United Nations system, it is not alone in the field. Beyond its national counterparts, UIS works in different types of collaborations and partnerships with various organizations. The Natural Sciences sector at UNESCO headquarters has been the most important partner, contributing to the strategic thinking, co-organizing numerous activities and establishing itself as a privileged stakeholder on this matter. The Organization for Economic Cooperation and Development (OECD), the Statistical Office of the European Communities (Eurostat) and the Ibero-American Network on Science and Technology Indicators) *i.e.* Red de Indicadores de Ciencia y Tecnología (RICYT) have been instrumental to lowering the burden on countries by establishing data-sharing agreements and collecting data from their countries of influence. Various projects and activities were also carried out with the participation of New Partnership for Africa's Development (NEPAD), African Union, African Technology Policy Studies Network (ATPS), Association of Southeast Asian Nations (ASEAN), International Development Research Centre, (IDRC) Canada, Institut national de la recherche scientifique (INRS) *i.e.* National Institute of Scientific Research, Québec, Islamic Educational, Scientific and Cultural Organization (ISESCO), the Inter-Academy Council, and the Centre for Science and Technology of the Non-Aligned and other Developing Countries (NAM S&T Centre).

R&D statistics

R&D statistics has always been the "bread and butter" of the S&T statistics programme. This area has also been the subject of immediate attention by the UIS, and constitutes the most demanded part of its outputs in terms of external data users. The S&T statistics programme includes a biennial global R&D data-collection exercise as well as specific training mechanisms (workshops), and an ongoing quest for the development of a methodological document providing more detailed guidelines for developing countries to produce R&D statistics.

Global survey on statistics of science and technology

The UIS survey on statistics of science and technology deals with R&D statistics and is conducted on a biennial basis, covering a total of 221 countries and territories. The 2004 S&T statistics survey was launched in May, 2004. The final results by region are given in Table 14.1. This was the first UNESCO statistical questionnaire to be available for completion on-line via the Internet. The latest resulting data were released on the UIS website in May 2006, providing statistics for 114 countries and territories. The 2006 S&T statistics survey was launched immediately after publication of the results of the 2004 survey in June 2006.

Table 14.1: UIS survey on statistics of science and technology 2004 and 2006: Results by region

Regions	*Questionnaire 2004*				*Questionnaire 2006*			
	Countries covered	*Responses*	*Response rate*		*Countries covered*	*Responses (May 2007)*	*Response rate*	
Sub-Saharan Africa	49	17	35%	35%	49	27	55%	57%
Arab States in Africa	5	2	40%		5	4	80%	
Asia (excluding Arab States and OECD)	32	23	72%	59%	32	21	66%	61%
Arab States in Asia	12	3	25%		12	6	50%	
Americas (excluding RICYT)	23	5	22%	23	3	13%		
Europe (excluding OECD and Eurostat)	15	9	60%	17	9	53%		
Oceania (excluding OECD)	19	2	11%	19	0	0%		
Sub-total	155	61	39%	157	70	45%		
Data from other sources:								
OECD + Eurostat	40	*Total coverage*		41	*Total coverage*			
RICYT	23			23				
Total	218	124	57%		221	134	61%	

The UIS questionnaire is sent out to 157 countries and territories. OECD and EUROSTAT (Statistics from European Union) provide UIS with data for their member states, and RICYT provides data for Latin American countries.

Tables included in UIS Questionnaire on statistics of science and technology 2006 are given below:

1. R&D personnel by occupation: Total; Researchers; Technicians and equivalent staff; Other supporting staff; in Headcounts (HC) and Full-time equivalents (FTE)
2. R&D personnel by sex: Total; Researchers; in HC and FTE
3. R&D personnel by sector of employment and occupation: Total; Researchers; Technicians and equivalent staff; Other supporting staff; in Business enterprise (BE); Government (GOV); Higher education (HE); Private non-profit (PNP) sector; in HC and FTE
4. R&D personnel by sector of employment and sex: Total; Researchers; in BE; GOV; HE; PNP sector; in HC and FTE
5. Researchers by formal qualification and sector of employment: Total, in BE; GOV; HE; PNP sector; International Standards Classification of Education, ISCED97 6; 5A; 5B qualifications; in HC and FTE
6. Researchers by formal qualification and sex; ISCED97 6; 5A; 5B qualifications; in HC and FTE
7. Researchers by fields of science and sector of employment: Total, in BE; GOV; HE; PNP sector; Natural Sciences, Engineering and technology; Medical sciences; Agricultural sciences; Social sciences; Humanities; in HC and FTE
8. Researchers by fields of science and sex: Natural Sciences, Engineering and technology; Medical sciences; Agricultural sciences; Social sciences; Humanities; in HC and FTE
9. Total expenditure in R&D by sector of performance: BE; GOV; HE; PNP
10. Total expenditure in R&D by source of funds: BE; GOV; HE; PNP; Abroad

The resulting statistics and indicators published at http://www.uis.unesco.org were

1. Number of total R&D personnel (Headcount-HC and Full-time equivalent-FTE) by sex
2. Total R&D personnel (FTE) by sector of employment
3. Researchers (HC and FTE) by sex
4. Researchers (FTE) by sector of employment
5. Technicians and equivalent staff (HC and FTE) by sex
6. Other supporting staff (HC and FTE) by sex
7. Researchers per million inhabitants (HC and FTE)
8. Technicians per million inhabitants (HC and FTE)

9. Women as a percentage of R&D personnel (HC and FTE)
10. Women as a percentage of Researchers (HC and FTE)
11. Total Gross Domestic Expenditure on R&D (GERD) in local currency and international Purchasing Power Parity (PPP) dollars
12. GERD by sector of performance (per cent)
13. GERD by source of funds (per cent)
14. GERD as a percentage of Gross Domestic Product (GDP)
15. GERD per capita

UIS contributes data to various international databases and reports produced by partner agencies, such as:

1. UNESCO Science Report 2005, UNESCO World Report, International Report on Science Technology and Gender 2006
2. UN Statistical Year Book (UN Statistical Division)
3. Human Development Report (UNDP)
4. World Development Indicators (World Bank)

Besides publishing the data on the UIS web site, a series of bulletins are published analysing the trends in specific areas such as Investment in R&D; Bibliometric Indicators; or Women in Science, all of which can be freely downloaded from the UIS website.

S&T statistics workshops

One of the main drawbacks in the collection of S&T statistics in developing countries is related to the low availability of these statistics at national level in many developing regions, particularly in Africa and Asia. This is an area where capacity building is urgently needed to provide countries the tools to develop their own S&T statistics.

A first response to this problem has been the establishment of a programme of regional workshops that would eventually address all developing countries, covering initially the regions with the lowest availability of data. The objectives of these workshops are

1. To increase the number of countries regularly producing quality S&T indicators
2. To create local capacities for the production of such indicators, with the final aim of establishing sustainable local S&T statistics systems
3. To promote the use of S&T indicators, seeking comprehension for evidence-based S&T policy making
4. To share experiences with other developing countries in the field of S&T indicators, and to address the problems that countries may have encountered in collecting S&T statistics

5. To gain knowledge about the particular characteristics of S&T statistics data collection and use in the context of countries in the same region
6. To generate initiatives that could be used as a demonstration for good practices in other countries of the region.

Participants in the workshops come from national S&T policy bodies and statistical offices, depending on where the responsibility for collecting S&T statistics lies at the country level. The workshops are organized in co-operation with the Division for Science Policy and Sustainable Development at UNESCO headquarters, as well as other local and international partners. The Islamic Educational, Scientific and Cultural Organization (ISESCO) has contributed with partial funding for some of the workshops.

Table 14.2 gives the details of the number of countries participating in S&T workshops.

Table 14.2: UIS S&T statistics at workshops 2004-2007: Number of countries and participants

Region	*Countries*
Africa (Sub-Saharan)	23: Botswana, Burundi, Cameroon, Cape-Verde, Kenya, Democratic Republic of Congo, Gabon, Guinea, Lesotho, Malawi, Mali, Mauritania, Mozambique, Namibia, Niger, Nigeria, Senegal, South Africa, Swaziland, Tanzania, Uganda, Zambia, Zimbabwe.
Arab States	8: Algeria, Egypt, Jordan, Lebanon, Morocco, Palestinian Authority, Tunisia.
Asia	21: Afghanistan, Armenia, Azerbaijan, Bangladesh, Bhutan, Georgia, India, Indonesia, Iran, Kazakhstan, Kyrgyzstan, Malaysia, Maldives, Mongolia, Myanmar, Nepal, Pakistan, Philippines, Sri Lanka, Uzbekistan, Vietnam.
Latin America and Caribbean	17: Argentina, Bolivia, Brazil, Chile, Colombia, Cuba, Ecuador, El Salvador, Guatemala, Honduras, Jamaica, Panama, Paraguay, Peru, Trinidad and Tobago, Uruguay, Venezuela.
South-East Europe	9: Albania, Bosnia and Herzegovina, Bulgaria, Croatia, FYR of Macedonia, Montenegro, Romania, Serbia, Turkey.

The workshops have had an impact on various fields of the S&T statistics programme. The first and most apparent one is the significant increase achieved both in timeliness and response rates to the UIS 2006 S&T Questionnaire vis-à-vis the 2004 survey. The workshops brought together the people responsible for collecting S&T data in the participating countries, and gave the opportunity for interaction within colleagues from countries with similar concerns and characteristics, leading hopefully in the long term to increased networking activities at sub-regional level.

As regarding UIS' quest for recovering UNESCO's leading position in the field, UIS received explicit support from participants, strengthening it significantly (Dakar workshop, 2006). UIS can now stand up in international fora such as the OECD's National Experts on Science and Technology Indicators (NESTI) meetings, and speak on behalf of developing countries after having discussed extensively with experts and practitioners in the concerned countries. Last but certainly not least, a substantial

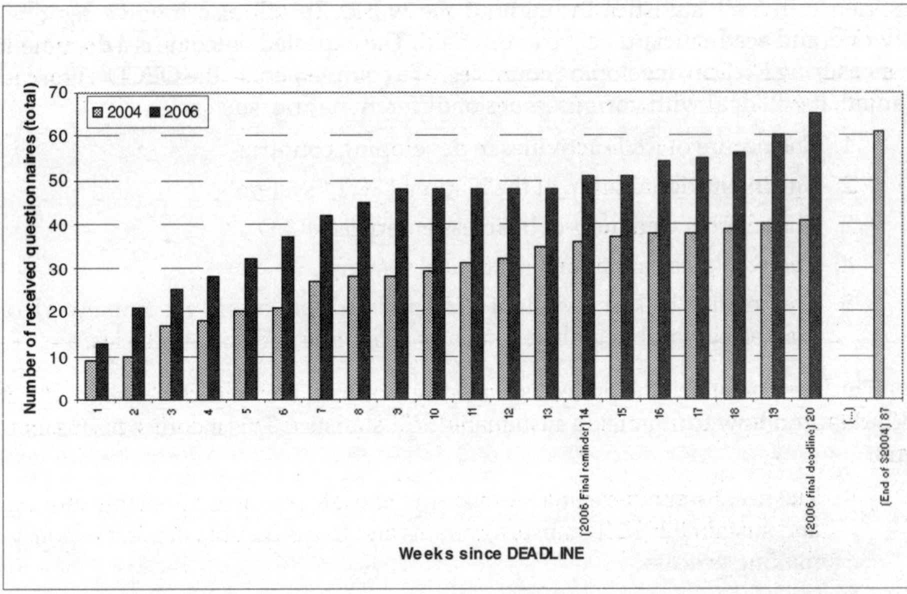

Figure 14.1: 2004 and 2006 surveys received questionnaires comparison

outcome of the workshops was the clear realisation that Frascati Manual methodologies are not directly applicable in many developing countries. Thus, a clear message was sent to UIS that new methodological guidelines were needed to collect S&T statistics in a more policy-relevant way for developing countries.

Standard setting/methodological developments

The UIS approach to R&D statistics has been up-to-now the one suggested by Rémi Barré in his assessment of the UNESCO science statistics programme in 1996, namely adapting the OECD's methodological framework and, therefore, applying and encouraging countries to apply the Frascati Manual (FM) (Barré, 1996). This has been effectively implemented, with consequences on different products. Mainly, the definitions underlying the UIS Questionnaire on S&T Statistics as well as the corresponding Instruction Manual are the ones proposed in the Frascati Manual. Complementarily, the definitions and methodologies proposed in the Frascati manual have been widely disseminated in the training workshops carried out.

The workshops were at the same time an opportunity to obtain feedback from the countries about the experience of applying FM in contexts very different from the original OECD one. The debates immediately surfaced the need for establishing complementary guidelines that could assist countries in collecting sound and quality R&D statistics. At the same time, it was stated explicitly that any move towards new guidelines should keep comparability, both backwards and across countries. The project of developing such guidelines for measuring R&D in developing countries is ongoing at the time of writing. The project involves experts and practitioners with experience in S&T statistics throughout the world. It will also involve regional networks and academicians active in this field. The expected outcome is a document on measuring R&D in developing countries, as a complement to the OECD's Frascati Manual. It will deal with various issues of different nature, such as:

1. The nature of R&D activities in developing countries
2. The institutional setup of the National S&T "system"
3. The relative weakness of business enterprise R&D
4. The weakness of national statistical systems
5. The relative lack of weight of R&D and S&T policies in the framework of mainstream public policies, such as economic and social policies.

The lines, on which the project is currently working, are related to how to collect R&D data and how to organize a sustainable S&T statistics. This includes discussions about

1. The need to convince policy-makers to allocate resources to establish sound and sustainable S&T statistics systems and to use the statistics in the policy-making process.
2. The importance of establishing widespread user and producer consultation mechanisms, involving all stakeholders, and potentially establishing national S&T statistics groups that promote a shared ownership of the resulting statistics and indicators
3. Countries should prioritize areas of work (R&D, innovation, CDH) based on the demand of the users. This can be accompanied by a "step-by-step approach" that enables countries to deal with the low degrees of resources frequently allocated to this area.
4. In order to produce statistics of the highest possible quality, it is important to establish training mechanisms for the producers of the primary information.
5. Countries might develop different strategies for different sectors, since once more in this area it is evident that "one size does not fit all". Sector-specific approaches and tools need to be developed to approach different realities in the government, higher education, business enterprise and other sectors. Examples of this could be:
    - o The use of combined R&D and innovation surveys to obtain business enterprise data.
    - o The use of administrative sources such as budgets and registers of government employees to obtain data from the public sector.
6. The establishment of sound S&T management information systems and databases capable of providing R&D data based on cross-nationally comparable methodologies, combining therefore S&T statistics and information systems.
7. The need for intensively interacting with colleagues from other countries with similar characteristics and problems.

Since this project is open for participation, it is foreseen that other issues will arise during its lifespan.

Human resources in S&T

The development of human resources for science and technology is one of the major priorities for developing countries in the field of S&T policy, if not the most important one. Two aspects of the human resources issue were addressed by UIS in the last years. On the one hand, the need to establish more balanced S&T workforces based on a higher participation of women. On the other hand, the need to create conditions for better making use of the highest qualified people in a society such as the doctorate holders. In this sense, the study of career and migration patterns should help to address phenomena such as the 'Brain Drain', which haunts many developing countries.

Women in science

In order to support decision making, the UIS has conducted a programme of work in this area based on:

1. Gender-mainstreaming of the S&T questionnaire
2. Raising awareness of the gender imbalance in the science and engineering professions
3. Disseminating statistics and indicators to inform the debate.

The results of a consultative questionnaire added to the 2004 UIS S&T survey showed that data with sex-breakdown that were up-to-then not demanded by UIS were available in many countries (Figure 14.2). At the same time, even more countries indicated their intention to collect such data in the near future.

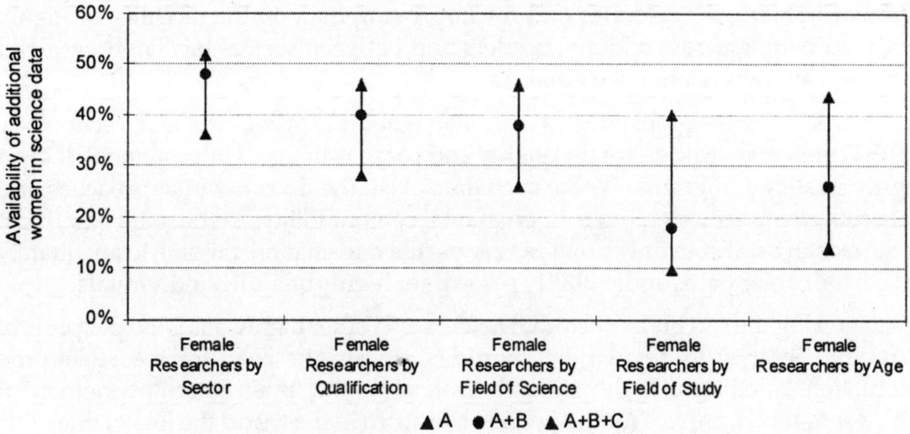

Note - A : Already available-by 2004, B : Available from 2005 on, and C: Possibly available before 2008

Source : UIS Questionnaire on women in science statistics

Figure 14.2: Availability of additional sex-disaggregated data on researchers, 2004

As a consequence, the 2006 UIS S&T survey questionnaire was redesigned with the goal of *gender-mainstreaming* it. Sex-breakdowns were included by adding tables crossing the sex variable with every other breakdown required in the human resources section. Therefore, it is expected that a more detailed picture of the participation of women in R&D will be available after the processing of the 2006 survey. Beyond establishing a more important presence of sex-disaggregated statistics in the UIS database, the UIS S&T statistics team has conducted various analyses combining R&D and Higher Education statistics to present the evidence establishing the gender issue as a significant concern for scientific and technological development. In this regard, the UIS highlighted the importance of the problem, both from the ethical point of view, and the economical one, *i.e.* the under-utilization of valuable human resources in a context of relative scarceness.

The results have been presented in various scientific conferences, workshops and policy debates, and published in the UIS bulletin on S&T statistics and as a chapter of the upcoming International Report on Science, Technology and Gender (IRSTG) to be published by UNESCO later in 2007. In this way, the UIS took a step towards promoting and informing evidence-based policy making in this field, seeking to incorporate statistics into policy debates.

Careers of Doctorate Holders (CDH)

Doctorate holders are an important part of the "highly qualified people" in a country and play a prominent role in scientific and technological development and stand for a critical part of the science and engineering (S&E) labour force worldwide. Doctorate holders represent the outcomes of tertiary education at ISCED 6, its highest level.

As one of the main axis in the measurement of human resources for science and technology (HRST), the need for collecting and exchanging information on the career-paths of holders of doctorates and for improving data on the mobility of highly qualified people across national borders and between sectors has lately received increased attention from policy-makers.

Statistics on the global trends in human resources for Science and Technology (HRST) are very weak, as are the quality and comparability of international data on highly qualified migrants. Where such data exist, the diversity of approaches and data collection methods hinders international comparability. At the same time, most of the existing statistics only provide very partial information, missing to adequately reflect the career paths and mobility patterns of highly qualified individuals.

Tracking the careers of doctorate holders and other highly qualified people is of particular interest to developing countries, due to the need for measuring the international mobility of highly skilled personnel (*i.e.* the "Brain Drain" phenomenon) which frequently implies for the poorest countries in the world the loss of one of its most precious resources: its most talented and skilled people. The 'Brain Drain' is one of the main priorities for UNESCO in relation to creating scientific and technological capabilities in Africa and other developing regions, but also in regions such as South East Europe.

The relevance of such measurement is also of significance to developed regions such as the European Union (EU), where current policies seek to enhance the region's attractiveness for research talents around the world through measures such as increased funding for training, mobility, and career development for researchers, as well as the improvement of the overall environment for researchers by enhancing mobility and removing obstacles in order to create a European Research Area (ERA). The United States face similar policy requirements regarding its S&E workforce and have been closely monitoring the changing patterns of academic recruitment. The OECD as a whole is also particularly interested in monitoring internal mobility patterns, *i.e.* mobility from university to industry.

Statistics related to career development of doctorates are being developed in a global framework, since the population of doctorate holders is highly mobile between countries and even continents. Hence the need for, and advantage of creating a unique set of tools that can be applicable at global level and meet both the requirements of developed countries, usually "absorbing" foreign doctorates, and developing countries, mostly "sending" doctorates abroad.

These statistics will in the future be extended from covering only doctorate holders to other groups of highly skilled people, depending on the interest of the countries actually carrying out surveys on this matter, where other highly qualified groups (such as teachers or nurses) might represent crucial parts of the skilled workforce.

Policy relevance and demand for CDH statistics

After discussing the policy relevance and the demand for statistics in this area with various stakeholders, four groups of research questions were identified:

1. Role of highly qualified people (HQP) in innovation and in the knowledge economy
2. Labour market supply and demand of HQP
3. Transition from higher education to work
4. International and national mobility of HQP.

Data needs were associated with each of the above-mentioned groups of policy research questions (Table 14.3).

The CDH project

The UIS, in co-operation with OECD and Eurostat, and with the support of the National Science Foundation of the United States, has undertaken the Careers of Doctorate Holders (CDH) project in order to meet the growing user demands. Each organization took responsibility for issues appropriate to their mandate and interest. Expertise and resources have been shared across the three partners.

The objectives of the CDH project are:

1. To design an internationally comparable tool for tracking the careers of doctorate holders and highly qualified people in different countries.

2. To conduct pilot tests of the developed instruments in volunteer countries.
3. To collect and exchange information on the career paths of holders of doctorates from existing data sources and the new survey tool.

Table 14.3: Possible links between policy research questions and data needs

Possible research questions	*Data needs*
Role in innovation, knowledge economy Where do doctorate holders work as compared to other tertiary graduates, what is their productivity?	Doctorate holders by type of employment (research, non-research), sector (industry, government, higher education), field of research; data on scientific productivity (publications, co-authorship, patents, citations)
Labour market supply and demand Do we train too many or too few doctorate holders? Are there mismatches on the labor market?	Employment status broken down by age, gender, country of origin, sector of employment, occupation, field of study vs. field of research.
Education to work	Time of transition to employment and postdoctoral experience statistics
Mobility How mobile are doctorate holders (among sectors and across countries)?	Doctorate holders' mobility between sectors (industry, government, higher education). Doctorate holders' mobility across countries: doctorate holders by country of birth, citizenship, residency status, type of visa. Data on permanent vs. non-permanent, length of stay, returns to country of origin.

Source: L. Auriol (2004), Why Do we Need Indicators on Careers of Doctorate Holders, paper presented at the OECD Workshop on User Needs for Indicators on Careers of Doctorate holders, (DSTI/EAS/STP/NESTI (2004 15), Paris, 27 September.

A set of indicators from various existing surveys on the careers of doctorate holders was drawn up. These indicators are divided into two categories, *viz.* "objective" and "subjective" (Table 14.4).

The outcome of this project (the "CDH toolkit") provides countries with an internationally comparable tool for tracking the careers of doctorate holders and highly qualified people, designed to furnish international statistics of high quality and timeliness on their mobility, career paths and employment destinations.

The CDH toolkit comprises an output tabulation programme, detailed definitions of the variables used in the tabulations, methodological guidelines, a core model questionnaire, and a manual for the questionnaire, as well as a detailed description of national data compilation methodologies currently applied by countries. The CDH toolkit has been developed by an expert group from more than 20 countries including Argentina, Australia, Belgium, Canada, China, Denmark, France, Germany, Greece, India, Italy, Japan, Korea, Malaysia, Norway, Portugal, Russian Federation, Spain, Switzerland, Ukraine, Uganda, and the United States.

Table 14.4: Indicators from existing surveys of careers of doctorate holders

"Objective" Indicators	*"Subjective" Indicators*
– Academic path in higher education	– Reason for seeking a doctorate
– Work while studying	– Employment preferences during studies
– Source of finance for doctoral studies	– Occupational ambition upon completion of doctorate
– Length of doctoral studies	– Links between qualifications and market-demanded skills
– Number of publications	– Correspondence between current job and thesis topic
– Location of postdoctoral experience	– Use of skills
– Migration abroad	– Collaboration between company and a university/ public
– Length of transition from doctorate to employment	– Research centre
– Unemployment rate	– Satisfaction in current job
– Profession/occupation	– Satisfaction with career progress
– Proportion of doctorate holders employed in higher education	
– Proportion of doctorate holders employed in R&D	
– Region of employment	
– Sector of employment	
– Starting salary	
– Current salary	
– Proportion employed part-time	
– Proportion employed full-time	
– Proportion on term contracts	
– Career advancement	
– Company has an R&D department	
– Company employs other doctorate holders	

Source: L. Auriol (2004), Conclusions on the workshop on user needs for indicators on careers of doctorate holders, held at the OECD, Paris-la Defense, 27 September 2004 (DSTI/EAS/STP/NESTI(2004)28).

The expert group was organized in taskforces that took responsibility on the different components of the project. These include taskforces on (*i*) Production of output tabulations, and development of common definitions (led by OECD and Canada), (*ii*) Drafting of methodological guidelines (led by Eurostat and Portugal), and Development of a model survey with a supporting methodology (led by UIS). UIS has been responsible for the development of the core model questionnaire and the corresponding instruction manual.

The CDH toolkit has been adopted by UNESCO, OECD and Eurostat and is expected to be implemented in a broad number of countries in the following years.

This will enable policy-makers, researchers and practitioners throughout the World to establish and evaluate the trends on the career of doctorate holders and the "Brain Drain", with a view to facilitating the implementation of evidence-based policies in the fields of human resources development for S&T, as well as for tertiary education.

Catering to different national approaches to CDH statistics

The CDH toolkit is flexible enough as to meet the options faced by the countries when taking decisions about the implementation of CDH statistics at national level:

1. "Use of existing data" vs. "new surveys"
2. "Snapshot" vs. "cohort" surveys.

The use of existing data ("Minimal approach") is favoured by an "output-oriented" design. This design method is based on an initial in-depth definition of the required output tabulations to meet the policy demands. Countries with existing CDH-type surveys, such as the US or Canada, or with extensive population registers (as is the case in Nordic countries), can use these sources to provide the information demanded in the output tabulations.

The creation of a new survey ("Maximal approach") is favoured by the development of a detailed "core model questionnaire" and a "model instruction manual". These tools are adaptable to national purposes. An intermediate approach for a country with a certain number of existing data could be to implement a reduced new survey to complement the data obtained from other sources. In practice, especially where there are constraints on resources (a particular issue for developing countries), most countries will adopt the intermediate approach; making use of existing data where available and seeking to fill the gaps with some form of dedicated survey. As a whole, the "methodological guidelines" provide the framework to guarantee cross-national comparability of the resulting data.

Two different types of survey approaches are used by countries collecting or planning to collect CDH-type statistics:

1. 'Snapshot' surveys that cover all doctorate holders in the country on a specific date
2. 'Cohort' surveys following a specific set of graduates from the date of their graduation into the future with repeated surveys of the same cohort.

The CDH toolkit can be used both to conduct surveys that cover the whole universe of doctorate holders in a given moment ("snapshot surveys")–which is the preferred option- and to establish a system to follow-up graduates by specific cohorts. Both approaches have pros and cons, and can certainly be combined, as they are in the USA. The 'snapshot' approach is likely to be most common in developing countries because resource constraints will limit their ability to carry out frequent or dedicated surveys. Cohort surveys can in some cases be used to follow-up doctorate holders who have left the country, but if enough countries collect CDH statistics, doctorate holders living abroad should be covered by the surveys taking place in their host country. This would in principle (and in the long term) allow data to be aggregated across national boundaries.

The CDH toolkit

The CDH toolkit comprises

1. Output tabulation programme with detailed definitions of the variables.
2. Methodological guidelines and description of national data compilation methodologies.
3. Core model questionnaire and instruction manual.

Output tabulations and definitions

The Output tabulation programme was conceived to be used by countries to deliver the requested data to the international organizations. It is also the base for an output-driven approach, guiding countries to design their data-collection activities based on the expected outputs. The output tabulation groups include personal characteristics, educational history, work history, scientific output, perception of current work situation, and international mobility (retrospective and future plans).

The definitions constitute a complement to the output tabulations, directed to allowing international comparability of the data. These include definition for doctorate, citizenship and residential status, recent doctorate recipients, employed persons, unemployed and inactive persons, researcher, temporary/permanent employment, full-time and part-time employment, and Postdoc. Most of the definitions used are already existing internationally agreed definitions and classifications. The development of an adequate definition of "POSTDOCS" is still a pending issue. It is expected that the application of the core model questionnaire in national CDH surveys will provide information on the characteristics of postdoc positions to assist in the development of an internationally acceptable definition.

Methodological guidelines

The methodological guidelines for CDH have been prepared under the lead of Eurostat, following highest standards for statistical practices. The main issue addressed in these guidelines is the adequate definition of the actual target population, as well as the way to construct an appropriate sample frame. They also present some details on aspects of data collection, data processing and results estimation.

The methodological guidelines are an orientation for countries that plan to launch a CDH survey as well as for those who need to improve and align their existing national surveys to international standards.

The target population, as defined in the guidelines, consists of individuals fulfilling the following criteria:

1. Having an education at ISCED 6 level (doctorates) obtained anywhere in the world, and
2. Being resident (permanent or non-permanent) within the national borders of the surveying country.

Target population structure

Target population, residents in country X (the surveying country) that are:

1. Citizens of country X with doctorate awarded within country X
2. Citizens of foreign countries with doctorate awarded in country X
3. Citizens of country X with doctorate awarded abroad
4. Citizens of foreign countries with doctorate awarded abroad

Non-target population, residents outside country X that are:

1. Citizens of country X, with doctorate awarded within country X that have left country X
2. Citizens of foreign countries, with doctorate awarded in country X that have left country X

For many countries focusing particularly on the mobility of doctorate holders, the question of how to study doctorates living abroad arises.

It is intended that those graduates that have left the country (*i.e.* groups E. and F.) are surveyed in the countries where these doctorate holders are resident. However, until the CDH does not significantly extend its coverage, these groups–a significant part of the "Brain Drain" in many countries- might get lost to the survey. Therefore, if

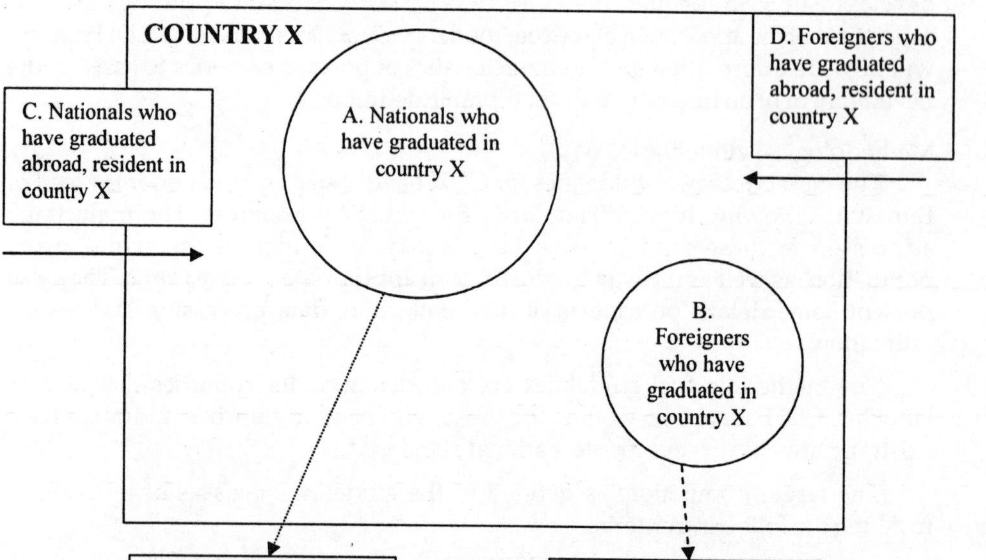

Figure 14.3: Structure of the target population

***Source*: Statistics on the Careers of Doctorate Holders (CDH): Methodological guidelines.**

countries are able to construct a sample frame to reach the national graduates who have left the country, the survey could be extended to this group, even if countries are not requested to report information to the international organizations on them.

Doctorate holders are not a large enough population to be adequately represented in sample-based household or labour-force surveys. Therefore, each country has to seek a combined approach based on existing sources of information on doctorate holders.

CDH statistics can be based on two different types of surveys, covering the whole current population of doctorate holders ("snapshot surveys") or targeting only a certain number of cohorts of recent graduates.

The sources for constructing sampling frames include:

1. National education registers
2. Universities
3. International and national foundations
4. National libraries
5. Population censuses
6. Research institutes
7. Professional organizations
8. Central registers of foreigners
9. Alumni organizations
10. Previous surveys of doctorate holders

The "snapshot" surveys are based on a cross-sectional retrospective sample of all doctorates in the country. Such a survey can be administered independently, on an occasional basis, and configures an appropriate tool for the collection of information related to the entire target population.

The "cohort" examines only recent graduates, has to be administered on graduation and should be repeated every few years. In such a survey, it is simpler and more straightforward to establish a sound sampling frame, but the coverage of the target population is limited. However, it can as well be used as a starting point for CDH statistics, and–as is the case in the US- be used to feed the sample frame for a periodic "snapshot survey". Therefore, both survey approaches should not be seen as mutually excluding but rather complementary tools to obtain CDH statistics.

Core model questionnaire:

The CDH core model questionnaire was designed by UIS as a tool for countries to obtain the necessary information for the completion of the output tabulations. It is mainly foreseen for countries which need to start CDH-type surveys without having concrete antecedents of surveying doctorate holders.

CDH-type survey questionnaires used in the CDH project include the following:

1. USA (Survey of Earned Doctorates–SED; and Survey of Doctorate Recipients–SDR)

2. Canada (Survey of Earned Doctorates–SED)
3. Switzerland (Survey of Tertiary Graduates, and Doctorates surveyed the year after they received their degree)
4. Portugal (Survey of former PhD Scholarships receivers concerning their professional situation)
5. India (Pilot study on the Career Profile and Professional Achievements of the PhD in science from the Selected Central Universities/Institutions)
6. Ukraine (Structure of database of Doctorate Holders in Ukraine)

The development of the questionnaire took into account existing questions featured in questionnaires applied in different contexts. The use wherever possible of existing survey questions means that most of the questions in the core model questionnaire have been tried and tested in a number of different countries and national contexts. Countries using the core model questionnaire will be able to access metadata on these questions as well as comparing responses to those from previous surveys.

Some questions however have required some rewriting in order to make them 'culturally neutral' with regard to language and content. The questions are organized in modules and the order of the questions maximizes user-friendliness and minimizes burden on the respondent. The questionnaire was tested in various countries, and discussed with numerous experts, resulting in a broad consensus around the questions. The questionnaire will be applied in 2007 in numerous countries, especially in Europe, where it is currently being translated and adapted to various country-requirements.

Several countries have launched specific CDH-type surveys, based on the core model questionnaire. Argentina has conducted a complete CDH survey, including a survey of the doctorates abroad. Eurostat has launched a grant programme to encourage EU-countries to conduct CDH surveys. In this framework, more than 20 countries have committed themselves to carrying-out such surveys. Both Korea and Japan will survey foreign doctorate holders in their countries. India is planning to adapt its next survey of doctorates to make it compatible with the CDH methodology. In North America, both the USA and Canada conduct such surveys regularly.

The core model questionnaire highlights *"core questions"* which provide the necessary information to fill-in the output tabulations. However, the modules of the core model questionnaire do not necessarily follow the logic of the output tabulation modules. The modules as selected are thematic and user-oriented. Countries can choose to exclude some of the proposed modules if the information can be made available from other sources, or to add extra modules to address local issues.

The module includes Doctoral Education (EDU), Recent graduates (REC), POSTDOCS (POS), Employment situation (EMP), Career-related experience and scientific productivity (CAR), International mobility (MOB), and Personal characteristics (PER).

The module on Doctoral Education is designed to collect information on the educational history of doctorate holders, such as the institution in which the doctorate was earned, the field of science, duration, sources of financial support of the doctorate, as well as the countries in which previous degrees were obtained.

The module on Recent Graduates is only applied to doctorate holders having received their degree in the last 2 years. It collects their complete educational history, including names and locations of secondary and postsecondary institutions, dates of attendance, field of study, and date of degrees. In order to obtain information about the transition from education to work, it enquires about the time elapsed before obtaining the first *career-path job.*

In order to get closer to a definition of "POSTDOCS", the corresponding module is build around the following question:

"Would your principal job be considered a POSTDOC position in your country of residence?"

This module provides a description of the title, field, and duration of the POSTDOC, as well as the activities involved (research, teaching, other) and the main source of financial support obtained. It is expected that the results of the application of this module in a certain number of countries will give enough information as to characterize a "postdoc" and permit the establishment of an international definition of "postdoc". The module further enquires about the reasons the doctorate holder had for taking this postdoc.

The model dealing with the employment situation of the doctorate holder inquires about employment status, working hours, principal employer, sector and type of employment (postdoc, temporary or permanent employment, part-time or full-time job), annual earnings, as well as about the retrospective career history (within past ten years). It also asks about the level of satisfaction with the job, and the degree of relationship with content of the doctorate degree. In order to accommodate the characteristics of many developing countries, a "multi-job" pattern is foreseen.

One of the core-modules of the questionnaire deals with international mobility, including the measurement of inflows (group D) and returning outflows (group C). Through the information to be collected, it will be possible to distinguish temporary from permanent mobility, as well as to better understand the reasons for departure and return. This module also provides data on intentions to move out of the country within the next year (including the destination planned) and enquires about linkages with country of origin for the foreign-born doctorate holders.

In order to better understand the position of the doctorate holder in her or his career, a specific module deals with career-related experience and scientific productivity. This includes data on scientific and technological output, as well as on teaching. It studies in particular those doctorate holders having pursued a career as researchers, and inquires about the reasons behind such decision. Linked to the international mobility module, international cooperation activities are surveyed in this module.

A module on personal characteristics will permit to cross the different variables surveyed with the most significant demographic variables, such as family characteristics, age, place of birth and citizenship and type of residential status. It also provides contact information for future surveys. The inclusion of demographic variables will open a whole new information source for studies in the field of "Women in Science", providing for the first time comparable data on salaries and productivity by sex.

The way forward

The CDH project is an ongoing activity, since the results of the initial methodological development are being widely applied for the first time in 2007, mainly in European countries. UIS is promoting the methodology and encouraging developing countries to conduct CDH-type surveys to produce cross-nationally comparable statistics on careers of doctorate holders. Argentina has already conducted such a survey, with promising results. The CDH toolkit is being tested worldwide, and might need to undergo a revision after its first wide application in 2007.

With the widening of the application of the CDH methodology to more countries both in the developed and developing world, in the medium term it will be possible to establish an international database on doctorate holders, which will be a preferred source of statistics on the international mobility of the highly skilled.

One of the principal characteristics of the CDH project is that the methodology was developed "from the scratch" by UIS together with OECD and Eurostat, and with the participation of experts from developed and developing countries, therefore taking into account the interests of both developed and developing countries from the beginning. This contrasts with the approach adopted in other circumstances, where methodologies developed with developed countries in mind have been more or less adapted to developing country situations.

Innovation

After re-establishing the UIS survey on S&T statistics, innovation is the upcoming priority for the UIS. A different approach is needed for innovation statistics than for R&D statistics. UIS has never before collected innovation statistics, nor have most developing countries.

The Oslo Manual (Guidelines for collecting and interpreting innovation data) is currently the most important reference document for the statistical definition of innovation and forms the basis for surveys of innovation throughout the world. This Manual has been put in place by OECD and Eurostat in 1992 and first revised in 1996. During 2003 and 2004, the Oslo Manual underwent its second revision process.

In the framework of this revision process, based on a proposal by RICYT the UIS co-ordinated the development of an annex to the Oslo Manual on "Measuring Innovation in Developing Countries". For this purpose, RICYT provided a base document that was complemented by an analysis of the existing experiences in innovation measurement in other developing countries. The resulting document was put forward for discussion to experts across the developing world. These experts were principally those responsible for conducting the innovation surveys in developing countries, as well as researchers involved in the analysis of the results of these surveys. It is important to highlight that most of the countries having conducted innovation surveys could be characterized as middle- or high-income economies.

Currently, the UIS is involved in the dissemination of the Oslo Manual and its Annex, while discussing its applicability to the context of other developing countries. This is being complemented by a pilot data-collection exercise to be conducted in 2007.

Measuring innovation in developing countries

After the publication of the 2nd Oslo Manual, developing countries also started conducting innovation surveys. The design of the surveys was intended to comply with Oslo Manual standards, with adaptations for capturing the particular characteristics of innovation processes. Adaptations were prepared by each country separately and with different approaches. In order to compile and organize these adaptations, RICYT elaborated the Bogotá Manual as a first effort to compile particularities and guide the design of cross-nationally comparable innovation surveys, for the Latin American context. However, countries in other regions (such as China, India, Malaysia, Philippines, Singapore, South Africa, and Thailand) also had started conducting innovation surveys.

Based on these experiences, the annex identifies the characteristics of innovation in developing countries, describes the difficulties in measuring innovation in these contexts, and proposes a certain number of adaptations and measurement priorities.

The annex to the Oslo manual on measuring innovation in developing countries

Innovation processes in developing countries' economies present differential characteristics due to a number of factors, including:

1. Limited size and structure of markets and firms
2. Instability, both in political and economic terms
3. High degree of informality in the economy
4. Particular economic and innovation environments, frequently involving a significant direct participation of the state in the economy
5. Reduced innovation decision-making powers, due to the overall weight of foreign multinational companies
6. Weak innovation systems, both in terms of institutions and linkages
7. Particular characteristics of innovation (*e.g.* reverse engineering, minor changes).

The recognition of the particularities of innovation in developing countries implies the need for adjusting the methodology for measurement. The OECD/Eurostat Oslo Manual has been conceived based on the experience of the Member States of both organizations, which are usually called 'developed countries'. In order to establish a truly global system of innovation statistics, it is paramount to maintain comparability with Oslo Manual data. Therefore, the basic definitions of the Oslo Manual should be adopted and remain unchanged. However, there are many issues in the Manual where additional definitions can be included. At the same time, it is particularly critical for developing countries to shift some of the measurement priorities towards more policy-relevant results.

Complementary to the existing definitions for "innovative" and "innovation-active firm", the concept of "potentially innovative firm" is brought forward. While the innovative firm is one that has introduced an innovation, an innovation active firm is one that has had innovation activities. A potentially innovative firm is one type of "innovation active firm", which has made innovation efforts but not achieved results. This is a key element in innovation policies in developing countries: other than benchmarking successes, the priority lies most frequently in helping firms to overcome the obstacles that prevent them from being innovative, therefore converting efforts into innovations. Potentially innovative firms have been analysed in many Latin American countries, as well as in the Tunisian innovation survey. Developing countries should shift their innovation measurement priorities in order to give broader attention to innovation capabilities of the firm (*i.e.* human resources, linkages, quality assurance systems, information and communication technologies), innovation efforts (*i.e.* expenditure on innovation activities), as well as to the measurement of organizational innovations.

In order to implement the proposed measurement priorities, developing countries should include in innovation surveys a focus on ICTs, particularly the strategic use of new technologies ("Back office"), give increased attention to linkages, as well as expand the list of innovation activities to include activities such as "reverse engineering", which are frequently a key element of innovation. Innovation surveys should also be strengthened in the area of human resources, and particularly in the availability of and expenditure in training activities. Quality and environmental management activities should also be surveyed.

It is important to highlight that measuring innovation in developing countries also requires the adaptation of the survey methodology to the particular characteristics of the countries. Information systems are frequently very weak, as are national statistical systems as a whole. This has consequences on the sampling, response rates and the amount of context information demanded to the firm in order to produce sound results. The way in which the survey is applied is related to infrastructural constraints, as well as to cost structures (telephone vs. mail vs. personal interviews).

Thinking ahead

The exercise of developing an annex to the Oslo Manual brought up a number of unsolved issues related to innovation measurement in this context, such as:

1. Measuring the role of entrepreneurs and their attitudes towards innovation
2. Capturing innovations driven by factors other than market forces, and in particular innovations conducted by the public sector
3. Measuring innovation in the primary sector (particularly in agriculture)

4. Better measuring minor or incremental changes, including innovative applications of existing products or processes, and the so-called 'backwards integration' of technological capability
5. Develop indicators reflecting sub-national (regional) innovation systems
6. Measuring innovation in the informal sector
7. Measuring innovation based on traditional knowledge

These issues are yet insufficiently studied and need further theoretical reflection, as well as pilot application by innovating developing countries. At the same time, they are setting an agenda for future methodological documents in this field.

Innovation data collection

In 2007-2008, the UIS will for the first time collect innovation data and metadata through a special survey. This survey will be conducted in co-ordination with OECD, and is expected to provide further knowledge on innovation in the developing world. While mainly countries with existing innovation surveys will be approached, this tool should also encourage countries to conduct innovation surveys and further disseminate the concept of innovation surveys worldwide.

Output and impact

Output and impact indicators are part of the longer term UIS priorities in the area of S&T. However, some activities were carried out to establish an initial framework, in partnership with INRS Québec. In the area of output, an analysis of the scientific output of developing countries with traditional bibliometric indicators was conducted, discussing pros and cons of existing methodologies.

On the other hand, in order to start addressing the problem of measuring the social impact of S&T, a workshop has been convened to discuss possible approaches. This area remains closer to a research programme than to actual statistics. However, the demand for such indicators in the developing world is very significant.

Conclusions

The UIS has been applying three significantly different approaches to the development of methodologies appropriate for the measurement of the various areas of S&T in developing countries. The different approaches are driven by historical factors, as well as the strength of UNESCO in the various fields. These approaches are also based in the role UNESCO and its partners play in the development of the corresponding methodology. These approaches could be classified as:

1. Development ex novo
2. Adaptation
3. Adoption

The development of statistics on the career of doctorate holders (CDH) is a clear example of an international methodology developed from the scratch (ex novo). This has been accomplished in an environment of intense co-operation between international organizations (UNESCO, OECD and Eurostat), and national counterparts (particularly the U.S.A., which provided the conceptual groundwork as well as funding for the international development). The results of this process are considered to be equally useful, pertinent and applicable in both developed and developing countries.

The ex novo model could in the future be applied to other new areas of S&T statistics, such as in the measurement of social impact, or in the quest for adequate statistics to measure science education.

In the field of innovation statistics, the UIS has opted for the adaptation of existing OECD/Eurostat methodologies, namely the Oslo Manual. The Oslo Manual is a well established methodology and has also been applied, although with a significant number of variations and adaptations, in developing countries' innovation surveys. With the methodological support and experience of RICYT, and the input from experts from the UIS's worldwide networks, guidelines were established to mainstream the adaptations in developing countries and therefore improve comparability between surveys.

In the case of R&D measurement, the focus has initially been put on keeping comparability with developed countries, therefore adopting the OECD's Frascati Manual, and setting aside the outdated UNESCO methodologies from the 70's and 80's. R&D statistics are well established, particularly in OECD countries, but also in developing countries where UNESCO has been collecting these data since the 1960s, and many countries have been widely applying or trying to apply the Frascati Manual.

However, this adoption of OECD methodologies is not without risk. Countries feel frequently forced to push their statistics and statistical methods and practices towards Frascati Manual standards, without taking into account their specificities. After thorough discussions with S&T statisticians all over the World it is clear that an alternative approach is at this point needed.

The process leading towards the preparation of the Annex to the Oslo Manual, as well as the underlying concept, could be used as an inspiration for further development in the area of R&D measurement. Since the experience of measuring R&D in developing countries is bigger than in the case of innovation, the resulting document will most probably be both more comprehensive and more detailed. The development of such a document has already been agreed by UIS and OECD and should start in 2007, with the participation as well of various regional organizations and networks. At that stage, in the field of R&D statistics UIS would be moving from "adoption" to "adaptation" in order to better meet the needs of developing countries.

References

Barré, R., 1996. UNESCO's Activities in the Field of Scientific and Technological Statistics, BPE-97/WS/1, UNESCO, Paris.

Dakar workshop, Final communiqué of the Dakar workshop available at http://www.uis.unesco.org/ev.php?ID=6759_201&ID2=DO_TOPIC.

Fernández Polcuch, Ernesto, 2006. Crunching numbers: Science and technology statistics at UNESCO, in "60 Years of Science at UNESCO 1945-2005", UNESCO, Paris.

UNESCO, Immediate, Medium and Longer-Term Strategy in Science and Technology Statistics: http://www.uis.unesco.org/ev.php?ID=5519_201&ID2=DO_TOPIC

UNESCO, International Review of S&T Statistics and Indicators; http://www.uis.unesco.org/ev.php?ID=3815_201&ID2=DO_TOPIC

Chapter 15

Interactive policy research for rural innovation

Rajeswari Sarala Raina

Abstract

S&T policy makers use policies and policy research inputs that are inadequate to enable innovation for poverty reduction in developing countries. This paper presents lessons on technological and institutional change, from recent research on rural innovation. It argues that policy relevant information for innovation can be generated through processes that entail interactive learning. The indicators of innovation that policy makers can use are highly context specific process indicators. In order to enable rural innovation, policy makers must acknowledge that S&T is only one among several other sources of knowledge. Like all other components of an innovation system S&T policy has to go through processes of institutional learning and change.

Keywords: Competency, Indicators, R&D, Rural innovation, S&T policy, Science, technology and innovation (STI), Technology.

Introduction

Rural innovation entails the generation and utilization of knowledge, including S&T, in socially progressive and economically productive ways. By definition, policy research and policy making for rural innovation demands capacities that go beyond policy research and policy making for scientific research and technology generation. To facilitate innovation policy makers must enable technological and institutional changes. The latter being changes in the rules or ways of working of organizations and individuals, is more difficult to comprehend, formulate and implement. S&T

policy research inputs which S&T policy makers have access to and actually use, are inadequate to address the complex contexts and processes of rural innovation. Using an innovation systems framework, this paper explores how policy relevant information can be generated and used to enable rural innovation. Recent research reveals how policy relevant information can be generated through alternative policy research processes. These entail interactive learning and policy research processes; a significant shift from current prescriptive policy research.

The nature and extent of rural distress in developing countries today demands serious policy attention. National governments, policy makers and scientists as major decision makers expect science and technology (S&T) to provide viable and sustainable solutions for many a problem. The relentless attempt to apply scientific and technological knowledge to solve rural problems, however, is yet to provide some payoffs in terms of enhanced rural incomes and poverty reduction.

While the population living below poverty line or at the edge of poverty is increasing steadily in Sub-Saharan Africa, the poverty afflicted population in South Asia is by far the greatest problem in the developing world–in terms of sheer numbers (living on less than 1 US$ per day) (World Bank, 2003; Figure 15.1). There is a strong likelihood of at least another 500 million more joining the former (currently living on 1-2 US $ per day) by the end of 2010. This is despite the fact that many developing countries have entered the globalization process and have over the past couple of decades, doubled their share in world trade and raised their per capita incomes by two-thirds (between 1980-1997) (IMF, 2002). There is a massive proportion of the Asian population still engaged in primary production (agriculture, mining, gathering etc.) and value addition activities (FAO, 2006). Any application of knowledge in this

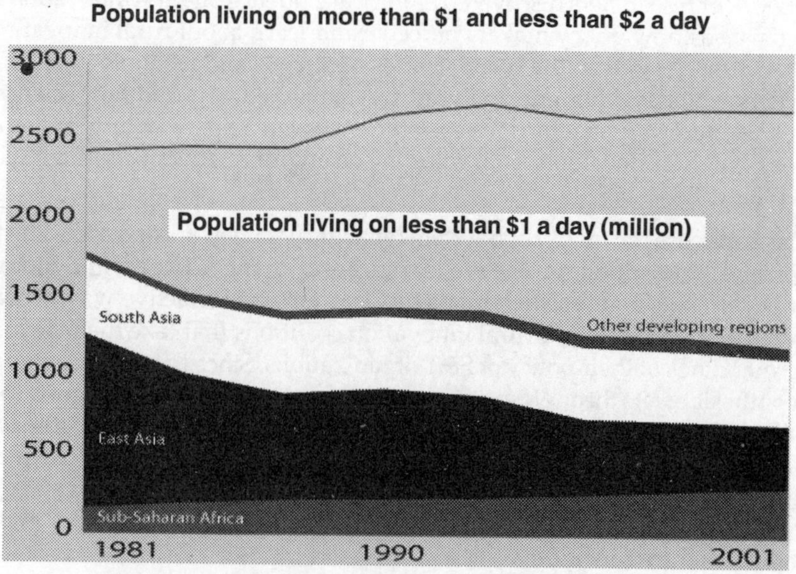

Figure 15.1: Rising poverty in Sub-Saharan Africa and status quo in South Asia

***Source*: World Bank (2003)**

context of poverty, largely generated by un- and under- employment of otherwise productive hands in the economy, has to address how these hands can be selectively withdrawn and converted into productive forces in other sectors in the economy, without adversely affecting the production levels, environmental quality and human health/well-being in the primary sector (Sen, 1967; Lewis, 1954; Schultz, 1964). In other words, knowledge has to play a role in transforming the structure of the primary, secondary (manufacturing) and tertiary (service) sectors, while ensuring that more hands get employed productively in secondary and tertiary sectors. Enhancing the innovation capacity of a range of actors in locally relevant and rural systems offers an answer.

Getting politicians and policy makers to enable widespread rural innovation is however, not an easy task. Addressing S&T policy alone in order to deliver rural innovation and developmental benefits is a highly limited or narrow agenda. Many of us (the author included) have spent a considerable number of years validating and verifying S&T statistics and correlating these statistics with performance measures (Rajeswari, 1992; Rajeswari 1995; Raina, 1999; Raina *et al.*, 2006). The measurements of research intensity, personnel performance, organizational performance etc. based on these statistics (research inputs–mainly expenditure and personnel, and research outputs–mainly publications) have however, not told us (as policy researchers) enough about (a) the S&T research processes and the presence/integration of S&T with other actors in society, (b) the kind of indicators and information that policy makers need to understand and foster the important role that S&T can play in meeting social/economic requirements. After much research, publications (research papers that policy makers never read, and policy briefs that some policy makers tend to read), seminar presentations, soul searching and consternation, some of us working as a network of researchers discovered empirical evidence on rural innovation which taught us to see how policy makers perceive and learn about rural innovation. We also drew from lessons in the social studies of science and technology (STS); from researchers who have thought, analysed and explained why and how science does what it does, and ways in which the social sciences can help the natural and physical sciences play their developmental roles in society more effectively (Elzinga *et al.*, 1990; Fuller, 1999).

This paper uses an innovation systems analytical framework and a few illustrative cases of rural innovation, pro-active learning and change. The main hypothesis is that policy makers and research managers learn in an interactive way, when they work with and within wider rural innovation coalitions that take them beyond the limited mandates of their policy or S&T organizations. Since the testing and proof of this hypothesis exists through specific projects (and their results) designed to study institutional learning and change (using empirical cases of innovation), the focus of this paper is not to test for the same again–to reinvent the wheel, but to explore the nature of policy research processes that can enable institutional learning and change. The main messages in this paper for S&T policy researchers and policy makers in developing countries include:

1. The urgency to make the transition from S&T policy to a broader set of policies for STI (Science, Technology and Innovation) focusing on the generation as well as utilization of knowledge in society,

2. The importance of the capacity for institutional learning and change within S&T organizations, and initiation of attempts to build these capacities, and
3. The need for new social science tools to conduct STI policy research and engage proactively with communities of natural science researchers, policy making bodies, public, private, voluntary and co-operative sectors and a range of relevant stakeholders.

Rural innovation: Some cases and questions for S&T policy

Rural innovation is defined as the changes that take place in rural areas when knowledge, technology and information are generated, accessed and used by a range of actors in socially progressive and economically productive ways (UNU-INTECH, 2005). Conceptually the innovation systems framework tells us that the generation and utilization of knowledge in society takes place when several actors (individuals and organizations) come together and learn to work together in mutually complementing ways in a conducive policy environment. The innovation systems concept has its origins in widespread disillusionment with conventional economics to explain technological change, innovation and economic development. The assumptions of linear technology generation, diffusion and adoption are rejected with the advent of the non-linear and systems thinking promoted by the innovation systems framework which draws theoretically from evolutionary institutional economics.

The innovation systems framework has not been widely applied to developing country problems and certainly not in agriculture and the rural sector, till recent times. Indeed, there are those who argue that innovation systems concepts are really a theory of how private company's behaviour evolves in market driven economic systems and therefore it has limited relevance to development issues. However, recent developments in the social sciences have applied the innovation systems framework to explore and explain agricultural and rural innovation, draw principles for replication and further development from such cases, and identify points of intervention and investment options to promote development through rural innovation. (Hall, *et al.*, 2000; Hall *et al.*, 2004; ARD, 2006). While there are clearly qualitative differences between the innovation context in developed and developing country setting, there is great attraction to using innovation systems concepts as a broad heuristic for thinking about innovation as complex systems phenomena–especially in poor rural economies where sheer survival demands a range of technological, economic and social skills (Hall *et al.*, 2004).

Clearly, the actors in locally adapted innovation systems in developing countries are different–the private sector is often much less developed, the poor are also caste differentiated and have limited economic opportunity, women face caste, class, and gender oppression and work just as much or more than men in rural areas, any technology whether it is Bt cotton seed, fertilizers, land levelers, or, breeds of cattle, is used by the poor in a judicious combination of traditional, local and modern practices and contexts, influenced by many actors, ranging from local traders, money lenders, extension agents, caste formations, etc. The innovation systems framework has been

Box 1: TIDE's energy efficient biomass based dryers

Stoves and dryers are the two fuel-efficient biomass based devices developed and diffused by TIDE (Technology Informatics Design Endeavour, a small NGO in Bangalore). Recognizing the crisis and wastefulness in rural energy in developing countries, and the need for a device for drying horticultural and agricultural produce at the source of their origin ASTRA (Application of Science and Technology to Rural Areas, in the Indian Institute of Science) had successfully developed and field tested fuel efficient biomass burning tray dryers (TIDE, 1998). The ASTRA cabinet dryer is a batch tray dryer consisting of a cabinet for housing trays, an efficient stove for burning the biomass fuel, ducts for conveying the flue gases through various positions in the dryer and ports at the bottom portion to suck in the drying air which flows up the dryers in an ordered fashion over the flue ducts and trays by natural convection.

The opportunity to develop and diffuse fuel efficient dryers in fishing villages presented itself in the form of a UNEP Global Environment Facility (GEF) project for using environment friendly technologies for income generating activities in rural areas. The first fish drying enterprise established by TIDE in Azhikode was a failure. TIDE did an assessment and selection of a second self help group (SHG) in Anapuzha, near Kodungalloor, during the last few months of 2003. This time TIDE had the active participation of a local NGO, who had evident strengths in working with the SHGs in the area. Besides confirmation from the Kudumbasree officers and the NGO about the willingness of this SHG to take up this dryer and experiment with the idea of a fish drying enterprise, TIDE staff made their own assessment, and had the IFP scientists come to the village to train and work with the SHG members. In addition, with the involvement of the local Kudumbasree unit, TIDE and the SHG also explored wholesale marketing of the dried fish. This unit made a profit of Rs. 10,471 over the first year (January 2004-December 2004), after paying for their wages (a total amount Rs. 14550) (See, George, 2004, Table 8). Given that 10 women were active in this enterprise it was an annual profit of Rs. 1047 per SHG member from this micro enterprise.

Within a few months, from the end of 2004 to March 2005, there was a group of SHGs drying and processing (value addition to dried) fish. Each of these dryers was constructed by (and is maintained by) local entrepreneurs, apprenticed with TIDE. As this production increased, a new marketing strategy evolved, with a common brand name that spoke of quality produce from the sea "SagarSree". As a micro-enterprise owned and managed by poor women in fishing villages, the dryer, drying technologies and protocols of drying for different products, marketing strategies, cost-benefit sharing mechanisms, and experiments with other biofuels, ducting material etc. have been successful. This is a rural innovation case that is still evolving to address new and more value added products, packaging technologies, and markets.

used for exploring partnerships, institutions and learning processes associated with innovation (CGIAR, 2003; Horton *et al.*, 2004; Hall *et al.*, 2004). The innovation systems framework has also been used to explore the presence and role of a unique actor (not mentioned in the literature thus far), the professional associations in science (Raina *et al.*, 2006). The purpose is to acknowledge and encourage the professional associations as actors in innovation systems, which can enable the natural sciences to interact pro-actively with other actors in society and lead to technological and institutional innovation, as we see in the cases discussed below.

S&T for pro-poor rural innovation

Here we discuss some cases of rural innovation that have transformed the livelihoods options and other opportunities open to rural people. We shall briefly illustrate one case, that of biomass based fish drying in Kerala and then discuss several others briefly to highlight the role of science and the research domain in the innovation system. The case illustrated here (Box1) and another one on solar lighting systems for street vendors and rural poor in Hassan, Karnataka have major research domains associated with them. The analysis focuses on the processes which enable innovation–*the conversion of this idea of clean technologies and locally accessible energy sources into economic products in poor communities* (Reddy, 2002). In both cases we find rural employment and enterprise development; the cases highlight the processes of access to and utilization of technology by the poor, and the capacities developed within communities to understand and demand more of information and other technological and market based knowledge.

What is evident in the TIDE case is a constant interaction between the personnel engaged in scientific research such as heat and mass transfer, materials research, in food / fish processing, fisheries, gasifiers and stoves, biomass, etc. with others engaged in packaging technology and protocols, technology development, technology and enterprise assessment, local NGOs, Government Departments, local actors / agencies such as traders, transporters, wholesale agents, input dealers, local banks, or local Governments (panchayats). There are different pieces of a messy puzzle that have come together- different types and sources of knowledge, each one learning and changing rules/ways of working, building new competencies and evolving. The technology (be it fuel efficient cabinet dryers, fish drying protocols, etc.) are important. But it is the innovations in ways of working that are critical driving forces that enable different pieces of knowledge and skills to come together, and to evolve into an innovation system.

When the Himachal Pradesh Government, in India, imposed a ban on tree felling and use of wooden boxes to transport tomatoes from Himachal to distant markets (like Delhi), there was an alignment of organizations and individuals to consciously seek an alternative packing material; this was an institutional innovation (Clark *et al.*, 2003). The technological innovation brought about by this coalition was a cardboard box for packing tomatoes (with institutional innovations/new ways of working that enabled this). The processes of design and testing, monitoring and evaluation and learning together have prevailed along with the technology in this innovation system. Another set of institutional innovations helped *char* dwellers in

Bangladesh access post harvest innovations and markets (Biggs and Matsaert, 2004). The nature of actor linkages and the intensity of interactions also enabled appropriate technology choices and designs/adaptations. The sorghum breeding community being involved directly in poultry feeding trials with sorghum farmers (small holder), poultry feed manufacturers, and animal nutritionists in a wide coalition of relevant knowledge partners, would have sounded scandalous a few years ago in ICRISAT in India. However, the coalition enabled several interesting experiments–changed the design of pot feeding trials based on the parameters of monitoring convenience suggested by feed manufacturers, and have evolved some viable institutional innovations for breeding and utilizing rainy season sorghum for poultry feed, in other words for pro-poor development (Crewe, 2004). The case of enabling cheap affordable micro irrigation and pomegranate cultivation in the arid lands of Solapur (Maharashtra state in India) is again one of a series of learning experiences and institutional innovations (Raina, 2004). Here, the pace of research, learning and change has been really rapid because of the opportunities offered by globalization and expanding agricultural/horticultural exports–this arid part of the Deccan plateau in which almost every household would migrate to neighbouring district every dry season (4-6 months) annually, now accounts for very little seasonal distress migrations and hosts the Agri Export Zone for pomegranates set up by the Ministry of Commerce of the Government of India (in 2003). In all these cases, the rural poor know where to go to, whom to contact and how to access the inputs (including technological ones) and services they need for the rural enterprise to work successfully and evolve further. This is a crucial part of the wider collective capacity for interaction and learning among a group of actors.

All these cases have used technologies generated by R&D organizations. In all these cases R&D actors have worked with others and not stopped at technology generation. Innovation–the generation and utilization of knowledge, technology and information is evident in each of these. The poor in rural areas–women in fishing villages, small farmers, geographically and economically vulnerable *char* dwellers, seasonal migrant households in the poor arid lands, have developed capacities for further innovation and development.

These cases reveal the features of an innovation system, namely, (*a*) interaction of diverse research and non-research actors, (*b*) combinations of technological and institutional innovations, (*c*) shifting roles for information producers, information users and (*d*) a need based exchange of knowledge, continuous evolutionary cycles of learning and innovation.

What do these cases of innovation tell S&T policy makers?

The conventional practice is to demand replication of these successes, increase funding for food processing research, or schemes for subsidising items like dryers/drip irrigation systems/treadle pumps. In this event, when replication does not take place, S&T is blamed and Government schemes are blamed. The key question here is–"replicate what?" S&T policy is equipped at best to replicate the R&D projects that generated the technologies, the extension/technology development and dissemination effort, the subsidies, more NGOs and grants for them to work as tail end implementers.

But S&T policy is not equipped to replicate or re-enact a series of socially embedded processes among a set of actors inhabiting different domains. It is replication or re-enactment of these linkages, interactions and learning processes that will enable widespread innovation. Policy research has to inform policy makers about how to replicate these processes. A simple illustration of the domains involved in a typical rural innovation case (a case of solar energy for the rural poor and urban poor) will help us see these relevant processes (See Figure 15.2). The innovation systems framework gives us a handle–an operational tool to see (a) whether all these domains are present in each context, (*b*) whether all the relevant actors are there (including which ones are relevant and which ones are absolutely necessary), (*c*) what are the kinds of interactions/relationships that S&T has with each of these actors and their domains, and (*d*) the overall capacity for interaction and learning among actors within and between domains.

Innovation systems analysis: Adding "some more" to S&T policy research

Theoretically, the innovation systems framework conceptualizes innovation as an interactive and evolutionary process, whereby several actors (organizations and individuals) work together as a coalition in appropriate policy and institutional arrangements to shape and use knowledge such as scientific, market related, or gender related in socially and economically productive ways. While the explanatory power

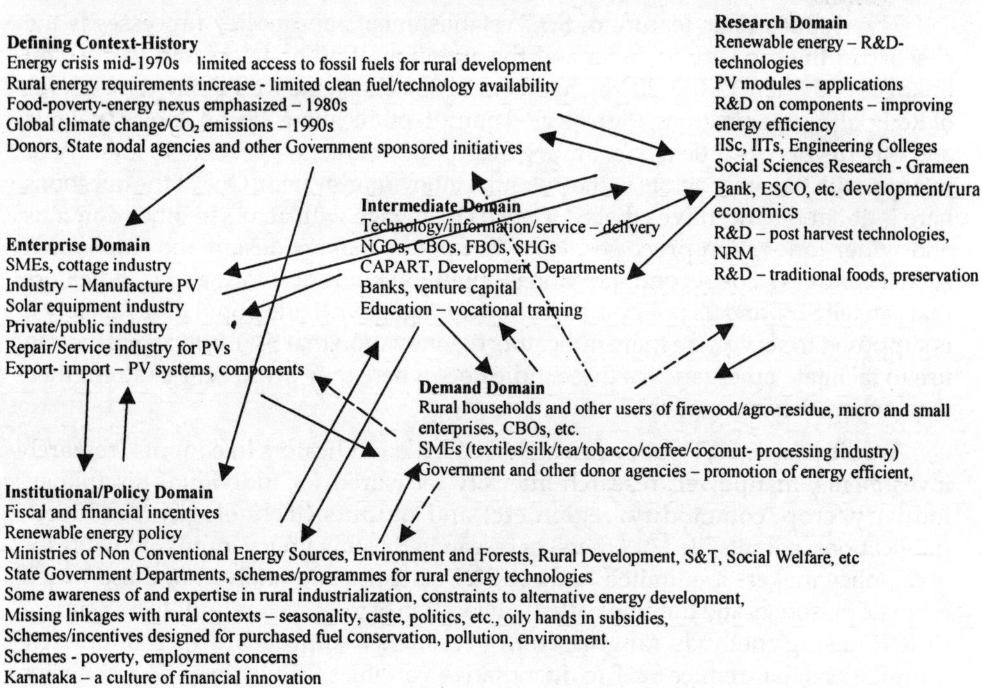

Figure 15.2: Solar lighting for the rural poor: A domain map

of the innovation systems framework is crucial, its application as a diagnostic tool to answer several pragmatic concerns of research decision makers, policy makers, donors, private traders, farmers, rural women, rural financial organizations (banks, venture capital) is increasing. In a study conducted for the World Bank, the innovation systems framework was applied to eight case studies of agricultural/rural innovation across four countries, to develop an intervention framework for assessing innovation systems and identifying potential interventions (ARD, 2006). Clearly, increasing funding or research organizations for agricultural S&T is perhaps, only one among several such potential interventions.

In all the cases we have discussed above, and in several other cases of rural innovation analysed and explained, *the main message for S&T policy makers is that innovation is not the conduct of science or generation of technology.* Even more important (and perhaps a little disheartening to those of us who, were taught and, firmly believe that science is "absolute truth", is "the source of all knowledge"), is that science is not the main trigger or driver of innovation. Scientific research is important–it is not the central component of innovation (ARD, 2006; Biggs, 2007). If the purpose of S&T policy is to enable better social and economic outcomes in the world then S&T policy has to engage with and integrate itself with other actors relevant to the innovation processes. To this end, S&T policymaking needs to command the attention of and develop concrete linkages with actors in other domains in an innovation system (Figure 15.2).

The characteristic feature of S&T establishments and policy processes is the absence of the capacity to command the attention of other actors and to develop linkages with them (ARD, 2006; Chambers, 2005). Given the well-defined mandates of R&D labs/organizations, the expected output–publications/technologies/patents and others, there is little desire among S&T policy makers or leaders to identify or interact with the other actors in the potential innovation system. One of the questions here is about the incentives that S&T needs to engage with actors in other domains and wider innovation processes. How can policy makers design and implement such incentives? The second question is about the kinds of information/indicators that can tell S&T and its policy makers how to engage with the population/society it is supposed to serve. Are there indicators of innovation that S&T policy makers can use to facilitate emergence of these indicators where they are absent or strengthen them where they are evident?

Traditionally, S&T policy research focused on indicators like inputs (research investments, manpower, research intensity measured for individual disciplines, industry/crop/commodity/region etc) and outputs (technologies produced, publications, patents filed or others) of research. Existing policy inputs available for S&T policy makers are limited to these S&T statistics and analysis of these which estimate personnel and funding patterns, allocation of resources, returns to investment in R&D using methods ranging from precedence approaches to econometric formulations that reduce S&T to quantitative variables. While these S&T statistics and information, mainly quantitative are important, they are only capable of providing information/recommendations about S&T or research intensity, location, allocation, or performance.

Currently available statistics are important and do play a critical role in telling us how the overall S&T sector fares; but they fail to provide any policy insights on the ways to improve the contribution of S&T to reducing poverty or to give a good picture on the content and ways of working of S&T. The S&T statistics throw little light on how S&T integrates with other forms of knowledge, technology and information. The traditional indicators offer little information about the way S&T organizations work in society (their institutions–habits and practices, competencies, flexibility, ability to report and learn from every experiment–successful or otherwise), the relationships among scientific knowledge and other forms of knowledge, say, markets, labour, finance, education, and the processes that enable effective generation and utilization of technologies/information in society.

The empirical evidence available from cases of rural innovation tells us that indicators that speak of the processes, relationships and ways of working are crucial for us to understand innovation capacity. The cases discussed here reveal that the indicators of STI strength or intensity are different from the S&T indicators. STI indicators are not limited to publications/patents, research expenditure, or number of personnel with Ph.D. or post-graduate level degrees. The indicators of STI performance are the patterns of collaboration/linkages, degree of and nature of response to socio-economic problems, response time, new rules/norms or behavioural changes, and problems in the innovation process in the short run. In the long run, the STI indicators are indicators of social and economic performance due to innovation. Technology adoption which is the accepted S&T indicator in the long term, assumes a linear flow of technology down to the beneficiary and does not tell policy makers about the range and nature of actors and interactions that lead to this technology adoption. Conversely, S&T policy makers are denied the opportunity to learn about ways in which the rural poor can gain from S&T.

Indicators of ST&I

Taking cue from the innovation systems theoretical framework and the empirical evidence that case studies of rural innovation provide, we now have some key indicators that can tell policy makers whether (*i*) the S&T establishment, individual organizations/programmes/projects and personnel within these S&T labs are or can be part of potential innovation systems, and (*ii*) how the S&T establishment can become part of or trigger/steer rural innovation.

While the specific sector we are concerned about here is rural, it must be emphasized that the rural sector is one that is intricately related to and shaped by the rate and patterns of urban growth and overall economic growth in the country. In other words, any intervention (S&T based or otherwise) must be (*a*) planned, designed and implemented with the ideal outcomes in mind, especially for the poorest, (*b*) flexible enough to respond to opportunities and constraints. For policy makers, then, developing and using science, technology and innovation for rural development demands some of the conventional sources of information plus some new indicators of innovation processes. The STI indicators listed below are applicable to regional rural sectors (say, within a district or State) or sector specific interventions (say, spices and plantation crops, dairy industry, rural energy).

Indicators of relative location

Sector maps: Domains and actors within domains

Sector maps are perhaps the most illustrative and handy among the indicators of STI. The main pieces of information that a sector map offers are given below:

1. Main domains visualized in the innovation systems framework and main domains present in the sector–with reasons for presence/absence, if any.
2. List of actors within each domain–identify major actors (organizations and individuals) who are currently involved in or can champion the innovation/sector.
3. Presence, absence or nature of engagement of domains like policy or demand domain with the research domain–identify winners and losers (with lessons from both treated equally valuable).
4. Evidence of linkages between actors in different domains–say a sector development report prepared by dairy development department, dairy exporters consortium, dairy research institute, food chemistry research, milk marketing federation and dairy equipment manufacturers, etc.
5. Evidence of intra-domain linkages–among research organizations (say university departments, private R&D and mandated research institutes in the research domain, small/medium scale enterprises, industry, individual households, other users in the demand domain).

Here, we must add that some parts or aspects of inter-sectoral collaboration or linkages have been developed and used in rural development. In India, these include the co-ordinated research projects of the ICAR, the co-ordinated projects/schemes of CAPART (Ministry of Rural Development), the research consortia on important crops or areas (say cotton or fermented foods of South Asia), the integrated sector support initiatives (for handlooms in the textile sector, or for spices development), major technology development and transfer programmes (the Technology Mission on Oilseeds and Pulses), industrial clusters or associations (UNIDO's industrial cluster programmes, other rural industrialization clusters by the Ministry of Industries). Other demands that are being made include sector development authorities to co-ordinate and guide the sector (say the National Drinking water Mission, National Watersheds Development Programmes, or proposed National Dry-land Development Authority). While the potential for all these inter-domain activities must be appreciated, it must be acknowledged that all of them stick to their given mandates, lacking a mandate to collaborate with or provide expertise or assistance to another domain/actor not listed in the current mandate.

The sector map gives the STI policy maker a handle on aspects like (who is who or who is present or absent) the usual sector statistics that are made available such as size of the sector, market structure, leaders and shares, share in agricultural GDP, growth rate, employment–male and female, or export contribution.

Indicators of patterns of interaction/partnership

Collaborating, complementary/substitutive and learning interactions

Several cases discussed above reveal collaborative interactions, especially among research and non-research actors, public and private sectors, civil society and individual households/beneficiaries and line departments, local middlemen/traders or SHGs and banks. Very often a contract between the rural development department and an NGO to implement a vaccination programme or vegetable production programme, or a contract between an agricultural research institute and an NGO to transfer technologies, is called a partnership. These unfortunately transgress all known definitions of partnerships and involve only an addition often very rigid and inflexible relationship, to conventional linear modes of knowledge development and transfer or awareness building. The collaborating interactions or partnerships involving shared pains and gains, intense and frequent interactions, and generally several partners (besides a one-to-one contract), evident in rural innovation cases are very different from these paternalistic and hierarchical contracts between S&T actors and non-research actors allegedly lower in the knowledge ladder.

Competitive interactions and learning

Almost all the innovation systems analysis pertaining to industrial or firm level innovation reveals competitive interactions and learning processes. These competitive interactions, crucial for the firm/industry are important determinants of rural innovation too. Market participation and control of market information of small producers, middlemen, local processors, and civil society organizations and donor agencies have been crucial factors shaping innovation. In most cases of pro-poor innovation in rural areas, it is the absence of or modification in competitive behaviour that appears striking. In these cases as in the fish drying micro-enterprises or the cheap drip irrigation in arid lands, even the private actors (dealers, or wholesale agents or others) had developed norms of profit margin (not more than 5-7 per cent), adjusted procurement or sale timings to suit the poorest, and through these had expanded/developed their own market.

For a group of STI policy makers it is important to maintain a checklist that will tell them whether the innovation system features of "diverse technological and institutional interactions" exist or can be created. Policy makers can use a matrix to asses existing partnerships and their nature and intensity, to create essential interactions (Table 15.1) among actors in the research domain and with other non-research actors/domains.

In each context, whether it is national, State, district or village level, there are certain ways in which the actors learn. These ways of learning are often closely related to the kind of interactions and relationships that the S&T organizations have with the others. A cursory tabulation of number of actors and the intensity of interactions, along with a grading of relative intensity of each of these interactions and learning, can help policy makers assess the magnitude and dynamism of interaction and learning among a group of actors.

Table 15.1: Types of interactions and learning

Types of learning	*Types of interactions*			
	Collaborative–partnerships	*Networks*	*Contract/ prescribed linkages*	*Competitive*
Training				
Research				
Interactions/discussions				
Searching/surveying				
Imitative behaviour				
Trial and error				
Etc. etc.				

There can be many other types of interactions and learning processes. As an STI policy input it is important to keep a tab on them to enable the sciences to learn from rural development actors, contexts and processes, and to enable the latter to interact with and learn from science. It is important to note here how existing indicators of research expenditure or publications in disciplines, or Ph.Ds among research personnel throw more light when used along with these interactions/learning indicators to understand innovation capacities and potential.

Indicators of competencies

Documenting and analyzing research and non-research skills that exist within the research domain is in itself an important indicator of innovation capacity. These include disciplinary competencies, inter-and multi-disciplinary competencies, learning competencies, collaboration skills, local trust/relationship building skills, communication skills, policy/other advocacy skills, and accountability/attribution skills and capacities. Other skills that S&T actors can acquire or share with partners in the innovation system are market know how, caste/social fabric, political positioning of technology, gender and ecological assessment skills.

While documenting disciplinary competencies is an important indicator of scientific strength, it is not an indicator of innovation capacity. In a research programme or organization with a STI orientation, the skills and competencies that the programme or organization possesses currently, has learnt or acquired recently, or has access to through collaboration with another non-research actor, can provide a picture of innovation. This is also a self-reflective exercise for the research organization, and helps it to keep a tab on what kind of research/non-research activity can help further innovation. The best examples among the cases cited above is the case of the cheap (pepsi) drip irrigation system–the local agricultural university, when asked to assess this technology acclaimed the success of the cheap (pepsi) drip system and said that it had over 200 irrigation technologies on shelves and publications too but lacked the competence to communicate with others like farmers, NGOs, Government Departments to even assess if these were the appropriate technologies they needed, or find out how they would perform in the field.

For STI policy making, it is important to see these indicators of competencies in the light of the indicators of interaction and learning listed above. It must be acknowledged both within scientific research and STI policy making that rural innovation is not a technology/artefact or a process or organization like a new marketing yard or e-commerce in agri-horti products. It is a continuously evolving process, and therefore calls for capacities for constant learning among STI actors and policy makers.

These indicators listed here are mainly for STI policy makers to check capacities for innovation in S&T organizations. Other organizations/individuals in the innovation system do have and bring a different set of capacities to the innovation system. This is what makes STI policy more than just one policy statement but a constantly evolving portfolio of policies and practices designed to cater to rural development. To plan, intervene in, establish and nurture pro-poor rural innovation systems, the STI policy portfolio has to be oriented to and draw from different domains and sectors.

Indicators of innovation potential and direction

A typology of innovation can be used to characterize the innovation capacity, the role of S&T in existing/potential innovation systems, and the appropriate phases and points where interventions from a portfolio of STI policies and practices can be effective (ARD 2006). An existing R&D system may be orchestrated with a strong public sector planned foundation and traditional R&D roles or may be opportunity driven with leading roles for private R&D and civil society actors, showing opportunistic behaviour. Both can be complemented by sector driven generic research and training services. For either to evolve into an innovation system it is important to recognize and cultivate the capacity to respond to emerging challenges and opportunities (*ibid* 2006).

To assess the innovation potential and direction of a system, the indicators have to be ones that help the R&D manager/policy maker see the specific features of a context specific innovation system. Thereby, the indicators are ones that show:

1. The number and nature of non-research partners who can lead the sector innovation system,
2. The processes of reflection and learning present in the system and in each actor (organization/individual) in the system,
3. The frequency, transparency and trust in the communication processes among the actors,
4. The acknowledgement, documentation and conscious internalization of behavioural changes made/acquired in the process(es) of innovation,
5. Diverse ways of funding research and reporting–highly context dependent and poverty oriented,

6. Number and nature of social goals and indicators incorporated into mainstream science, and
7. Changes in overall knowledge generation and utilization policies–from strengthening R&D to developing, intervening and strengthening other components of the innovation system.

The last has received special attention in recent times from major donor agencies (the World Bank, IDRC and DFID). The idea of looking beyond conventional 'strengthening R&D' is finding increasing acceptance, and is to become a sector strategy within these donor agencies who are keen on operationalising the innovation systems framework to achieve poverty reduction and other social goals. In the world of S&T today, these attempts to shift from conventional S&T to a STI policy framework and implementation process is a noteworthy development.

Implications for policy research and policy making

We must note that the importance of learning and interactive processes in conducting scientific research as well as S&T policy research has been recommended before. Whether it is Miles and Huberman (1994) or Funtowicz and Ravetz (1990), there have been scholarly studies on and recommendations for research and policy making to interact closely and to learn from each other. The advantages noted are that communication of research results to policy makers is much more focused and direct, and the capacity of science to verify and validate its research questions, processes and results is enhanced with the inputs from practitioners and policy makers. These recommendations though, have not had much of an impact on prevailing mainstream modes of prescriptive policy research and socially isolated S&T activities. They disturb existing institutions or ways of working in both science and S&T policy making.

Our concern here is to see how policy makers can use the process indicators of science, technology and innovation. It is unlikely that any policy/practice will change because of a set of recommendations made by an expert committee on rural innovation. For one, the ways in which policy makers are assessed and held accountable for their policy formulation or implementation, are outdated and hardly counts even as a routine evaluation exercise. Secondly, it is likely that conventional S&T policy making organizations, research managers and organizations perceive this move from S&T policy to a wider portfolio of STI policies as a major threat. The move is partly a threat to the conventional complacency of S&T–it disturbs S&T into acknowledging social realities, and realizing that technology is only one among many sources of knowledge that can lead to rural innovation.

More importantly, it gives research managers the tools, *viz.* indicators and measures, methods, collaborators–partners and leaders, to do so. It also disturbs the conventional neo-classical economics dominated theories and practices of social science research that engages with investment and impact assessment of the agricultural and rural development sciences. These economists have a general apathy to increasing evidence of pro-poor institutional changes, and new methods or frameworks for assessing impacts of science (special issue of *Agricultural Systems*,

2003). The social studies of the agricultural sciences, as Fuller would argue about SSS in general, have dodged normative commitments in order to maintain a user-friendly presence (Fuller, 1999). Social science is channelled largely into a service field working on problems identified by biological and technical agriculture, not by the ultimate clients of that effort (Rhoades, 2000).

A brief overview of the evolution of S&T policy and rural development policy

An overview of the way the two estates of development policy and S&T policy have evolved over time, is essential to clear a few cobwebs and visualize where we stand with respect to our ability to develop pro-active S&T organizations and interactive policy research processes. It will help us move in the direction of a portfolio of STI policies for rural innovation. Given that rural development departments, financial organizations, or extension agencies are major intermediary agencies in an innovation systems domain map, (Figure 15.2), we use this as a foothold to see how S&T policy makers who are the major actors within the research domain and within the institutional/policy domain, can engage with and learn to work towards STI policies and practices. This section will also bring all of us–S&T policy analysts, research managers, scientists, development economists and several other actors, to a common platform and shared understanding about the nature and contents of a portfolio of STI policies for rural innovation.

There have been broadly three phases of S&T policy making (Gibbons *et al.*, 1994). During the first phase, the issue at stake was to formulate a policy for the growth of science or the scientific enterprise. Policy, based on the writings of Weinberg and others in the 1950s and 1960s was mainly concerned with choosing the right projects within disciplines, or formulating guidelines of cost-effectiveness. In India and in the developing world in general, this phase came as a State-led phase of establishment of R&D laboratories and Departments in Universities, often discipline-based (Raina and Jain, 1994). There was also a sense of national pride among the scientific elite–their newly acquired status as formal researchers, (which they had to fight for during the British period) and possession of well-equipped labs were important.

The second phase in S&T policy began in Europe with the Rothschild Report (1971), where the emphasis was on shifting from policy for science to science for policy. The main contention was that S&T should play a major role in shaping and achieving major policy objectives and not merely focus on the development of science (Gibbons *et al.*, 1994). For the developed and developing countries this marks the expansion phase in R&D–State-led in the latter and private sector led or shared in the former. In the West the decisions about investing, public or private in different research councils/institutes, industrial, agricultural, medical, space, defence, or others, was mainly shaped by the commercial saleability of the final product. In countries like India the State decided to invest in all these sectors with the private sector often procuring technology from the West or adapting a few home grown ones. It also marks the beginning of the emergence of the educated middle class professional as a selected not elected decision-maker in democratic contexts (Raina, 1997).

The third phase of contemporary S&T policy has been again grouped into two– one that focused on policy for technological innovation, and the other that focuses on science, technology and innovation for development. The former, beginning somewhere in the early-mid 1980s focused on using S&T for building industrial competitiveness. Emphasis was on strategic and then generic technologies, using technological advances to support national industry (Gibbons *et al.*, 1994). This private-sector led phase of generation and application of technology in industry as well as a renewed technological innovation policy seems to have helped the developed and developing countries only to a limited extent, despite some industrial growth rates and international trade growth rates in the developed world, in the 1980s and 1990s. Tucked into the major features of sluggish economic growth rates the world over, except in newly industrialized economies and the global IT players, are major development problems of over a third of humanity going to bed hungry every day.

Gibbons *et al.* (1994) argue that under the new conditions of knowledge production and utilization, marked by non-linear, simultaneous, and evolutionary processes and involving a whole range of research and non-research actors, the traditional phases of S&T policy (phase 1 and 2) and innovation policy (phase 3) seem inadequate. In reality, the three phases seem functionally inseparable. It is only when regarded and treated as a co-evolutionary process involving several actors and learning capacities that science, technology and innovation can cater to development– whether industrial, rural, agricultural or overall social and economic development.

Though the evolution of rural development policy and practice is significantly different, the current status, or phase in evolution emphasizes the same 'people and their capacities' as the critical resource. The different phases of rural development thinking/policy and practice (Box 2) begins with the 1960s and on to the 1970s, marked by the absolute confidence in the ability of agricultural/rural S&T to deliver the rural development public goods. All that was needed were the mechanisms for getting the rational farmers to adopt the technologies being generated by scientific research. Though this was an improvement over the 1950s idea of lazy peasants and their inability (lack), even in the 1970s the questions about the wider implications of green revolution had not been translated into appropriate schemes/policies for rural development. The institutional arrangements that led to the green revolution in the Asian countries seem to have been guided by intuitive and focused policy makers, than by academics in the social or natural sciences.

In the rural development time line (Box 2), we find that almost till the mid-late 1970s, there was little awareness or explicit mention of the linkages and mutual dependence between technologies and institutional arrangements. Then on, however, there has been an explosion of institutional issues and social science analysis for the same, to facilitate rural development. The basic needs approach, integrated rural development, price policy/marketing arrangements and infrastructure, finding NGO partners in development programmes, and so on were being analysed in the social sciences and implemented in government schemes/programmes for rural development. Some research methods in and evolved by the social sciences such as farming systems research (FSR), actor oriented research and development systems, rapid rural appraisal and participatory rural appraisal, were also incorporated into

Box 2: Rural development ideas timeline

1950s

modernisation
dual economy model
'backward' agriculture
community development
lazy peasants

1960s

transformation approach
technology transfer
mechanisation
agricultural extension
growth role of agriculture
green revolution (start)
rational peasants

1970s

redistribution with growth
basic needs
integrated rural development
state agric. policies
state-led credit
urban bias
induced innovation
green revolution (cont.)
rural growth linkages

1980s

structural adjustment
free markets
'getting prices right'
retreat of the state
rise of NGOs
rapid rural appraisal (RRA)
farming systems research (FSR)
food security & famine analysis
RD as process not product
women in devt (WID)
poverty alleviation

1990s

microcredit
participatory rural appraisal (PRA)
actor-oriented RD
stakeholder analysis
rural safety nets
gender & development (GAD)
environment & sustainability
poverty reduction

2000s

sustainable livelihoods
good governance
decentralisation
critique of participation
sector-wide approaches
social protection
poverty eradication

Source: Ellis & Biggs (2001)

rural development ideas and specific schemes. The Government of India and several State Governments for instance invested heavily in specific schemes for agro-forestry, integrated farming systems, and integrated pest management.

Rural development policies have evolved in a more inclusive fashion compared to the agricultural sciences and S&T policies, revealing a relatively open-ended social and technological determinism of development. In terms of utilizing scientific and technological knowledge for rural development, the approach even within the rural development schemes is rather linear paying obeisance to the superiority and generalized scientific knowledge inputs that are transferred down the line to the farmers/rural artisans/women. There have been some attempts at linking the rural development actors and processes to the sources of knowledge *e.g.*, S&T, extension, farmer-based/indigenous know-how, cottage industry clusters, and other actors through several experiments that have suggested the usefulness of such interactions (Chambers, 2005).

Current thinking in rural development emphasises the need to learn from cases of successful innovation and positive institutional change, to build on positive capacities of people in order to generate growth engines using S, T and I. This also demands investments in a wide range of relevant actors other than R&D actors who are key components of rural innovation and development (Biggs, 2007; ARD, 2006). The key concern here is that little of this rural development policy/practice seems to have touched mainstream S&T policy or even policy research processes which inform S&T. The most important message is that in order to enable innovation, S&T has to engage with other development actors more pro-actively. That this demands institutional change, seems to have eluded policy makers.

Conclusion: Interactive policy research processes for a portfolio of STI policies and practices

There are two main empirical findings that rural innovation offers for STI policy and practice, *viz.*

1. S&T is only one part of several sources of knowledge and technology evident in rural innovation, and therefore, to enable innovation for rural development and poverty eradication, S&T must learn to work with and become part of several other socially embedded processes, and
2. S&T like all the other actors has to go through processes of institutional learning and change to enable rural innovation.

Institutional innovations, or new ways of working such as new rules for managing a collective natural resource, new travel norms in research organizations enabling scientists to collaborate more effectively with non-research partners, legal reforms granting access to land and water resources or new markets, an agreement from a bank permitting a flexible rate of repayment to 10 per cent of the poorest households, or a norm for extracting payment from each household for equipment maintenance services, are important if the rural poor are to gain from S&T. Today, the social sciences have led us to a better understanding of rural innovation–the processes of formation of coalitions of actors and the evolution of innovation.

Involving wide networks of people and organizations, their rules, habits and practices, institutional innovations are always slow, more complex and painful to arrive at, compared to the technological innovations. The major breakthroughs in the adaptation and utilization of a technology, be it the HYVs of the green revolution, electricity or railways take place only after the institutional innovations have picked up, and gained momentum. The history of S&T teaches us that these institutional innovations were made through gradual and sometimes drastic changes in society–more specifically in the social organization of activities. S&T policy makers, managers and scientists must appreciate the need for institutional changes to deliver development. It must be emphasized that public sector has a major role to play in rural innovation and development, and therefore has a significant responsibility in shaping the kind of science, technology and innovation efforts in each country.

The problem is that few scientific research committees or programmes are even cognizant of the mutually complementing role of technological and institutional innovations. It follows then, that there are few who appreciate the need for research that tells policy makers and scientists about these institutional changes–especially about the changes in the rules that govern the structure and conduct of S&T organizations, and their capacity to learn from and with other rural innovation actors. Within S&T policy circles, whatever has been facilitated as institutional changes and capacity building for the same has come from lessons learnt through individual work experience, and often appears as ad hoc decisions. The innovation systems framework offers rigorous analytical tools that can be applied to understand the actors, their interactions, and their collective capacities for innovation. The framework can be used at different levels- national, regional or local, and for specific sectors, because it enables inclusion of relevant actors and their knowledge (market-based, location specific, etc.) to be incorporated in the analysis. Using the innovation systems framework it is possible for S&T policy making to learn interactively and enable widespread innovation and progress in rural areas. Empirical evidence and some recent international initiatives in this direction emphasise the need for interactive policy research for S&T to benefit the rural poor in developing countries.

Current S&T policy research is conducted in a prescriptive mode. It is no secret that it is not for dearth of recommendations that an S&T policy is not formulated clearly or is not changed to meet new or emerging challenges in society. In an interactive learning mode, policy makers and other actors involved (scientists and research organizations in the public/private/voluntary sectors, other non-research organizations and other rural stakeholders including local governments) participate in the policy research process, constantly communicate and learn from each other, as well as change their habits and practices to facilitate innovation. Ways of funding and promoting interactive policy research are rare–and even when organizations like the DST do fund new and development oriented research projects designed to build innovation capacity among NGOs, there is little attempt to learn from them and promote the lessons within DST and other NGOs/actors.

A portfolio of STI policies involving and relating to changes in several currently compartmentalized areas/departments seems imperative. An interactive STI policy portfolio will involve regular and continuous interactions and learning among policy makers and implementers involved in rural development, infrastructure, credit, employment, human resources, health, or education, through effective inter-ministerial/departmental communication. The dynamism of these interactions and learning processes will be the most crucial indicator of innovation capacity in the system. It is difficult to unsettle the complacency of policy making organizations, even in the face of increasing poverty and the pockets of evidence of poverty reduction through innovation. Allowing or forcing policy making organizations to learn through interactive policy research processes seems to be a critical requirement for widespread rural innovation.

References

ARD 2006. Enhancing Agricultural Innovation: How To Go Beyond Strengthening Research System. Washington, DC: Agriculture and Rural Development Department, The World Bank.

Biggs, S., 2007. Building on the positive: an actor innovation systems approach to finding and promoting pro-poor natural resources institutional and technical innovations, International Journal of Agricultural Resources, Governance and Ecology, Vol. 6(2): 144-164.

Biggs, S. and Matsaert, H., 2004. Strengthening Poverty Reduction Programmes Using an Actor Oriented Approach: Examples from Natural Resources Innovation System. AgREN Network Paper. No. 134. London: Overseas Development Institute (available at www. Odi.org.uk).

Chambers, R., 2005. Ideas for Development, Earthscan: London.

Clark, NG, Hall AJ, Rasheed Sulaiman V and Guru Naik., 2003. Research as Capacity Building: The Case of an NGO facilitated post-harvest innovation system for the Himalayan Hills. World Development 31 (11): 1845-1863.

Crewe, E., (2004) (mimeo) "Institutional History of the CPHP- South Asia project 'Exploring Marketing Opportunities through a Research, Industry and Users Coalition: Sorghum Poultry Feed'." CRISP: Patancheru India.

Douthwaite, Boru, 2002. Enabling Innovation–A Practical Guide to Understanding and Fostering Technological Change, Zed Books: London with Cambia: Canberra.

Ellis, F., and Biggs, S., 2001. Evolving Themes in Rural Development 1950s-2000s. Development Policy Review 19 (4) pp. 437-448.

Elzinga, A., Nolin, J., Pranger, R., Sunesson, S. (1990) In Science We Trust? Sigurdson, J. (ed.) Science and Technology Policy Studies 2. Lund University Press: Sweden.

FAO, 2006. The State of Food and Agriculture in Asia and the Pacific FAO, Regional Office for Asia and the Pacific: Bangkok.

Freeman, C., 1987. Technology and Economic Performance: Lessons from Japan. Pinter: London.

Fullerm S., 1999. Why Science Studies has never been critical of Science–Some recent lessons on how to be a helpful nuisance and a harmless radical. Philosophy of the Social Sciences 30, pp. 5-32.

Funtowicz, S.O. and Ravetz, J. R., 1990. Uncertainty and Quality in Science for Policy, Kluwer Academic Publishers: Dordrecht, the Netherlands.

George, S., 2004. (mimeo) Process documentation and analysis of TIDE's fish drying micro-enterprises in Kerala. TIDE: Bangalore.

Gibbons, M., Limoges, C., Nowotny, H., Schwartzman, S., Scott, P., and Trow, M., 1994. The New Production of Knowledge–The dynamics of science and research in contemporary societies. SAGE: London and New Delhi

Hall, A. J., Clark, NG., Sulaiman, R.V., Sivamohan, M.V.K., and Yoganand, B., 2000. "New Agendas for Agricultural Research in Developing Countries: Policy Analysis and Institutional Implications," Knowledge, Technology and Policy, 13 (1) pp. 70-91.

Hall, A. J., Sulaiman, R.V., Clark, N.G., and Yoganand, B., 2003a. From measuring impact to learning institutional lessons: an innovative systems perspective on improving the management of international agricultural research. Agricultural Systems 78 (2) pp. 213-214.

Hall, A.J., Yoganand, B., Sulaiman, R.V., Raina, R. S., Prasad, C. S., Naik, C.G., and Clark, N.G., 2004. Innovations in innovation: reflections on partnership, institutions and learning. CPHP, South Asia, ICRISAT, and NCAP: Patancheru and New Delhi.

Horton, D. L. Dupleich, and A. Andersen (Eds.) (1998) Assessing Organizational Impact–Report of a Review and Synthesis Workshop, 4-6 August 1998. The Hague: ISNAR.

Miles, M.B. and Huberman, A. M., 1994. Qualitative Data Analysis. Thousand Oaks: Sage.

Lewis, W. A. 1954. Economic Development with unlimited supplies of labour, Manchester School of Economics and Social Studies, No.22, pp. 139-191.

Lundvall, B.A., 1992. National Systems of Innovation: Towards a Theory of Innovation and Interactive Learning, London: Pinter.

Nelson, R. and S. Winter, 1977. In Search of a Useful Theory of Innovation, Research Policy, Vol. pp.36-76.

Raina, Dhruv and Ashok Jain, 1994. Interdisciplinary studies at NISTADS. Interdisciplinary Science Reviews 19 (3), pp. 211 -217.

Raina, R.S., 1999. Professionalisation and Evaluation: The case of Indian agricultural research", Knowledge, Technology and Policy, 11 (4), pp. 69- 96.

Raina, R. S., 2004. Technological and institutional innovations: the pomegranate innovation system in Maharashtra. In: Innovations in innovation: reflections on partnership, institutions and learning. (Eds. A.J. Hall, B. Yoganand, R.V. Sulaiman, Rajeswari S. Raina, C. Shambu Prasad, Guru C Naik and N.G. Clark), CPHP, South Asia, ICRISAT, and NCAP: Patancheru and New Delhi., pp. 81-103.

Raina, R.S., 1997. Public Patronage and Political Neutrality in Agricultural Research: Lessons from British experience. Economic and Political Weekly, Vol.32 (39) pp. 2473-2485.

Raina, R.S., Sangar, S., Sulaiman, R.V., and Hall, A.J., 2006. The Soil Sciences in India: policy lessons for agricultural innovation. Research Policy, Vol. 35 (5) pp. 6691-714.

Rajeswari, S., 1992. The Organisation of Agricultural Research in India: An Economic Analysis of Technology Generation, 1860-1990. Unpublished Ph.D Dissertation, Centre for Development Studies. Kerala University, Trivandrum.

Rajeswari, S., 1995. Agricultural Research Effort: Conceptual Clarity and Measurement. World Development 23, pp. 617-635.

Reddy, A.K.N., 2002. Reddy, A. K.N. (2002) "The Evolution of an Energy Analyst: Some Personal Reflections," Annual Review of Energy and Environment, Vol. 27: 23-56.

Rhoades, R.E., 2000. Agricultural Anthropology: New Disciplinary Blood in an International Research organization. In: Classics of Practicing Anthropology 1978-1998. (Eds. P.J. Higgins and J.A. Paredas) Society for Applied Anthropology: Oklahoma City, pp. 159-162.

Schultz, T.W., 1964 Transforming Traditional Agriculture, Yale University Press: New Haven.

Sen, A.K., 1967. Surplus labour in India: A Critique of Schultz's statistical test, the Economic Journal, Vol.: 154-160.

UNU-INTECH, 2005. RIPWiG Reporter, December 2005. UNU-INTECH: Maastricht.

World Bank, 2003. World Development Indicators. Washington, D.C.: World Bank.

Notes:

[1] This last one was however more than publication statistics. It relates research publications in several sub-disciplines of the soil sciences to the ways of working – or institutional features of the ICAR system, and through this, addresses the issue of changes in these rules to enhance the capacity of the soil sciences to play their role in natural resources innovation.

[2] The project on "Institutional Learning and Change in post-harvest innovation systems" was conducted by a coalition of researchers and practitioners to document and analyse ways of learning, behavioural/institutional changes within and among coalitions of actors involved in rural/ post-harvest innovations (under the CPHP South Asia). The project analysed four on-going CPHP projects and three non-CPHP projects, and covered institutional learning practices and processes in innovation systems ranging from rain-fed sorghum for poultry feed, NTFP and tribal livelihoods (mahua and lac), pineapple production and processing in tribal areas, herbal medicine and traditional health care, rural energy efficient devices, etc.

[3] Pioneers in innovation systems analyses in the developed countries include Nelson and Winter (1977), Freeman (1987), and Lundvall (1992).

[4] Both were case studies within a portfolio of rural innovation cases analysed in the UNU-INTECH/CPR/CRISP/XIIM B/ACTS project on "New Insights into Promoting Pro-poor Rural Innovation" in 2005-06.

[5] We shall not dwell upon the benefits or cost effectiveness of clean energy sources/devices such as biomass/biogas or solar power or wind energy. Innumerable authors and the Ministry (MNES) highlight these.

[6] Since 1999 a number of policy analysts have started to explicitly use the innovation concept in relation to developing country agriculture (See Hall *et al.*, 1998, 2000, 2001, Clark *et al.*, 2003). Douthwaite's important and widely cited book *Enabling Innovation,*(2002) demonstrates that the underpinning innovation principle that he refers to as 'learning selection', can be observed to be at work in technology contexts ranging from rice harvesters to computer software.

[7] For a brief analysis of DST's STARD Scheme, its unique features and achievements see UNU/INTECH 2005 (RIPWiG reporter).

Chapter 16

Collaborative links between academic and research institutions and industry for stimulating technological innovation and economic development: Need for science and technology policy initiatives and networking amongst NAM and other developing countries

M. Bandyopadhyay

Abstract

The capability for innovation and its diffusion can be developed through a dynamic process involving governments, academic and research institutions and industry with the help of a combination of schemes and programmes for research and development in priority areas of technologies, for the transfer and adaptation of technologies, and for the development of new technology-based industry, backed up by adequate support facilities and suitable institutional mechanisms.

Development of institution-industry links is now widely recognized to be of increasing significance in the overall technological and industrial competitiveness. Besides their obvious importance of being the repository of technical manpower, academic and research institutions also have their indirect impacts on the economy through technology transfer, creation of high technology spin-off firms, employment generation, continuing education and providing research and technical advice to existing industrial units.

It has been exhibited from the experiences of various advanced nations that it is possible to develop a knowledge based economic infrastructure within a country by creating a suitable environment for an organic relationship between the institutions and industry and for the utilization of the outputs of the higher education and research sector for industrial use. The NAM and other developing countries greatly suffer from the absence of a meaningful Science and Technology Policy that may stimulate the process of economic development by appropriate use of the S&T resources. The S&T Policy in the NAM and other developing countries should be oriented in such a manner that it should be able to motivate the academic and research institutions and the S&T persons to establish collaborative linkages with industry and create suitable infrastructure for technology transfer and stimulate technological innovation. A network of institutions and technology transfer professionals in the NAM and other developing countries should be created for exchanging information and sharing of knowledge and expertise in this regard.

Keywords: Academic and research institutions, Economic development, Institution-industry links, Gross expenditure on R&D (GERD), NAM and other developing countries, S&T policy, Technological innovation, Technology transfer.

Introduction

The key to economic development in a country lies in its ability to innovate and to effectively exploit innovations. In the competitive world, only those countries that possess high-class technologically skilled manpower, indigenous capability in state-of-the-art technologies, and a strong climate of innovation and entrepreneurship can be successful in governing their economy.

Technological innovation makes a significant contribution to develop competitive strength in the international markets. Technological change results in a change in production function or in the availability of new products. Since there is no satisfactory way to measure the rate of technological change directly, the economists often measure it by its effects. The rate of technological change depends to a large extent on the amount of resources devoted to the improvement of technology.

There have been serious concern and distrust about the cooperation between the academic and research institutions and industry. There were fears that in case of industry's involvement, they would tend to direct or interfere in scientific research in an institution. It was felt that universities would neglect fundamental research and education in case of a greater association with industry, while industry was cynical about the capability of scientists in dealing with practical techno-commercial problems.

The recent experience in the developed countries has shown that these issues have not come in way of improving relations between institutions and industry for their mutual benefit. They have been able to agree on undertaking collaborative projects without distorting each other's values. In the recent past, the universities, other higher educational institutions and research laboratories in many developed countries have taken a leading part by changing their role and strategy for research and by interacting actively with industry and thereby have been able to contribute greatly in stimulating technological innovation and economic development. For instance, US Federal laboratories are participating in several collaborative programmes with industry and under one of such initiatives *viz.* Cooperative Research and Development Agreements (CRADAs), 3,500 and 2,936 agreements were executed in the years 1996 and 2003 respectively (NSF, 2006). Similarly, the EU's 2010 goals in R&D, as set by the Lisbon Summit strategy is to achieve an R&D intensity of 3 per cent for the EU as a whole as well as having two thirds of R&D expenditure financed by the Business Enterprise Sector (Frank, 2005).

In contrast, most of the NAM and other developing countries greatly suffer from the absence of a meaningful Science and Technology Policy that may stimulate the process of economic development by appropriate use of the S&T resources. In order to correct this imbalance, the S&T policy in a developing country should be determined through consultation between the governments, institutions and industry, taking into account the specific technological needs and resources of the country.

Technological innovation and economic development

The concept of innovation

The term Innovation has a broad connotation, which refers to an idea perceived as new by the individuals. Thus it relates to all spheres of life. For industry, Innovation can be regarded as the total process of the inception of an idea into the manufacture of a new or improved product or a process, and finally of its commercial distribution in the market. Innovation is of importance not only for increasing the wealth of nations, but also in the more fundamental sense of enabling men to do things which have never been done before (Freeman, 1974).

The concept of innovation originates from the work of several classical economists who viewed that adoption of innovation leads to the generation of extra profit for surplus value. According to the *'theory of creative destruction'* advanced by Schumpeter, *"the capitalist economies are highly unstable and there is continuing change as a result of a succession of innovations, leading to ...... continual reorganization of the economic system...", in which the re-establishment of equilibrium is pre-empted by further rounds of innovation* (Schumpeter 1934). The Schumpeterian model gives a credible interpretation of competition in several knowledge-based industrial sectors *e.g.* microelectronics and information technology, chemicals, pharmaceuticals and biotechnology.

From an organizational perspective, Luecke and Katz give the definition: *"Innovation... is generally understood as the introduction of a new thing or method... Innovation is the embodiment, combination, or synthesis of knowledge in original, relevant, valued new products, processes, or services."* (Luecke and Katz 2003).

Invention and innovation

A clear distinction between 'Invention' and 'Innovation' is necessary in order to understand the processes involved in innovation fully. An invention is only an early part, a precursor of the entire process. An idea or the conception of a new approach to a problem or opportunity becomes an invention when the inventor gives it a physical form, such as, drawing, formula, prototype. Thus, invention is a scientific activity, which is not necessarily motivated by economic advancement. When such an invention is successfully put to use by society as a new commercial product or technological process, it becomes an innovation. The inventor develops a technique, which the innovator seeks to exploit for the creation of wealth. As such, innovation includes invention followed by various stages of its implementation such as research, development, production and marketing. Freeman distinguishes between invention and innovation in the following words:

" .... *An invention is an idea, a sketch or model for a new or improved device, product, process or system. While.... an innovation in the economic sense is accomplished only with the first commercial transaction involving the new device, product, process or system, although the word is used to describe the whole process....*" (Freeman 1974).

Technological innovation and R&D

Technological innovation is a dynamic and complex process of converting a research result into a commercial enterprise leading to a positive change in the existing economy. Success of technological innovation always requires the existence of three factors, namely, S&T capability, market demand, and an agent, which transforms this capability into goods and services to satisfy the demand. An entrepreneur is that agent with his motivation and skills to carry out the process of innovation. The OECD describes Technological Innovation as: *"Technological Product and Process (TPP) innovations comprise technologically implemented new products and processes and significant technological improvements in products and processes. A TPP innovation is considered implemented if it has been introduced in the market (product innovation) or used within a production process (process innovation)"* (OECD 1995).

Outcome of scientific and technological research positively influences innovation process. There are many examples of scientific discoveries or research results having been converted into commercially viable products and processes leading to industrial development. The economists have designated this effect of S&T research as *'technology push'* to innovation, whereas, influence of changing consumer preferences and market economies on the system forcing technological development has been characterized as *'market pull'* (Figure 16.1).

There are basically four stages involved in the cycle of activities that occur from the identification of a market need to its realization as a product. After the initial concept as an idea, an investigative stage of research may be necessary to prove its scientific validity. Product development, prototyping and business plans then follow, after which financing, full-scale production and marketing can be established. The feed back of consumer preferences often leads to an improved product for which the whole process may have to be repeated.

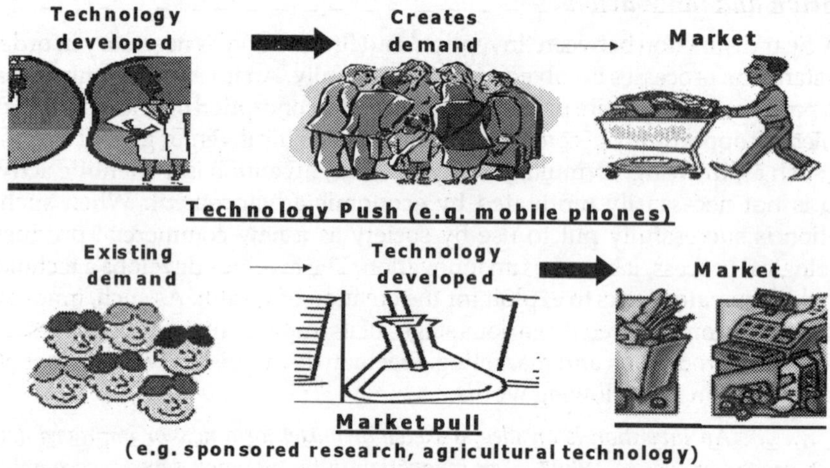

Figure 16.1: Technological innovation

Various models of innovation have been suggested in order to explain the complexity involved in the linkage between technology and market, and emphasis laid on different aspects of the process. There is no professionally agreed standard model for innovation. However, the primary events in the process of innovation can be schematically represented as given in Figure 16.2.

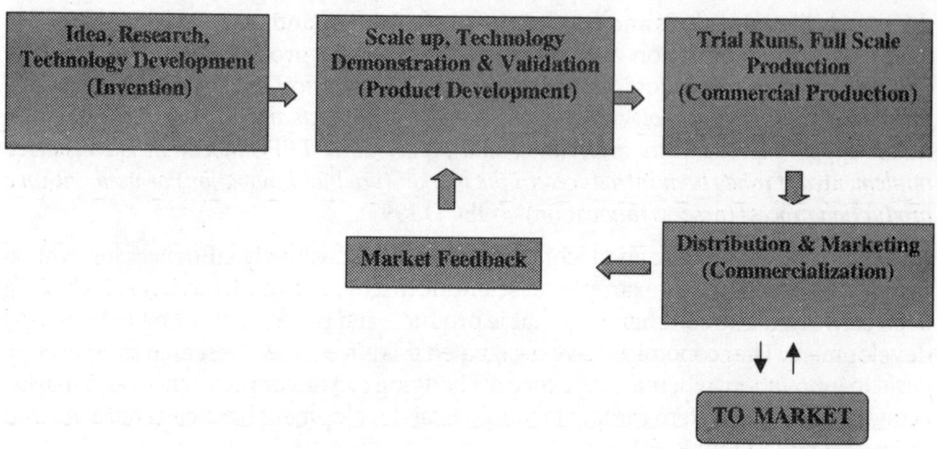

Figure 16.2: Suggested model of technological innovation

The role of technology in economic development

The role of technology in economic development has been extensively studied after R M Solow announced that nearly 90 per cent of the growth of per capita income in the United States during 1909 to 1949 was due to technological progress (Solow 1957). Kutznets concluded that large proportion of the rate of growth in per capita product must be assigned to an increase in efficiency in the productive resources, or to the effects of changing arrangements, or to the impact of technological changes, or to all the three (UNCTAD, 1990). The realization that the qualitative improvements or innovation in various elements of production function can bring about important changes in the output and productivity emphasizes the importance of the economic aspects of technological changes and innovation. However, it is difficult to find a clear-cut relationship between technological development and economic growth because quantitative measurement of these two factors is not possible. Efforts have been made by economists to relate technological progress and increase in productivity, taken as one of the measures of economic growth.

R&D expenditure and economic development

Several studies have attempted to relate technological advances to the investment made in research and development. Some empirical studies have been attempted considering R&D expenditure as an indirect measurement of technological development (Goddard *et al.*, 1987). R&D expenditures indicate both the relative importance of promoting S&T as compared with other objectives as well as the importance given to technological innovations in a country. R&D must compete for funding with other activities supported through government budget *e.g.* education, health, social security, defence etc. The allocation made by a government for R&D activities thus reflects its commitment to R&D as compared with other national programmes.

Though such an assumption may not always be valid, the higher rate of economic growth in the developed countries can be attributed to the higher R&D expenditure incurred by them. Thus it would be relevant to note that R&D expenditure of the European Union, Japan and the US has increased more than twofold between 1990 and 2003. The majority of EU member states R&D expenditure underwent sustained growth over the years and in the four years 1998-2002, the EU's R&D expenditure rose by 3.98 per cent a year on the average (Frank, 2005).

It may be noted from the data presented in Table 16.1 (UIS, 2005) that the industrially advanced countries spend much more on R&D as compared with the developing countries. The share of developed countries was more than 3/4th of the world's Gross Expenditure on R&D (GERD) in 2002 as against less than 1/4th only by the developing countries. The ratio of R&D expenditures to GDP is a useful indicator of the intensity of R&D activity in relation to other economic activities and can be used to gauge a nation's commitment to R&D at different points of time. The GERD/GDP ratio has been in the range of 2-4 per cent by most of the developed countries while the same was less than 1 per cent by almost all the developing countries in 2002.

Despite our limited knowledge on the exact contribution made by R&D to economic growth and the increase of productivity, there is undoubtedly a widespread conviction that it is large. The emphasis on growth as an objective of public policy has thus led to the advancement of science and technology and the ways of organising and stimulating it in both public and private sectors thus becomes an important focal point for studies.

Table 16.1: R&D expenditure in developed and developing countries

Country groups/ selected countries	*% World GDP*	*% World GERD*	*% GERD/GDP*
Developed countries	59.4	77.8	2.3
North America	23.8	37.0	2.7
Europe	27.9	27.3	1.7
USA	21.9	35.0	2.8
UK	3.3	3.5	1.8
Germany	4.7	6.7	2.5
France	3.4	4.2	2.2
Japan	7.3	12.8	3.8
Developing countries	39.1	22.1	1.0
China	12.2	8.7	1.2
India*	5.8	2.5	0.7
Brazil*	2.7	1.6	1.0
South Africa	0.9	0.4	0.7

* Data for 2000. All other data for 2002.

Collaborative links between institutions and industry

Institutions-industry linkage

With the realization of the importance of knowledge as a key source of economic growth and increasing competitiveness of individual firms and the need for the emergence of knowledge-intensive industries, the university-industry relationship has become an important global issue for the consideration of the policy makers. In view of the positive impact of technological development on economic progress as discussed above, it is imperative for the academic and research institutions to establish collaborative links with industry. Many authors (*e.g.* Leydesdorff and Etzkowitz, 1996) have emphasized on the "Triple Helix" theory underlying the need for the universities to form direct links with industry to maximize capitalization of knowledge, under which an institutional set up, comprised of university, industry and government is proposed to be set up.

An institution-industry link is an organizational structure *e.g.*, a department or a unit created to support collaborative activities between an institution and industry. The function of such a structure may widely vary from performing a simple role *e.g.*, liaison with an outside industrial unit or to work as a full-fledged organization with its own infrastructure, objectives and strategy housed in or around the parent institution. An institution-industry linkage structure, in fact, is a bridge between academics and business, in which the staff is multi-disciplinary and is competent to carry out applied, task-oriented projects, which could never be handled by an academic department alone.

An institution-industry link is a kind of public-private partnership which may be engaged in joint activities relating to development of technology or R&D, technology transfer, technology assistance etc. The partnership may also involve financial support through grant or joint funding. It may be in the form of a cooperative agreement, professional or student internship or exchange, technology business incubator, or science / technology parks.

Constraints in institution-industry linkage

As the expenditure on science and technology continued to rise, critics started probing whether the output of such a large Research and Development support system created by various countries has been commensurate with investment and expectation. There has been a persistent criticism that universities and higher academic institutions are closed-loop systems, which are driven by internal incentives of publications, awards and other kinds of recognition for high science activities with little or no relation to wider economic needs of the society (Smith, 1992). Universities have been regarded as *'Ivory Towers of Knowledge'* and government research laboratories have been found by many critics as *'black holes of public funding'*, because academics and researchers have generally failed to develop a workable link with industry for the commercial exploitation of indigenous research results and the resources of institutions.

There is some sort of a chicken-and-egg problem in the development of technological links between institutions and industry. The enthusiasm of many institutions is constrained by internal fears of threats to academic freedom, by uncertainty about how to manage relationship with industry, and by a lack of financial resources for developing necessary services. At the same time, much of industry continues to doubt the ability of academics to provide practical problem-solving assistance, and so holds back from helping to develop higher education's technology transfer activities.

Unfortunately, in many cases, there is some truth in these apprehensions. Academic and research institutions, particularly in the NAM and other developing countries, have not generally sought, or played an effective role in technological innovation and economic development-related programmes

Motivations of industry to collaborate with institutions

There is now a growing awareness among industry of the benefits, which they might get by collaborating with academic and research institutions. In the competitive market, industry needs more and more access to technological information and resources, which they can get from institutions at a much lower opportunity cost than by attempting on entirely in-house development. The formation of links with institutions is related to the technological needs of the industry at different stages of the innovation process. These links are also a function of the competitive concerns of the participating companies, and are related to the technological opportunities and competitive strategies. Industry now recognizes that innovation can occur more efficiently in some areas outside the company so it reallocates its finances and manpower to pay for and then absorbs advances made elsewhere, and thereby achieves economies of scale. A number of studies have shown that there is a relationship between the collaboration of a company with a University and the ability of the former to innovate (Laursen and Salter, 2004). Lööf and Broström analyze the benefits that cooperating firms could have by collaborating with universities and undertaking innovations by increased propensity to patent and higher sales (Lööf and Broström, 2006).

Lee presents three most important reasons for a firm to seek partnership with Universities *viz.* (*i*) access to new research (*ii*) development of new products and (*iii*) maintaining a relationship with the University (Lee, 2000). The important motives, which a firm may have for establishing collaborative links with institutions, are listed at Table 16.2 (Charles and Howells, 1992 and others).

Table 16.2: Motivations of industry to collaborate with institutions

Sl.No.	*Motivation of industry*
1.	Saving of cost
2.	Inadequate scale of in-house R&D
3.	Development of standards
4.	Access to new areas of expertise
5.	Strategic benefits and opportunities
6.	Participation in public programmes
7.	Access to public sector markets
8.	Internal problems
9.	Carrying out trials
10.	Access to equipment
11.	Testing and calibration
12.	Student projects
13.	Use of facilities

Motivations of institutions to collaborate with industry

Various motivations for the academic and research institutions to interact with industry can be summed up as listed in Table 16.3.

Table 16.3: Motivations of institutions to collaborate with industry

Sl.No.	*Motivation of institutions*
1.	Commercial exploitation of the outputs of research carried out in the institution
2.	Additional income
3.	Improving the image of the institution before the society
4.	Improving the employment opportunity for the outgoing graduates
5.	Providing financial incentives to the researchers/inventors
6.	Promoting S&T based entrepreneurship

Howells *et al.* in their survey, rank following motives of universities in the UK to collaborate with industry (Table 16.4) (Howells *et al.*, 1996).

Table 16.4: Motivations of UK universities to collaborate with industry

Sl.No.	*Motive*
1.	To access industrial funding
2.	Collaboration with industry is a strategic institutional policy
3.	To find an exploitation outlet for research capabilities
4.	To access complementary expertise
5.	To provide an outlet for research results
6.	To access state-of-the-art equipment and facilities
7.	To contribute to local economy
8.	Government policy and/or political pressure
9.	To contribute to UK economy

Classification of institution-industry linkage mechanisms

A large number of studies have been carried out to identify the nature and variety of institution-industry links, which present many common themes and emphasize various aspects of collaborative mechanisms. Various institution-industry linkage systems can be classified into following types as given in Table 16.5 (Stankiewicz, 1986).

Table 16.5: Types of institution-industry linkage systems

Sl.No.	*Types of Links*	*Examples*
1.	General support link	– Non-research sponsorship, – Links with business clubs – Industrial membership of advisory boards for Research Centres
2.	Education and training links	– Course development for commercial and industrial organizations – Sponsorship of students – Endowment of professorial and lecturing posts.
3.	Research and development links	– Sponsored research – Joint research with external organizations
4.	Commercialization of institution infrastructure	– Provision of consultancy services, – Conference facilities – Patenting and licensing of research ideas – Creation of spin-off companies

S&T institutions and industry in various countries have established a variety of linkages depending upon the specific requirements of the partners of the links, government policies on promoting such facilities and investments made by the partners and/or government in developing the same. Different typologies of university-industry relationship have been suggested by many experts, such as, Howells *et al.*, to identify four main forms of university-industry linkages in the United Kingdom, *viz.*: i) through informal contacts and spin-outs from university departments; ii) from contract and collaborative research performed by universities on behalf of industry; iii) as a result of property-led initiatives in the form of science parks; and iv) via the commercial exploitation of university research through the management and licensing of intellectual property rights (IPR) (Howells *et al.*, 1996). Based on various mechanisms currently operating successfully in different countries, institution-industry links may be classified into following three distinct groups (Table 16.6) (Bandyopadhyay, 1995; Monck *et al.*, 1988).

Table 16.6: Classification of linkage mechanisms

Sl.No.	*Type of mechanisms*	*Examples of mechanisms*
1.	Institution-based linkage mechanisms: linkages as extensions of institutions	– Industrial Liaison Office (ILO) – Technology Transfer Office (TTO) – Research Contract – Consulting agreement – Affiliate programme – Research consortia – Industry cooperative – Technology brokerage company – Teaching company scheme – Exchange of personnel – University-based institutes serving industrial needs
2.	Linkages through the creation of additional organizational infrastructure	– Cooperative research centre – Centre of excellence – Incubation centre – Technology incubator/business incubator
3.	Property based initiatives in/around institutions with investment by government bodies, institutions, industry, venture capital funding agency, either solely or jointly	– Science park – Innovation centre – Technology park – Research park – Industrial park – High technology developments – Technopoles – University company

Need for S&T policy initiatives in the NAM and other developing countries

Problems and constraints in S&T policy formulation

It is important to have a strategically designed science and technology policy at the national level, commensurate with the challenges of the globalizing society. But, in many NAM and other developing countries, science and technology policies are rather focused on research in basic sciences, and do not address to the needs of the local industry and economic conditions. Moreover, most developing countries do not have proper human resource to design their own national S&T policies and thus they lack experience of S&T policy formulation and implementation.

For a variety of reasons, the Science and Technology infrastructure and capabilities created in the NAM and other developing countries have, in most cases, failed to fulfil their potential. In these countries, the real technological requirements of the productive sector have not been adequately appreciated in the formulation and implementation of Science and Technology policies. The reasons may stem from financing, management problems, economic factors and policy orientations.

The need for S&T policy orientation

The distinction between the current state of economy in a developing country and that in an industrially developed country is obvious. The problems to which solutions are to be found in these countries are also different. As such, the mechanisms, which have been found successful in the countries like the USA and UK, may not be entirely replicable for the developing countries. However, there were certain basic considerations under which various programmes on institution-industry interaction were undertaken in the developed countries and such considerations are applicable to the developing countries as well.

Innovation theories contain insights into how and why technological capabilities are developed in the industrial sectors of the advanced countries. In effect they give some new dimensions of meaning to the concept of accumulation of local technological capabilities, which has come to play an important part in technology policy in developing countries. The theoretical framework that has been developed from empirical studies of innovation and behaviour of industry in the developed countries could give useful guidelines for policy studies in the developing countries.

Under the present circumstances, the future of S&T institutions in the NAM and other developing countries lies increasingly with strategic collaboration with industry for stimulating technological innovation and accelerating economic development. However, there is a great diversity among different institutions in terms of their size and facilities and among different developing countries in terms of their level of S&T capabilities and infrastructure. As such, any attempt to devise a uniform model for technology management and linkage activities would not be advisable. Different approaches may be appropriate for different institutions and different countries. Whatever may be the approach, it requires a clear perspective planning, discipline, motivation, commitment, energy and resources and most importantly, the right kind of people. Finally, all such actions must be supported through a set of policy initiatives.

Suggested policy initiatives

Some of the important issues which are required to be seriously considered while formulating the S&T Policy of a developing country are summarized below (Bandyopadhyay, 1992; Jain and Bandyopadhyay, 2001).

Increase in investment on S&T

1. The most important step that should be taken by the NAM and other developing countries is to make adequate allocation of funds required for S&T capacity building, research and development and human resource development.
2. Public sector funding plays an important role in encouraging the emergence of a technology-based economic infrastructure. An organizational infrastructure in or around an institution such as innovation centre, business incubator or science park is an important tool towards achieving this goal. Though such a facility should be driven by a 'for profit' attitude, it is essentially a 'non-profit' making development oriented organization. Thus, liberal funding from government and government owned developmental agencies for establishing such infrastructure is essential.
3. Till the technology transfer activities of the institutions become self-sufficient, the respective governments should make specific funds available to assist the institutions in the planning, launching and operation of the same.
4. Abundant venture capital funding with liberal lending terms should be made available for the creation of new technology based enterprises and spin-off firms arising out of institution-industry linkage. Venture capital support may be in the form of participation in equity, purchase of debentures, soft term or low/no interest loans and technology development loans.

Motivating the institutions

1. Governments in the NAM and other developing countries should encourage academic and research institutions to undertake innovative activities and provide technology transfer services to industry through appropriate policy decisions with greater flexibility wherever required.
2. In addition to maintaining the broader educational objectives and supporting basic research, S&T institutions must be supported for facilitating the transfer of technology and knowledge into the wider economy. Higher education and research system must take a fuller role in providing technological assistance to industry. A credible focus and entrepreneurial confidence on the part of the institutions are needed to drive the phenomenon to succeed. For this, a multi-pronged approach by undertaking a number of programmes simultaneously in an institution would be effective. Each institution should adopt one or more forms of in-house mechanisms for technology transfer and providing services to industry. Business incubators and science parks should also be established in or around the institutions to complement such activities and provide infrastructural and support facilities to entrepreneurs.
3. Institutions should be allowed to take part in commercial activities, to launch fully owned or partnership companies or to participate in equity of other companies.

4. From the beginning, explicit and equitable rules should be framed about how technology transfer income is to be shared between the individual scientists, departments and the central institution.

Motivating the industry

1. Presently, investment by industry on R&D is very poor in the NAM and other developing countries. Private sector investment in institutions on research and technology transfer programmes should be encouraged by appropriate policy measures.
2. Financial support may be provided by way of subsidy on the expenditure incurred by industry for their participation in government R&D projects, and also for in-house work on prototype development, pilot plant trials, testing and evaluation etc
3. Tax rebates and incentives may be given to companies for sponsoring R&D projects in institutions, investments made in in-house R&D and expenditure on manufacturing activities in selected high technology areas.
4. Customs duty exemption may be given for the import of equipment, instruments and consumables required for carrying our R&D projects.
5. Awards and recognition would have a positive impact on indigenous technology development, innovation and entrepreneurship.

Motivating institute personnel and human resource development

1. Research scientists and members of the staff in an institution must be suitably motivated to accept assignments from industry. Institutions should not tie up their scientists with rules about the amount of outside work they can undertake or limit their earnings from such work. Internal institutional rules on academic employment, regulations regarding owning property, on taking equity in companies, on receiving commercial income; patent law, conditions for intellectual property rights of the inventor via-a-vis institution should be made flexible or problems sorted out.
2. Careful thoughts must be given to the staff requirements for technology transfer management. The special skills needed for effective services in technology evaluation, marketing, and intellectual property protection and negotiation merit more attention. Suitable programmes on training and retraining of staff should be undertaken.
3. Developing countries should be committed to retaining high-level scientists, stimulating them, and providing funds and other support to encourage and maintain their productivity. Proactive policies to reverse the 'Brain Drain' should be adopted so as to utilize the expertise of the non-resident scientists and professionals in high technology areas.

Choice of technology and promotion of technology based enterprises

1. Developing countries should lay emphasis on making use of a mix of intermediate and appropriate technologies on the one hand and high and

leading edge technologies on the other. While efforts must be made to build indigenous technological capability by strengthening science and technology infrastructure, modern technologies may be imported selectively from developed countries to keep pace with the world wide development in various fields.

2. Science and Technology-oriented small business development training programmes should be organized and supported for converting the higher education and research system into a breeding ground for new technology-based small industries.
3. Adequate policy measures should be adopted for promoting small technology based enterprises and for giving them preference to multinationals and big industry houses. Small enterprises should be helped to build up long term research and technical collaborations with academic and public funded research establishments by providing incentives, funds and other promotional assistance.

Other policy measures

1. In order to facilitate rapid exchange of information and expedite technology transfer from R&D institutes and universities to industry, the national S&T information system including technology data collection, application and distribution must be strengthened.
2. A national level system on standardization, quality control and accreditation should be established to promote the quality of industrial products, standards and measurements, and inspection and testing of products.
3. Appropriate Intellectual Property Rights Protection law should be put in place to encourage new invention and innovation activities by protecting the rights of the inventors.
4. A network of academic and research institutions and S&T Policy experts of the NAM and other developing countries should be established for sharing of expertise, information and data on institution-industry linkage, technological innovation and related issues and to fill the gaps that are existing in the S&T policy formulation in different countries. The platform of the NAM S&T Centre could be useful for the developing countries for such an exercise.

Networking amongst NAM and other developing countries on S&T policy and technological innovation

Most of the NAM and other Developing Countries suffer from serious constraints of the lack of adequate expertise and non-availability of suitably trained manpower on the formulation and implementation of science and technology and innovation policies. Since developing these capabilities in the respective countries independently would be cost intensive and take a considerable time, it is imperative that the developing countries should cooperate amongst themselves in sharing of information and expertise that are presently available with them and in undertaking training programmes and joint projects on S&T policy formulation, technology transfer and technological innovation. For this, a network of S&T policy makers and technology transfer professionals, government agencies engaged in S&T policy and the S&T institutions in general should be created. The vast pool of S&T institutions and the experts, trainers and resource persons presently associated with the NAM S&T Centre should be made use of for this purpose.

Networking through the NAM S&T Centre

Consequent upon the decisions taken in various summits of the heads of states and governments of the countries of the Non Aligned Movement (NAM), the Centre for Science and Technology of the Non-aligned and Other Developing Countries (NAM S&T Centre) was established in 1989 as an inter-governmental organization with its headquarters in New Delhi. The main objectives and functions of the Centre are: promotion of mutually beneficial collaboration among scientists and technologists and scientific organisations from the non-aligned and other developing countries, helping in the establishment of links between national and regional centres, acting as a clearing house of information regarding technological capabilities of individual countries, maintaining a registry of S&T experts of high calibre, stimulating and promoting joint R&D projects etc. The Centre provides a very good platform for networking amongst the NAM and other developing countries through its following initiatives and activities.

1. The Centre has a vast network of S&T ministries, departments, agencies and institutions and individual experts and professionals through its Focal Points in its 40 member countries and also in 53 other countries, which are not yet its members. About 4000 entries pertaining to such organizations and individuals are already available in the database of the Centre.
2. The Centre closely cooperates with several other international organizations and non-governmental organizations for the formulation and implementation of international collaborative projects.
3. In order to encourage participation of the scientific agencies and institutions and industry/industries associations in the activities of the Centre and also to promote public-private partnership in S&T, a 'NAM S&T–Industry Network' has been set up by the Centre, which has been joined by several S&T organizations and institutions as its members.
4. The Centre has initiated a scheme to encourage affiliation of scientists and researchers of the developing countries with the Centres of Excellence in other countries. Under this scheme, a cooperation programme with the Academy of Scientific Research and Technology (ASRT) of Egypt and a joint fellowship programme with the International Centre for Chemical and Biological Sciences (ICCBS), HEJ Research Institute of Chemistry, University of Karachi, Pakistan have been initiated.
5. The Centre has instituted a NAM S&T Research Fellowship Award, which provides scientists and scientific institutions in member countries an

opportunity to establish closer cooperation with fellow scientists in various fields of science and technology.

6. Finally, the Centre provides ample opportunities to the developing countries for capacity building, human resource development and interaction among scientists and professionals by organizing international seminars, workshops, roundtables and training programmes on various priority areas. So far, about 2500 scientists from 90 countries have benefited from such activities.
7. The Centre has in its credit an impressive list of technical publications in various S&T areas including those related to S&T policy issues *e.g.* IPR, S&T popularisation, disaster management, technology transfer etc., which are widely disseminated to the institutions and scientists in the developing countries besides their availability in the open market.

It is proposed that a collaborative project may be taken up under the aegis of the NAM S&T Centre to establish a 'Technological Innovation Network' or expand and enlarge the scope of its existing 'S&T-Industry Network' in order to promote collaborative linkages between academic and research institutions and industry for stimulating technological innovation in the NAM and other developing countries.

Conclusions

Science and technology in NAM and other developing countries suffer from the weak institutional infrastructures, inadequate financial resources, and the absence of an appropriate S&T human resource development programme. One of the major factors for the marginal growth of science and technology in most of the developing countries is the lack of planning and management of appropriate schemes and programmes on research and development in priority areas. In fact, so far, only a limited number of developing countries have attempted to formulate and adopt a meaningful national policy on science and technology. Further, the industry in the developing countries is faced with the problem of high inflation, sluggish production, and above all, lack of international competitiveness due to serious technological backwardness resulting in low production efficiency and poor quality of products.

Unlike the advanced nations, the concept of linking technological resources in academic and research institutions with industry has not percolated down to the NAM and other developing countries which still lack technological capability and largely depend on obsolete technologies imported from abroad long back. In some of these countries, even though there is a good degree of S&T base, transfer of technology from S&T institutions to industry is poor for want of suitable support mechanisms and motivation. Whatever the industrial base these countries might have, they suffer from the disadvantages of technological inadequacies. The governments in these countries are yet to take a worthwhile policy initiative to harness the potential of science and technology for the promotion of innovation in industry.

It is necessary for the decision makers in the NAM and other developing countries to undertake appropriate policy measures and reorient their existing S&T policies with a view to motivating the S&T institutions and industry to collaborate with each other for mutual benefit and for facilitating technological innovation and economic development and to create necessary facilities and infrastructure on institution-industry linkage. The lack of expertise and the prevailing disparities amongst the developing countries must be minimized through sharing of expertise and exchange of information, for which an active network of S&T institutions, S&T persons and policy professionals should be established. The vast pool of individual scientists, experts and professionals, academic and research institutions, and government ministries, departments and agencies dealing with S&T policy and management in various countries, which are closely associated with the NAM S&T Centre should be utilized for this purpose.

References

Bandyopadhyay, M., 1992. The Role of Academic and Research Institutions for introducing Technological Innovation in Small Scale Industry; Queen Elizabeth House Senior Visiting Fellowship Dissertation, University of Oxford, England.

Bandyopadhyay, M., 1995. Economic Development by Utilizing the Expertise of Academic and Research Institutions in Britain: A Role Model for India; in the book "Science, Technology and Industrial Development in India", Ed. By S.C.Pakrashi and G.P.Phondke; Wiley Eastern Ltd., New Delhi

Charles, D. and Howells, J., 1992. Technology Transfer in Europe: Public and Private Networks; Pinter Publishers, London.

Freeman, C., 1974. The Economics of Industrial Innovation; 2nd Edition; Frances Printers, London.

Goddard, J.B., Charles, D., Howells, J. and Thwaites, A., 1987. Research and Technological Development in the less favoured regions of the Community (STRIDE); Final Report; Office of the Official Publications of the European Commission, Luxembourg.

Howells, J., Nedeva, M., and Georghiou L., 1998. Industry-Academic Links in the UK, HEFCE ref 98/70, Higher Education Funding Council for England, Policy Research in Engineering, Science and Technology, Scottish Higher Education Funding Council and Higher Education Funding Council for Wales, December 1998.

Jain, B. K., and Bandyopadhyay, M., 2001. Emerging Challenges before Higher Educational Institutions–Need for Reorientation of Activities; VISION-The Journal of Business Perspectives, Gurgaon, India; Vol. 5, No. 2 (July-December 2001).

Laursen, K. and Salter, A., 2004. Searching low and high: what types of firms use universities as a source of innovation? Research Policy 33, 1201–1215.

Lee, Y.S., 2000. The Sustainability of University-Industry Research Collaboration: An Empirical Assessment, The Journal of Technology Transfer, Vol 25, No 2, pp. 111-133.

Leydesdorff, L., and Etzkowitz, H., 1996. Emergence of a Triple Helix of university-industry-government relations, Science and Public Policy Vol 23151.

Lööf, H. and Broström, A., 2006. Does Knowledge Diffusion between University and Industry Increase Innovativeness? Working Paper presented at the World Bank Workshop in Cambridge, September 2005, Journal of Technology Transfer, 2006.

Luecke, Richard, Ralph Katz, 2003. Managing Creativity and Innovation. Boston, MA: Harvard Business School Press.

Monck, C., Quintas, P., Storey, D., and Wynarczyck, P., 1988. Science Parks and the Growth of High Technology Firms, Croom Helm, Beckenham.

NAM S&T Centre Website: http://www.namstct.org.

NSF, 2006. Science and Engineering Indicators 2006; National Science Foundation, February 2006 (http://www.nsf.gov/statistics/seind06/).

OECD, 1995. The Measurement of Scientific and Technological Activities. Proposed Guidelines for Collecting and Interpreting Technological Innovation Data. Oslo Manual. 2nd edition, DSTI, OECD/European Commission Eurostat, Paris 31 Dec 1995.

Schumpeter, J., 1934. The Theory of Economic Development. Cambridge, MA: Harvard University Press.

Segal, Quince and Wicksteed, S.Q.W., 1988. Universities, Enterprises and Local Economic Development: An Exploration of Links; A Report for the Manpower Service Commission; HMSO, London.

Frank, S., 2005. R&D Expenditure in the European Union; Statistics in Focus, Science and Technology 2/2005, R&D Statistics; European Communities (http://www.europa.eu.int/comm/eurostat/).

Smith, B.L.R., 1992. Strengthening the US University Research System in the book "Research and Higher Education: The United Kingdom and the United States" Ed. by T.G. Whiston and R.L. Geiger, The Society for Research into Higher Education, Open University Press, Bristol.

Solow, R.M., 1957. Technical Change and the Aggregate Production Function; Review of Economics and Statistics 39 (August 1957): pp. 312-20.

Stankiewicz, R., 1986. Academics and Entrepreneurs: Developing University-Industry Relations; Frances Pinters, London.

UIS, 2005. UNESCO Science Report 2005; UNESCO Institute of Statistics (UIS). Montreal, Canada.

UNCTAD, 1990. Transfer and Development of Technology in Developing Countries; A Compendium of Policy Issues; The UN Conference on Trade and Development (UNCTAD), New York.

Website of wikipedia (http://en.wikipedia.org/wiki/Innovation# Scholarly_ definitions).

Chapter 17

Protection of intellectual property: Technology acquisition, adaptation and diffusion

Vinod Kumar Gupta

Abstract

Given the centrality of technology to economic and social development, and the need for technology acquisition by developing countries as a means of furthering development, it is essential that countries should be able to benefit from the transfer, acquisition and diffusion of technology, particularly, in strengthening national endogenous technological capacities. In the emerging era of globalization and liberalization of investment climate, it would require abilities to acquire and adapt technologies from both external and internal sources as well as to create new technology in order to successfully compete in the international markets. The issues of intellectual property protection are significant both in acquiring technology through indigenous R&D efforts as well as from abroad, particularly, since the signing of agreement on Trade–related aspects of Intellectual Property Rights under WTO. This paper describes the basic concepts of intellectual property (IP) and delves upon the salient aspects of intellectual property protection in the specific context of technology acquisition from external sources, internal sources and in-house R&D and technology development. The critical aspects of negotiating intellectual property in finalizing the technology licenses have been described. In case of acquisition from internal sources, it points out the importance of protection of the results of research, and sharing of IP during S&T collaborative projects. Some of the aspects relevant to taking advantage of IP in building technological capabilities have been indicated.

Keywords: Intellectual property, Technology acquisition, Technology adaptation, Technology diffusion, Technology transfer.

Introduction

Technological competence is critical to, both development of an economy and its competitiveness in the marketplace. Most developing countries acquire technology in ways that include the import of capital goods and components; design specifications and technical assistance from foreign buyers; foreign technology licensing; foreign direct investment and strategic alliances; foreign education and training; mergers and acquisitions of firms, and local research and development (R&D) efforts. The provision of high-technology services – such as testing, design, calibration, and quality control for manufacturing – is another channel for the acquisition of foreign technology. The abilities of countries to acquire and absorb knowledge from elsewhere and then make use and adapt it for their own purposes is of importance, and to a large extent depends upon the capabilities of the enterprises in the economy. In particular, the small and the medium enterprises have an important role to play in the process of technology acquisition, adaptation and diffusion.

The issues of intellectual property protection are significant both in acquiring technology through indigenous R&D efforts as well as from abroad. In recent times, specially, since the signing of agreement on Trade–related aspects of Intellectual Property rights under WTO, the awareness about the protection of intellectual property is growing amongst the firms, R&D institutions, and policy makers of most developing countries, which has further required necessary skills in order to take advantage of IP during the process of technology acquisition, adaptation and diffusion. The present paper delves upon some of these issues, particularly, while negotiating technology acquisition or transfer agreements.

Basic concepts of intellectual property rights

The central objective of intellectual property protection is to encourage creative, inventive and innovative activities resulting from the process of research and development by legally providing exclusive rights for a limited period within a designated market. The rights are given to the inventors and creators to produce and sell a product, service or technology those results from such intellectual creations (Alikhan, 2000). An inventor will be encouraged to share his invention with the society if he is assured that his rights are protected.

In principle, the intellectual property is defined as a "product of the mind". It is similar to the property consisting of movable or immovable things like a house or a car, wherein the proprietor or owner might use his property as he wishes and nobody else can lawfully use his property without his permission. The Convention establishing the World Intellectual Property Organization (WIPO) in 1967, one of the specialised agencies of the United Nations System defined the "intellectual property" to include rights relating to (WIPO, 1997):

1. Literary, artistic, and scientific works,
2. Performance of performing artists, phonograms and broadcasts,
3. Inventions in all fields of human endeavour,
4. Scientific discoveries,

5. Industrial designs,
6. Trademarks, service marks and commercial names and designations, and
7. Protection against unfair competition and all other rights resulting from intellectual activity in the industrial, scientific, literary or artistic fields.

The main elements of intellectual property, thus, include patent, copyright, trademark, industrial design, geographical indication, layout design (topographies) of integrated circuit, protection of undisclosed information and the control of anti-competitive practices. These concepts are briefly defined below.

Patent

The patent relates to creations borne out of inventions, which are solutions to technical problems. A patent is a government granted and secured legal right to prevent others from 'practicing' *i.e.* making, using, selling or importing the inventions covered by the patent. An invention is patentable if it is new, involves an inventive step, *i.e.* it is not obvious in the sense that it will not occur to any specialist, if such a specialist was asked to find a solution to the particular problem, and is industrially applicable in the sense that it can be industrially manufactured and used (GOI, 1970, GOI, 1994). Examples are a new ballpoint pen or a new drug molecule or a new fire-fighting device.

Trademark

A trademark is an identification symbol, which is used in the course of trade to enable the purchasing public to distinguish one trader's goods from the similar goods of other traders. The public make use of trademarks in order to choose whose goods they can purchase. If they are satisfied with the purchase they can simply repeat their order by using the trademark (GOI, 1999a). In order to be eligible for protection a mark is to be distinctive of the proprietor so as to identify the proprietor's good, for example, KODAK for photographic films or COMPAQ for personal computers. Where a trademark is used in connection with services, it may be called 'service mark'. For example, service marks are used by hotels, airlines or travel agencies.

Design

The expression 'Design' means only the features of shape, configuration, pattern or ornament applied to any article by any industrial process or means whether manual, mechanical or chemical, separate or combined which in the finished article appeal to and are judged solely by the eye (GOI, 2000a). The design means the features applied to an article and not the article itself. The features are conceived in the author's intellect who gives those ideas conceived by him a material (visual) form as a pictorial illustration, or as a specimen, prototype, or model. By registration under the Act, the features are protected as design. The protection is given for independently created industrial designs that are new or original, for example, the distinctive shape of a coke bottle or a pen or a textile design.

Know-how/undisclosed information

Know-how is the confidential information that may be in the form of an aggregation of known processes/procedures, an accumulation of data, a secret formulation, or a combination of any of these. Know-how is often transferred together with licensing of patents and under transfer of technology arrangements. Individuals should be able to prevent information lawfully within their control from being disclosed to, acquired by, or used by others without their consent in a manner contrary to honest commercial practices (Gupta, 1999).

Geographical indication

A geographical indication is an indication that identifies a good as originating in a territory where a given quality, reputation or other characteristics of the good is essentially attributable to its geographical origin. The geographical indication may be simple and need not be associated with any obligation other than of production in a specified area, *e.g.* Darjeeling tea, French wine, Basmati rice, etc. It prevents unauthorised persons from using the protected geographical indication for products not from that region or from misleading the public as to the true origin of the product. The geographical indication the use of which is likely to deceive or cause confusion or contrary to law cannot be registered (GOI, 1999b).

Layout design in integrated circuit

Large-scale integration of a number of electrical functions in a very small component becomes possible due to advances in semi-conductor technology. Integrated circuits are creations of human mind. There is a need for the creation of new layout designs, which reduce the dimensions of existing integrated circuits as also increase their functions. For example, the law in India provides protection for designs that are original, not commercially exploited in the country or any convention country and is inherently distinctive. The law prohibits reproducing, importing, selling or otherwise distribution for commercial purposes of a protected layout design or an integrated circuit in which a protected layout design was incorporated or an article incorporating such an integrated circuit only insofar as it continues to contain an unlawfully reproduced layout design (GOI, 2000b).

Copyright

The copyright protects only the form of expression of ideas, not the ideas themselves. The creativity protected by copyright is creativity in the choice and arrangement of words, musical notes, colours, shapes etc. only for original works. The copyright protects the exclusive rights to authorise others to use the protected works *e.g.* reproduction, performing, recording, broadcasting, and translation (GOI, 1995). The examples for protection include books, research publications, pamphlets, lectures, dramatic or musical works, cinematographic works, paintings, photographic works, works of applied arts, and maps. In laws of many countries, the computer programmes and databases are also protected under copyright.

Unfair competition

In the context of industrial property protection, acts of unfair competition are those that create confusion with the goods or the industrial or commercial activities of a competitor, false allegations in the course of trade that discredit the goods or the industrial or commercial activities of a competitor, or yet, indications or aggregations, the use of which in trade is likely to mislead the public as to the characteristics of the goods. The law prohibits such acts (GOI, 2001a).

Protection of plant varieties

There is another kind of intellectual property in the development of new plant variety that is protected under Plant breeders' rights (PBRs) as a *sui generis* system. The laws for protection may vary across countries. In India, the PBRs are granted to breeders of new, distinct, uniform and stable plant varieties. The novelty is not essential for the protection of an extant or farmer's variety. The farmers are entitled to save, use, sow, resow, exchange, share or sell his farm produce including seed of a variety protected under the Act in the same manner as he was entitled before except that he cannot sell branded seed of such registered variety (GOI, 2001b).

International regime for IP protection

The international agreement on Trade related aspects of Intellectual Property (TRIPs), under World Trade Organisation (WTO) in 1995 established the minimum international standards for the protection of intellectual property by mandating the extension of patentability to virtually all fields of technology, prolonging the term of patent protection for a uniform term of twenty years, and by providing legal recognition of the patentee's exclusive rights to import the patented products (WTO, 1995). The patent rights are enjoyable without discrimination as to the place of invention, the field of technology and whether products are imported or locally produced. The coverage of protection is also extended to micro-organisms and the plant varieties. All the member countries are required to harmonise their intellectual property regimes to these minimum standards of protection. The agreement tends to strike a balance between enforcing intellectual property rights and meeting the technological needs of developing countries. For example, the agreement stipulates that countries may, in formulating or amending their laws and regulations, adopt measures necessary to protect public health and nutrition, and to promote the public interest in sectors of vital importance to their socioeconomic and technological development, provided that such measures are consistent with the provisions of this Agreement. Also, the agreement provides countries with the freedom to adopt measures that may be needed to prevent the abuse of intellectual property rights by the rights holders or resort to practices which unreasonably restrain trade. On enforcement, it lays down detailed provisions on civil, administrative and criminal procedures and remedies. This development has considerable implications for the intellectual property issues relating to the international technology transfer and the process of acquisition of local technological capability by developing countries.

Protection of intellectual property

Most scholars argue that a stronger intellectual property system is one of the cornerstones of modern economic policy (David, 1996, Gupta, 2004, Maskus, 2000, World Bank, 1999). In contrast, other scholars argue that intellectual property rights are constraints on development (Martin, 1995, Patel, 1993). It has, however, been difficult to empirically prove that these rights make a direct contribution to economic development. The report of the UK Commission on Intellectual Property Rights (CIPR) for developing countries (CIPR, 2002) concludes that in most developing countries, with a weak scientific and technological infrastructure, intellectual property protection at the levels mandated by TRIPS is not a significant determinant of growth. Weak IPRs restrain the development of innovative local enterprises and discourage the acquisition of high-quality foreign technologies. However, the developing countries are increasingly recognizing that strengthening of IPRs promises substantial net benefits over the long term (Maskus, 1999).

Some of the aspects in which protection of IP is likely to benefit the process of technology acquisition are indicated below.

1. The strength of patent regime is likely to ensure greater contract certainty between supplier and the receiver and by providing more protection for commercializing technologies in local market. It is likely to enhance the flow of high technology goods and the superior technologies in the developing countries.
2. The effective deployment of technology and improved operational techniques invariably require dissemination of technical information, which can be acquired through training including on-the job training. The adequate protection of confidential information or trade or training manuals or guidelines which might be protected under copyright would enable the firms acquiring technology to receive relevant IP in confidence which might enhance their absorptive and adaptive capacities.
3. Quality assurance is significant to safeguard the interests of consumers. Also, the widespread sale of counterfeit goods can ruin the reputations of enterprises, which is achieved at high costs. In principle, the effective trademark enforcement both raises the average quality of products over time and provides a wider range of qualities from which consumers may choose. This is particularly important in food products, beverages, cosmetics and medicines where counterfeit goods can be hazardous.
4. Similarly, inadequate copyrights may encourage piracy in creative industries like music, films, software. Counterfeit production and piracy is likely to be reduced over time by trademark and copyright enforcement.

Technology acquisition, adaptation and diffusion

Technology transfer is the movement of technology from the holders or developers of the technology to the recipients of it. Given the centrality of technology to economic and social development, and the need for technology acquisition by developing countries as a means of furthering development, it is essential that countries should be able to benefit from the transfer, acquisition and diffusion of technology, particularly, in strengthening national endogenous technological capacities. In the emerging era of globalization and liberalization of investment climate, both the public and private sectors have to play an important role and would require abilities to acquire and adapt technologies from both external and internal sources as well as to create new technology in order to successfully compete in international markets. Salient aspects of intellectual property protection in each of these respective contexts are discussed below.

Acquisition from external sources

There is a subtle difference between the terms "transfer" and "acquisition". Simply, transferring a technology from external sources is likely to be one time relationship while during the process of technology acquisition; a long-term relationship is established between the supplier of the technology and the firms acquiring the technology. The later enables these firms to build capabilities. The acquisition may include disembodied technology in the form of patent or know-how, trademarks, design, or services with a technological content or embodied technology in the form of machinery and equipment connected to either product or process innovations.

During late 1980s, the UN-APCTT described the process of acquiring technology to include several steps like analyzing and planning, identifying the need for a technology, finding the right technology, search strategies for finding sources of technology, evaluating sources of technology, negotiating a contract and writing a formal contract. Based on the successful experiences of the Republic of Korea, it identified key ideas in the chain of acquiring technology, *viz.* (*i*) development required technology, (*ii*) technology was know-how, (*iii*) know–how was transferred by documentation, training, and technical assistance, (*iv*) foreign investment was an important source of technology, and (*v*) all foreign investment agreements contained strong technology transfer provisions. The essential message was that the foreign investment was a source of technology (of skills and know-how), and not simply a source of machinery and projects. It was necessary to create a favourable climate for acquiring technology and building up of capabilities in designing, producing and marketing of goods (ESCAP, ST/ESCAP/862). The process of acquisition, essentially, involves a cumulative process of learning encompassing the full range of activities–R&D, design, production, logistics, marketing, distribution, support services–that bring products from conception to end use and beyond (UNCTAD, 2002).

More recently, the Korean experience in technology transfer has been examined based on an integrated analytical framework including: (*i*) technological trajectory (mature technology, intermediate technology, and emerging technology), (*ii*) technological complexity (small batch and unit production, large batch and mass production, and continuous batch production), (*iii*) absorptive capacity (existing knowledge base, intensity of learning effort), and iv) mechanisms of technology transfer (formal and informal). It indicates that the effects of IPRs on technology transfer to and local innovations in developing countries vary according to countries' level of economic development and to the technological nature of the economic activities. Benefits of strong IPRs can only be reaped only after they reach a certain level of threshold in their industrialization (Kim, 2003).

The firms in the developing countries needed to acquire new scientific and technological competencies and gradually move up these chains to perform higher value-added activities (McEvily, 2004).

The experience of most developing countries including India has shown that the technology acquisition depends largely on the capabilities of the firms seeking to acquire technology. In most cases, the entrepreneurs, who do not have adequate R&D or technological capabilities, would prefer proven technologies that could be easily adapted in the system. Such firms may also resort to reassembling of components or reverse engineering. However, in case of the firms with innovative capabilities, the choice may be to obtain the know-how, designs, drawings etc. by paying the royalty while taking care of the rest based on its own capabilities.

The purpose of technology acquisition is to obtain the best available technology in the market place. The technology sought to be acquired may be protected by one or more patents, subject to copyright and/or may have been kept trade secret. There may be other intellectual property rights surrounding the technology, such as trademarks protecting the brand or name of the company, copyright protecting the documentation, trade secrets protecting a whole host of confidential information including the know-how. One of the ways to acquire such IP is to obtain technology through licenses, the key determinant of which include the capabilities of the firm in undertaking the due diligence exercise and effectively negotiating the terms and conditions of the contract including those relating to intellectual property. If the technology is not protected, it is in public domain, in which case one should avoid negotiating for a license or pay for it. If the technology is protected by a patent it is important to check whether the patent is still valid in the country or the region where it is to be acquired. Since the IP rights are territorial, their validity may be limited to the national or regional jurisdiction for which they may have been granted. It is possible that a patent may not have been applied for in the country in which the invention is to be exploited.

Due diligence exercise

In undertaking the due diligence exercise, information is sought with respect to (*i*) the ownership of technology, (*ii*) whether technology is proprietary and have all proper procedures been followed to ensure its protection in the relevant markets, (*iii*) whether any third party is claiming rights over the intellectual property, and (*iv*) whether intellectual property rights other than patents or know-how need to be acquired to fully implement the technology (WIPO, 2005).

Negotiating intellectual property: Technology license

There may be several aspects relevant to intellectual property that may form part of negotiations in agreements on technology licenses. Some of these aspects are described below (WIPO, 2005, Gupta, 1997a, Gupta, 1997b).

Non-disclosure agreement

Prior to and during negotiations for obtaining a technology, it may be necessary to obtain confidential information, without which one may not be able to assess the feasibility of the technology being acquired. In such cases, generally, confidentiality or secrecy agreement or a non-disclosure agreement is entered with the licensor of the technology. The information is disclosed to the receiver under certain terms and conditions, which essentially require the receiver to treat all such information confidential and not to use the information in any way other than for the purposes stated in the agreement. The receiver does not have right on the information so acquired and is required to seek the permission of the discloser if the information is to be shared with other related or unrelated party. The information may have to be protected for certain period even after termination of the agreement.

The receiver needs to ensure that the supplier has commercially proven technology or know-how, as well as the right and the capability to transfer this know-how to the receiver. The invention should not have been published or publicly used by the technology supplier or by anyone under its control.

Scope of the agreement

Once it is agreed to acquire the technology, the agreement should clearly identify the technology necessary for the manufacture of the product or the application of the process or for some other given purpose.

All technology contracts should include a provision whereby the recipient enterprises acquires explicit rights for the use and exploitation of the technology in question, and the period covering these rights should be clearly specified. In cases where the enterprise is acquiring the right to practice a process, the concept of know-how should be clearly expressed and defined. In this connection, concepts such as "technical information" or "technical services" are treated as complementary to the know-how. The geographical areas in which the receiver has the right to exploit the IP may be designated. To ensure a continuous flow of information between licensor and licensee, the agreement should provide for access to licensor's plants and related R&D facilities

Working of the patented invention

It is expected that the receiver will fully exploit the rights granted and the technology supplied.

Technological advances/improvements

The agreement should include provision to safeguard the interests of the receiver resulting due to subsequent technological advances made or acquired by one or both of the parties. The receiver should have access to improvements on technology including both patented and non-patented developments. Moreover, if the improvements are made by the receiver, the agreement should provide flexibility for its appropriate protection and benefits to the receiver.

Research and development

Acquisition of technology may involve cooperation in research and development, between the technology supplier and the recipient. The agreement should permit one or the other parties to take out the patent and oblige that party to grant a license free of charge or against payment depending on the respective contributions to the inventive activity which each has made. Another possibility is to provide that both the parties shall take out a joint patent.

Maintenance and prosecution of the patents

The supplier usually bears responsibility for the cost of filing, prosecuting and maintaining licensed patents. In the event that the supplier believes that a given patent does not justify the expense of maintenance, the receiver should be so informed and might be given the opportunity to take an assignment of the patent, normally for a nominal consideration. This includes future patent applications on improvements. Typically the following cost categories are associated with a patent:

1. Patentability search
2. Preparation of a Patent application
3. Filing fees with the Patent Office
4. Fees chargeable by patent attorneys/agents.

It is, therefore, imperative that only ideas with good commercial potential should be patented world-wide. A lot of thinking is necessary before filing in other countries.

The production phase

The terms and conditions should cover the manufacture or assembly of the product or of its components by a subcontractor, and necessary measures to safeguard against the disclosure of technical information in the event of subcontracting.

Cost of patent license

One of the important issues to be negotiated is the "price" or the "cost" of the intellectual property rights to be acquired. The "price" or the "cost" is dependent upon a number of factors, including the nature of the intellectual property rights, the technology and the relative bargaining power of the two parties. Several methods can be used to value a technology, however, it is to be noted that the value is likely to be higher if the technology is well protected by patents and other forms of IP.

Payment for the exploitation right

The principles and the basis of the calculation of benefits from the exploitation of patent rights are to be simple. These may include lump sum payments or payments for transfer of know-how or prototype. Royalties are normally calculated as a percentage of the turnover or sale. Provision should be made for the reduction of royalty in case of discontinuation of the intellectual property rights.

Cross-licensing of patents

There could be situations where both parties own intellectual property of interest to the other and have the legal right to prohibit the other from using it. In such a case, they might enter into a cross license agreement through which they would license each other the right to use and exploit their respective intellectual properties. There may or may not be royalty payments, depending upon the value attributed to the intellectual property owned by each party.

Infringement

Licensor and licensee rights concerning infringement suits may vary from country to country owing to differences in the applicable laws. It is necessary to check the infringement clauses carefully against what is allowable for the countries covered by the agreement. The licensor will want the licensee to advise it promptly, in writing, of any infringement the licensee discovers. The licensor will also want the right to sue the infringing party to recover damages after assessing the facts of the alleged infringement. The licensor usually claims the right to choose and control the counsel selected to prosecute the suit and seek the full cooperation by the licensee. The licensor may also want to control legal action by the licensee against an alleged infringer. Damages collected in such a suit usually pay for the cost of litigation. The parties then divide the remainder evenly between them if the infringement caused significant damages to the licensee. Alternatively, the rights may be given to the licensee to file the suit and claim the damages. The licensor may extend legal and technical assistance to the licensee. The licensee should attempt to have royalty payment relief written into the agreement for situations in which the licensor will not allow the licensee to take legal action against an alleged infringer.

The supplier may retain the power to license industrial property rights and to transfer the know–how to more than one technology recipient or also to exercise those rights and use that know-how by himself within a given territory.

Agreement for the acquisition of know-how

Know-how agreements ensure the disclosure of inventions to third parties through legal channels. Such agreements for the communication of know-how should (Gupta, 1997b):

1. Include a statement by the supplier on the secret nature of know-how to be transferred which also specify the items of information to be deemed confidential
2. Stipulate the effect of disclosure of the know-how on the validity of the agreement, including termination where the know-how has lost its secret character through no act of the recipient
3. Determine the duration of confidentiality obligations, which should not extend, in principle, beyond the lifetime of the agreement or a reasonable period thereafter
4. Explicitly state the right of the recipient to continue in the use of transferred know-how, after the expiration of the agreement
5. Authorize the use of know-how by the recipient in any possible field of application, without prejudice to corresponding payment obligations emerging from the uses not specifically agreed upon.

Mergers and acquisitions

The driving force behind a majority of mergers is the acquirer's desire to obtain target's intellectual property assets. In addition to conventional forms of intellectual property, the mask-works and internet domain names have also become important. In the event of merger, the acquiring party should obtain equitable and record ownership of these intangible assets or else acquire appropriate license to use such property. Typically, the acquirer should ensure that the schedule of intangible assets is complete and accurate, the seller is the rightful owner, no liens or encumbrances exist, the IP do not infringe rights of third party, and the buyer is indemnified. The ownership of the intangible assets can be put in a holding company, which then licenses back the assets for use by the operating company. It is necessary to obtain worldwide transfer of the intellectual property rights promptly soon after the merger (Bryer and Lebson, Accessed 15.10.2006).

Credibility in acquisition through technology exports

Firms in some of the scientifically advanced countries like India have also developed capabilities for export of technology, which may also help in technology acquisition as the credibility of such firms in the international market is likely to be high. These firms may also have greater awareness about the international market and thus not only identify credible suppliers but also be in a better position to negotiate (Agarwal, 2004).

Importance of patent information in technology acquisition

The availability of computerization of patent databases has made patent documents as well as their bibliographic apparatus easily accessible. The ways in which patent information can be used in the technology development, transfer and acquisition include locating business partners for collaboration, identifying products for marketing, avoiding possible infringement problems, locating areas of investment, assessing technology related proposals for financial support, and monitoring viability and progress of projects. From a technical point of view patent information can help in avoiding duplication of research, assessing state-of-the-art, finding ready solution to technical problems, and identify and evaluate technology for transfer or acquisition. From the legal point of view, patent information can be used to assess patentability of inventions, oppose grant of patents, and move for revocation of patents. The patent information has, thus, the importance in technology acquisition, the necessary skills for which should be acquired by the firms seeking technology from external sources.

Acquisition from internal sources

Firms can also acquire technology through internal sources either by in-house R&D and technology development or through collaboration with publicly funded R&D institutes or universities with in the country. In such cases, a choice is to be made between in-house development or acquisition from other R&D institutions. In case of former, it is necessary to exploit R&D results as intellectual property and profit from such exploitation. In case of later, it will be necessary to establish linkages either by sponsoring R&D or collaborating in R&D and development of technology or acquiring technologies developed by the publicly funded institutions or universities. The decisions would depend on several factors including the nature and advancements in technologies, and the need to produce higher value-added products.

The salient aspects of protection of intellectual property in respective cases are indicated below.

Protection of IP during in-house R&D

In case of in-house R&D or technology development, the essential features of protection of intellectual property may include (Gupta, 1995):

1. Strengthening awareness about IP at the bench level scientists
2. Well established system for identification of results of research for protection
3. Ability to decide in which countries the protection is to be sought

4. Determining patentability of the invention is an important decision. Two aspects which should be carefully considered in this context are (a) extent of prior art and prior public use, (b) anticipating any possibility of infringement of patent by others
5. Taking decision about patenting or other appropriate means of protection
6. Taking decision as to when to obtain patent
7. Making use of technical information contained in the patent documents for identifying opportunities for R&D and technology development
8. Ensuring protection of the intellectual property possessed by employees through appropriate terms and conditions in the employment agreement. Generally, an employee holds the results of all investigations made by him while in service and within three years of the termination of his services in trust for the employer. The employee is also required to disclose the full and complete description of the nature of his invention and the mode of performing the same to a party identified by the employer (Gupta, 1997c, Gupta, 2000a).
9. Protection of confidential information possessed by employees. A contract of employment may expressly prevent an employee from disclosing information learnt in the course of their employment. Such a provision is more likely if the employee has access to secret information whether about R&D, technology or trade practices. The employee's duty of confidence lasts for the whole period of employment. Even after the end of employment, the obligation of confidence continues for a period designated in the employment agreement (Gupta, 2000a). However, adequate flexibility needs to be given to the employee in using his knowledge and skills to earn a livelihood after termination of his employment.
10. Instituting appropriate measures for the disclosure of confidential information during the course of R&D work. While disclosing such information, the employees need to put it in writing or some other permanent form of record. Keep a copy of what is disclosed, and a record of when and to whom it was disclosed. If an oral disclosure is made in confidence confirm the same in writing what was disclosed and that it was given in confidence (Gupta, 2000a).
11. Establishing a good system for keeping records of research and experiments.
12. Rationalizing decision-making while naming team of inventors. Generally, disputes arise on deciding the team of inventors while taking patents or sharing benefits arising out of exploitation of IP. In such cases, the basis of naming inventors is that all those persons who contribute towards the development of patentable features of an invention should be named as inventor (s)/co-inventor (s)
13. Managing IPR during commercialization of results of research and transfer of technology: exploit the results of research by outright sale of technology or in-house production or through licensing, royalty and benefit sharing.

Managing IP during S&T cooperation

Networking and collaboration are new mechanisms of technology acquisition (Agarwal, 2004). Technological collaborations occur when two companies establish joint ventures or formally agree to develop technical knowledge and products, while maintaining their respective ownership. Companies may also enter into collaboration with R&D institutions for the development and exploitation of results of research or the technologies. The important aspect of the management of IP during R&D collaborations with organisations within or outside the country is the identification of the principles for sharing the ownership and benefits of the intellectual property (Gupta, 1996, Gupta, 2000b).

The cooperation in R&D may result in either of the three ways *viz.* contract research, grant from a funding body, and collaborative or joint R&D projects. The principal purpose of a contract is to undertake research as a direct input in well–defined programmes and missions of the agency giving the contract. Normally, for contract research both tangible and intangible intellectual property are to be wholly-owned by the donor/supplier of the funds; but the recipient of the funds may have royalty-free licence for his non–commercial use.

One of the purposes of a grant is to accomplish the given objective through stimulating or supporting the acquisition of knowledge or understanding of the subject or phenomena under study. Both tangible and intangible property are to be wholly–owned by the recipient of money; but the donor of money may have royalty-free license for his own use.

In case of collaborative R&D projects, there is a substantial involvement of both the collaborators in making contribution to the technical aspects of the effort. The project would not be possible without extensive technical collaboration. In principle, the intellectual property is to be jointly owned by the collaborating S&T institutions and the resulting benefits are to be equally shared. Alternatively, the IP and the benefits are shared in proportion to the respective inputs of the collaborating parties.

The recognition of the confidentiality obligations is essential for the R&D institutions while dealing with industrial firms as one of the collaborators. For example, the confidential and proprietary (*i.e.*, trade secret) information owned by the industry and transferred to the scientists in university or S&T institution should be guarded.

The confidential and proprietary information including the patent held or developed prior to the beginning of the collaborative research programme by the respective collaborators should be defined in the collaborating research agreement.

R&D Outsourcing: In some instances, it may be essential to give an R&D contract to a party/institution outside the country. In such a case, one should ensure that no application for patents should be filed, by the foreign institution receiving the contract, for inventions resulting from work undertaken under the contract without prior approval of the R&D agency giving out the contract. The agency should be notified of the patent application number, filing date, the applicant's name and be supplied with copies of the description and drawings filed with the application within a short span of time (say three months) from the date of filing of the patent application. The agency that gives a contract should keep with itself the rights without payment to an irrevocable non-exclusive licence to exploit the invention for its purposes and to grant sub-licence for these purposes to anybody in the country.

Adaptation and diffusion

Once a technology is identified and obtained, it is required to be adopted, which means acquiring capabilities to practice the technology as such without modifications. It may need learning through training, and sharing of confidential information. The next step is the adaptation of technology wherein the firm acquires capabilities by way of incremental modifications depending upon its local needs. It envisages necessary in-house R&D capabilities or associating publicly funded R&D institutes to help in the process of adaptation. The next step in acquiring technology is its assimilation and absorption, implying adequate capabilities with the firm to understand why of the technology.

Protection of IP has several implications in this context, primarily, relating to protection of results of research from in-house R&D and those resulting from co-operation with other R&D organizations, the details of which have already been given above. In particular, the protection of IP is likely to enhance endogenous technological capabilities of the firms.

The opportunities for adaptation may vary across industries depending, among other factors, upon the nature of technology, the gap between local and global standards, and the nature of intellectual property protection. The opportunities for adaptation were higher in capital goods industries, which required minor modifications, for example, in design to adapt them to local conditions (Nagesh and Siddharthan, 1997).

The ultimate aim is then to improve upon the technology so acquired and further diffuse them in the market, which will require extensive innovative capabilities. The capability in protecting and managing IP is likely to play a supplementary role. One company, however, may not have all the competence to effectively acquire technology and in the process may take the help of consultants, publicly funded R&D institutions and other stakeholders in the system. Technology diffusion needs strong R&D support, in house within the firms, and in publicly funded R&D laboratories.

Diffusion through patent applications: the availability of technological information in the vast pool of patent applications filed by inventors within one's own countries and or in other countries provides opportunities for improvements in technology and its diffusion across firms and users in developing countries. Necessary skill should be cultivated to exploit such information.

Taking advantage of IP

Following are some of the relevant dimensions for taking advantage of IP in developing endogenous technological capabilities, which need appropriate attention of the policy-makers in developing countries:

The role of the Government in funding R&D is gradually diminishing. Major resources are to be generated from user orientation of R&D. This would call for bold steps for commercialisation of R&D and transfer of technology. Effective enforcement of protection of IP may be essential to tap resources from private sources.

Only a very small percentage of all patents (say 5-7 per cent) granted reach the commercialisation phase of the innovation process. The reasons for failure may range from the quality of the invention to factors such as the high investment cost for a relatively small effect or improvement. One should note that in spite of such constraints, the inventive activity has continued to grow. Scientists and entrepreneurs should be encouraged and motivated to generate new knowledge and inventions and developing capabilities to acquire existing knowledge and technologies and suitably adapt them.

Encourage development of skills in technology prospecting, which entails the process of searching, identifying, adapting, and diffusing imported technology. In this process research institutions, the firms and expert organizations have a role to play, particularly, in undertaking global searches for patents and technologies and finding ways of adapting them to local and international markets.

One of the significant developments in structuring the economy in the context of globalization is the use of concepts like national innovation system and formation of innovation networks, by shaping several dimensions involving technology and intellectual property. National innovation system creates and distributes knowledge, and utilizes it in the form of diffusion of innovations in the economy. Network involves partnering activities designed to reduce the risks associated with the development of new products or innovations. These concepts should be effectively utilized in the implementation of policies and mechanisms of protection and management of IP and in structuring linkages in the economy.

The aim should be to strengthen own absorptive capacity, which would enable to identify relevant technology available elsewhere, add to bargaining power during negotiations, help in quick assimilation of acquired technology, in carrying out improvements in acquired products or processes building innovative strategy around IPRs, and generate own IPRs.

Conclusions

The issues of intellectual property protection are significant both in acquiring technology through indigenous R&D efforts as well as from abroad. In the emerging era of globalization and liberalization, the firms in developing countries would need to develop abilities to acquire and adapt technologies from both external and internal sources as well as to create new technology in order to successfully compete with the international markets. The signing of agreement on Trade related aspects of intellectual property rights tends to strike a balance between enforcing intellectual property rights and meeting the technological needs of developing countries. This has considerable implications for the intellectual property issues relating to the process of acquisition of local technological capability by developing countries.

The strength of patent regime is likely to ensure greater contract certainty between supplier and the receiver and is likely to enhance the flow of high technology goods and the superior technologies in the developing countries. The adequate protection of confidential information or trade or training manuals or guidelines which might be protected under copyright would enable the firms acquiring technology to receive relevant IP in confidence which might enhance their absorptive and adaptive capacities. The experience of most developing countries including India has shown that the technology acquisition depends largely on the capabilities of the firms seeking to acquire technology.

One of the ways to acquire IP is to obtain technology through licenses, the key determinant of which include the capabilities of the firm in undertaking the due diligence exercise and effectively negotiating the terms and conditions of the contract including those relating to intellectual property. Prior to and during negotiations for obtaining a technology, it may be necessary to obtain confidential information, without which one may not be able to assess the feasibility of the technology being acquired. The receiver needs to ensure that the supplier has commercially proven technology or know-how, as well as the right and the capability to transfer this know-how to the receiver. The invention should not have been published or publicly used by the technology supplier or by anyone under its control. The receiver should have access to improvements on technology including either patented or non-patented developments. Acquisition of technology may involve cooperation in research and development, between the technology supplier and the recipient. The agreement should permit one or the other of the parties to take out the patent and oblige that party to grant a license free of charge or against payment depending on the respective contributions to the inventive activity which each has made. Another possibility is to provide that a joint patent shall be taken out by both the parties. The patent information has importance in technology acquisition, the necessary skills for which should be acquired by the firms seeking technology from external sources.

In case of in-house R&D or technology development or acquisition from internal sources, the essential features of protection of intellectual property may include strengthening awareness about IP at the bench level scientists, making use of technical information contained in the patent documents for identifying opportunities for R&D and technology development, and ensuring that the intellectual property is protected by employees through appropriate terms and conditions in the employment agreement. A contract of employment may expressly prevent an employee from disclosing information learned in the course of their employment. Even after the end of employment, the obligation of confidence continues for a period designated in the employment agreement. Collaboration is one of the means for technology acquisition. An important aspect of the management of IPR during collaborations between R&D institutions and industrial firms as well as collaboration with similar organisations within or outside the country is the identification of the key principles for sharing the ownership and benefits of the intellectual property.

Acknowledgement

Author gratefully acknowledges his discussions with Professor S.P. Agarwal, Head, Centre of International Trade in Technology, Indian Institute of Foreign Trade, New Delhi, which greatly facilitated the completion of this paper. Prompt and timely help and encouragement by Dr. A. Lahiri, presently Office Incharge, UN Centre for

Asia and Pacific for Transfer of Technology, New Delhi is also thankfully acknowledged.

References

Agarwal, S.P., Gupta, A., and Gandhi, G.P., 2004. Technology transfer trends. Tech Monitor, May–June, pp. 36-41.

Alikhan, S., 2000. Socio-economic benefits of intellectual property protection in developing countries. WIPO, Geneva.

Bryer, L.G., Lebson, S. J., 2002. Intellectual Property Assets in Mergers and Acquisitions. John Wiley and Sons.

http://www.wipo.int/sme/en/documents/pdf/mergers.pdf (Accessed 15.10.2006)

CIPR, 2002. Report of the Commission on IPR (UK Commission on Intellectual Property Rights). London.

David, M. G., and William. C. G., 1996. The role of intellectual property rights in economic growth. Journal of Development Economics, 48, pp. 323-350.

ESCAP, 1992. Technology transfer: An ESCAP training manual, Booklet 3: The Process of Technology Acquisition. UN–ESCAP (ST/ESCAP/862).

GOI, 1970. Indian Patent Act 1970.

GOI, 1994. The Patents Act, 1970 and Patents Rules, 1972, Short notes. Sixth Edition, Eastern Book Company, Lucknow, India.

GOI, 1995. The Copyright Act, 1957, The Copyright Amendment Act, 1994 along with the Copyright Rules, 1958 and International Copyright Order, 1991, Short notes. Universal Book Traders, New Delhi, India.

GOI, 1999a. The Trademarks Act, 1999.

GOI, 1999b. The Geographical Indications of Goods (registration and protection) Act, 1999. Gazette of India, Extraordinary, Part II-Section I.

GOI, 2000a. The Designs Act, 2000. Gazette of India, Extraordinary part II-Section I.

GOI, 2000b. The Semiconductor Integrated Circuits Layout-Design Act, 2000.

GOI, 2001a.The Competition Law, 2001.

GOI, 2001b. The Protection of Plant Variety and Farmers' Rights Act, 2001.

Gupta, V. K., 2000a. Employer vs Employee Inventions: IPR Issues in R&D Organizations, Journal of Intellectual Property Rights 5, pp. 239-250.

Gupta, V K., 2000b. Negotiating IPR in international science and technology cooperation. Journal of Intellectual Property Rights 5, pp. 61-71.

Gupta, V.K., 1997a. Technical Report/Guide on Commercialisation of R&D and Transfer of Technology-IPR related Insights", prepared for TIFAC, New Delhi.

Gupta, V.K., 1997b. Intellectual Property Rights in Commercialisation of R&D and Transfer of Technology, Journal of Intellectual Property Rights 2, pp. 181-190.

Gupta, V.K., 1997c. Sharing of Intellectual Property–Relationship between R&D Organisations as Employer and Scientists as Employees. Prepared for Patent Facilitating Cell, TIFAC, DST, GOI.

Gupta, V.K., 1995. Patenting: Practice and Management in a R&D Organisation. Prepared for Patent Facilitating Cell, TIFAC, DST, GOI.

Gupta, V.K., 1996. Intellectual Property Rights and International Cooperation in R&D. Prepared for Patent Facilitating Cell, TIFAC, DST, GOI.

Gupta, V.K., 1999. Protecting confidential R&D information. Asia and Pacific Tech Monitor 17, pp. 37-42.

Gupta, V.K., 2004. Multi-disciplinary Studies on IPR in R&D: A Review, Journal of Intellectual Property Rights 9, pp. 34-42.

Kim, L., 2003. Technology transfer and intellectual property rights: the Korean experience. Issue paper no.2, UNCTAD-ICSD Project on IPRs and sustainable development, pp1-31.

Martin. B., 1995. Against intellectual property. Philosophy and Social Action, 21, pp.7-19.

Maskus, K., 2000. Intellectual Property Rights in the Global Economy. Institute for International Economics, Washington DC.

Maskus, K., 1999. Competition Policy and Intellectual Property Rights in Developing Countries: Interests in Unilateral Initiatives and a WTO Agreement, World Bank Global Conference on Developing Countries and the Millennium Round, Geneva, September 20-21.

McEvily, S.K., *et. al.*, 2004. The Global acquisition, leverage, and protection of technological competencies. Strategic Management Journal 25, pp. 713-722.

Nagesh K., and Siddharthan N.S., 1997. Technology, Market Structure and Internationalization: Issues and Policies for Developing Countries, The United Nations University, Routledge in association with UNU Press., London, pp 73-75.

Patel, S. J., 1993. Intellectual property rights and national development. Journal of Scientific and Industrial Research 52, pp. 212-218.

Rao, M.K.D., and Gupta, V.K., 2003. IPR and sharing of biological research materials in R&D, Journal of Intellectual Property Rights 8, pp. 112-122.

UNCTAD, 2002. Financing Technology for SMEs: Issues Note by the UNCTAD Secretariat. Commission on Enterprises, Business Facilitation and Development Expert Meeting on Improving the Competitiveness of SMEs through Enhancing Productive Capacity: Financing Technology, Geneva, TD/B/COM.3/EM.16/2.

United Nations Conference on Trade and Development (www.unctad.org/en/docs/c3em16d2_en.pdf (Accessed 19.10.2006).

World Trade Organization (www.wto.org/english/docs_e/legal_e/legal_e.htm#TRIPs Accessed 15.10.2006).

WIPO, 1997. General information. WIPO Publication no. 400 (E), Geneva.

WIPO, 2005. Exchange value: negotiating Technology Licensing Agreements, A Training Manual, pp 1-177.

World Bank, 1999. World Development Report–Knowledge for development. Oxford University Press, New York.

WTO, 1995. Text of the Agreement on Trade Related Intellectual Property Rights. Geneva.

Chapter 18

Innovation and the role of IP system in Egypt

Janet Ibrahim Youseef

Abstract

Innovation and Invention are considered to be the essential factors in the advancement of social and economic status of Egypt, for which scientists and technicians play a significant role. In this context, the Innovation and Invention Development Agency (IIDA) was established for updating and improving the basic structure, supporting the agencies with modern technological equipments and devices, and for initial selection of patents, that could be commercialized, through examination by the Patents Committee. The academy has an effective role in developing technology via certain mechanisms in order to reach the distinctive Egyptian inventions into quantitative stage and commercialization so as to encourage creativeness, innovations and inventions. The paper presents the activities of IIDA in achieving the said objectives.

Keywords: ASRT, Egyptian inventions, Innovation and Invention Development Agency (IIDA), Patent.

Introduction

It is clear that the advancement and development for Egypt exist in what its technicians and scientists achieve in order to accomplish technological development, which is a prerequisite for the country's social and economical development. This is simply because scientific research plays a main role in developing human future and pushing scientific improvements forward through the ability of creation and innovation technology. Thus, the global concern of innovation and invention issues has increased according to the advancement in various fields of science and arts. Moreover, innovations and inventions have become one of the indicators through which the development of a nation could be estimated along with interaction with modern science movement. Innovations and inventions are considered essential factors and main steps for enhancing economic status, which is consequently reflected in the social status. These inventions may lead to creation of new industries and renewal of the old ones by using more efficient means to produce new products. These inventions and innovations also provide new job offers, which may further enhance the rate of the national income.

The development of a nation is related to self-ability of improving technological means convoyed with its environmental, social and economical status. The support extended for inventors and innovators may be considered as an investment from which the whole community may benefit. Accordingly, the Academy of Scientific Research and Technology (ASRT) supports distinctive researches, and patenting Egyptian innovations and inventions. It supports ideas that could be converted to technological projects in order to continue such research, their improvement, and the invention and innovation processes. Therefore, a resolution, in Egypt, was issued to establish the Innovation and Invention Development Agency (IIDA) affiliated to the technology development sector in the ASRT. IIDA depends on a certain strategy in order to accomplish its objectives such as updating, and improving the basic structure, supporting the agency with modern technological equipments and devices, and commercializing the patents.

Objectives and the mechanisms of IIDA strategy

The objectives of IIDA can be achieved through a central administration via three general departments that have been established for the purpose. A group of specialized Egyptian experts and scientists in all fields of technology helps the Agency in accomplishing its goals of generating and exploiting technical knowledge resulting from research projects and the protected inventions of the technological products.

The innovations should have relative advantages and should be able to be competitive in the local, regional and international markets, in terms of the outputs. These should be based on the recommendation of the Committees, namely, (*a*) Study and examination by the Patents Committee for initial selection of patents that could be commercialized, and (*b*) Management Cooperation Activity Committee for "Preparing prototypes" projects, within the framework of the Cooperation Agreement between the Academy and developing social fund.

The IIDA mechanism includes supporting innovators and inventors who aim at creating the marketable and distinctive invention outputs that are expected to reach the markets, accessible to quantitative stage, and commercialization via the following steps:

1. Supplying the Agency with periodical copies of the Egyptian patent abstract and the patent gazette,
2. Sending the applications of inventors to specialized committees for studying them,

3. Computing and classifying the protected Egyptian patents and the patents those fall in the public domain in various fields,
4. Computing, classifying, studying and selecting the patents expected to reach markets and sending them to concerned bodies to study how far these could be beneficial; by conducting a feasibility study to be followed by semi-industrial experiments as a primary stage, quantitative production stage, and finally, to commercialization,
5. Manufacturing prototypes for some distinctive inventions after carrying out the initial market studies, and coordinating with the owners of inventions in order to take the required executive steps towards manufacturing such as preparation of geometrical drawings, identifying manufacturing bodies, preparing technical schedules, and financial plans,
6. Promoting and marketing the patents that are industrially applicable via different medias (visual/audio/press) and preparing media bulletins, workshops, scientific seminars, industrial, productive and field visits, taking part in exhibitions and local, regional and international competitions,
7. Preparing feasibility studies for distinctive inventions,
8. Prototypes proved to be of industrially applicable nature are introduced to investors or to be resided in technological business incubators in an attempt to reach the quantitative production stage and commercialization,
9. Prepare the following marketing agreements to protect the rights of the inventor, the Academy and the beneficiary under the rule of the intellectual property protection:
   (*a*) A marketing agreement among the Academy, the inventor and the manufacturer of the device,
   (*b*) Sale agreement of the "know-how" among the Academy, the inventor and the beneficiary, and
   (*c*) A marketing agreement between the Academy and the beneficiary.

Achievements of the IIDA

Since its establishment in July 2001, the activities of IIDA resulted in a number of achievements. Some of these include the achievements such as computing, classifying, studying and selecting distinctive protected patents that are suitable for production to reach markets, and reaching the quantitative production stage and then commercialization. There are some basics and criteria for choice such as:

1. The product should be important to Egyptian community and contribute to the national income, general health, local environment and should be required locally.
2. The invention project cost should be met by the revenue generated from its outputs.
3. The invention project outputs should be compatible with the product lines. It is important that the sponsoring organizations also comprehend labour requirement of its technological production.

4. The project outputs, *i.e.* production should not depend on the imported materials, but on the local material as much as possible.

The Agency followed executive procedures within the framework of the commitment of the previous criteria, *viz.*

1. All the 118 Egyptian patented inventions issued in 2002 were computed, classified and studied. Several distinctive Egyptian inventions were selected. Some of these have already been commercialized, and others were expected to be marketed. In addition, a number of inventions were being examined by the specified committee for developing manufacturing prototypes.
2. The Egyptian patented inventions issued from 2003-2006 are being computed, and currently, there is a study to examine how far these outputs can be beneficial via computing, classifying, studying, clarifying and the selection process.

Presently, the prototypes for a number of distinguished protected Egyptian inventions are being manufactured. These include an instrument used for ejecting the air brake without bleeding air by the brake blade, connected piece between car clutch and pull wire with variable axel and length, regulating automatic valve for controlling fresh water, wall cleaning machine, paste machine, device scaling driving circle, obliquity from start movement position at naval land carriages and machines based upon gravity, a compass for triangular functions, multifunctional instrument used in the production of difference forms of wood and pastes, a moving chair (manual/automatic), dough divider machine, a spray of insecticides, manual scraper, petrol electronic injectors (cleaner machine), a car seat belt against sinking, solar cooker, and the elliptical solar energy heater.

These prototypes of the patented Egyptian inventions are being manufactured via classifying, and studying them as well as holding discussions with inventors for marketing. Besides, it involves presenting price offers, preparing technical schedules and financial plans, and identifying the industrial bodies for following the industrial steps. This gives an opportunity for testing the first design of the prototype and identifying its performance efficiency through specialized bodies while carrying out the executive procedures for marketing invention outputs.

Marketing and promoting

The following steps are involved:

1. Media promotion and highlighting the role of IIDA and some of the properties of the patented distinctive Egyptian inventions and innovations, which are expected to reach markets. This is to encourage industrial authorities and investment organizations to adopt different technologies and providing them with proper markets.
2. Prepare brochures, which include a summary about the role of the IIDA, its most important characteristics authenticated by specialized committees, as well as information about the new inventions and innovations.

Examples of such inventions include petrol electronic injectors (cleaner machine), an alarm before the occurrence of fires, a guiding apparatus to take aim at insertion of penetrating object from outside the human body to a small target inside it, machine that protects breaks, interactive intelligent- remotely controlled electronic system for air conditioners, feeder digital over current and earth fault relay, alarm set and automatic fire fighting in cars, electronic water neutralizer, and the dynamic brace.

IIDA actively participates and contributes to the local, regional and international exhibitions where prototypes are displayed, and takes an opportunity at such events in explaining its role in helping the inventors. As a result, the Academy (ASRT) has been honoured. Many such inventors have been granted financial prizes and recognition certificates for their inventions.

Examples of these prizes include:

1. The youth and sports exhibition in its 21st round – Eng. Brigadier General Abdl Flamed Sharaf, the inventor of electrostatic water neutralizer (patent no.20702) was awarded the second prize at an amount of 2000 L.E. and a recognition certificate,
2. The youth exhibition in its 24th round – Dr. Mohamed Gamal Hassan Ammar, the inventor of the dynamic brace (patent no. 20431) was awarded the third prize at an amount of 1000 L.E. and a recognition certificate,
3. The second exhibition for the small sized enter prizes technology – Eng. Brigadier General Abdl Hamid Sharaf, the inventor of electrostatic water neutralizer (patent no. 20702) was awarded an honour shield and a recognition certificate,
5. Other Egyptian inventors were awarded recognition certificates for their distinctive inventions,
6. Mobile science museum mobile exhibition – IIDA participated in the mobile science museum that aimed at spreading scientific knowledge in all governorates in Egypt where some prototypes were displayed to enlighten young inventors with the value of granting patents for their inventions. Examples of these inventions include paste machine, a wall cleaning machine, plastic anal, connection piece between car clutch and pall wire with variable axel and length, phenol formaldehyde product (improved bakelite) by adding urea and melamen formaldehyde, and non flammable substance,
7. The second youth innovations exhibition – participating in the 2nd youth innovations held from 31st March to 1st April 2004 in Al Azhar centre for conferences. More than 21 prototypes were displayed,
8. The Galactica exhibition held in Spain – Dr. Mohanlmcd Galnal Hassan Ammar, the inventor of the dynamic brace (patent no. 20431) was awarded a certificate of recognition,
9. International exhibition for new ideas, inventions and products,
10. Some of the distinctive Egyptian exhibits recommended for participation at these exhibitions, and the providing of participation requirements to inventors if their inventions were accepted,

11. International Exhibition for young inventors – facilitated participation of young inventors from 6 to 19 years old in the International Exhibition, organized by the Japanese Invention and Innovation Institution (JIII) in Tokyo/Japan, and
12. Recommendation of four inventors to travel to Japan for the display of their inventions such as: rehearing of sound using a reflecting laser beam, damage protecting circuit for electrical devices, a simple apparatus used in watering the plants during the absence of the home owner for a short or long time period, and electrical spoon.

IIDA's participation at conferences, symposium, seminars and workshops

Conferences

1. General Conference of the Academy of Scientific Research and Technology in the 12th round, under the title of the Academy and technological scientific development at the beginning of the 21st century
2. The conference, which took place along with the first international small enterprises technology exhibition held from 28th February to 3rd March 2002

Seminars

1. The first Arab seminar about industrial projects incubator
2. A seminar about intellectual property and its means of education (Tepri)

Symposiums

1. First symposium with the Arab intellection market and exhibition from 23rd to 26th January 2002
2. Second scientific symposium of Arab intellection market "Ideas for better future"
3. Round table conference about technological innovations and intellectual property within the framework of the new participations for the sake of developing Africa *viz.* What is the African strategy?

Workshops

1. A workshop about innovation in Egypt was organized by the industrial updating centre

Prizes and competitions

IIDA participated in number of competitions. As a result, a lot of distinctive Egyptian inventions were awarded certificates of recognition and financial prizes such as:

1. Invention and innovation development competition in the industrial and the technological development organized by the prize sector of the Academy

of Scientific Research and Technology. In 2001, two inventors were awarded this prize at the amount of 5000 L.E, shared between them. These two inventors are Mr. Mohammed Abd Alla, the inventor of multifunctional instrument used in the production of different forms of wood and pastes (patent no. 21277) and Mr. Fayez Ahmed Ibrahim Rageh, the inventor of El-meshabek machine (patent no. 21332),

2. Two inventors were awarded a prize of 5000 L.E, shared between them, namely, Eng. Heba Abd Rahman Ahmed Fayez, the inventor of improving the properties of fabricated PMMA matrix composite by argon and ND-YAGLASERS ablation (patent no. 20841) and Mr. Awad Mohamed Abdl Ghafar, the inventor of a car seat against sinking (patent no. 21 965),
3. Creativeness day competition 2000 organized by the incubator of technological projects in Tibin. Two inventors were awarded financial prizes and recognition certificates for inventing, *viz.* the dynamic brace device that was awarded the 1st prize at the amount of 2000 L.E., a medal and a certificate of recognition, and water keeping combination box device, which was awarded the 2nd prize of 2000 L.E, a medal and a certificate of recognition. Moreover, 4 distinctive inventors were awarded recognition certificates.

Efforts have been initiated in contacting the industrial, research and executive organizations besides renewing open communications channels with the production and service units in both the public and private sector. This is to keep in touch with the distinctive Egyptian invention outputs in various fields and the possibility of benefiting from them.

Its most important results are:

1. The Academy adopted a project for conducting clinical experiment on the invention outputs of accelerated hemodialysis lines with two options, *viz.* accelerated hemodialysis programme and accelerated hemodialysis valve; which cost 50,000 L.E of the total budget. It can be successfully used in the blood washing unit in hospitals.
2. 822 feeder digital over current and earth fault relay devices were commercialized for the Egyptian electricity company and for another middle level electricity distribution company in Egypt that were marketed in some Arab countries, for example, Saudi Arabia, Syria, and Libya.
3. 60,000 units of the interactive intelligent–remotely controlled electronic system for air-conditioner inventions were commercialized by the Union Air Company that were marketed in some Arab countries for example, Saudi Arabia, Syria, Libya, and Tunisia.
4. 148 units of electrostatic water neutralizer inventions were marketed for 57 farms. In addition, the inventor was awarded number of prizes, for example, a prize from the Health and Youth Exhibition in its 21st round, and an honour shield and recognition certificate from the 2nd small technological enterprises international exhibition.

5. 51 Egyptian KH-Y lithoclast devices were marketed for public and private hospitals. The inventor of the device was awarded number of prizes and a certificate of recognition including "A" degree from Dr. Mohamed Ahmed of the Nephritis and Urinary Paths Centre, Al Mansoura University, and another recognition certificate and a silver award at the Budapest Inventors Olympic Round in 2000.
6. 5 to 25 different diameter circular machine for sieving, classifying and sizing of all materials by vibration that contributed in solving the problems of sonic companies. This machine was exported to Sudan, Saudi Arabia, Palestine and Jordan.
7. The economic returns resulted in marketing a new mycoticides compound outputs and their methods of preparation, at the amount of 600.000 L.E.
8. 1000 plastic antenna were marketed to different bodies such as radio and T.V. Unions, and administrative prosecution and control authorities.
9. Anti-flammable material invention outputs were being marketed to Egyptian organizations.
10. Prototypes were being manufactured for two inventions, *viz.* eight–nozzle hydraulic servo value of low sensitivity to low force, and spool value of low sensitivity to flow forces. These two inventions were financed by technological creativity support company investors. They would be marketed via the same company, in case these prototypes proved to be qualified.
11. A scientific experiment was being carried on alarm set and automatic fire lighting in cars, and invention output on a bus. Besides, market executive procedures were followed via an agreement between the Academy, the inventor and the 99 military factory investors. This was to generalize invention output benefits on a number of C T A buses.
12. Another scientific experiment was being carried out to verify the tools and methods for queen honeybee rearing in the faculty of Agriculture, Cairo University, upon the inventor's request in order to take it into the commercialization stage.
13. A marketing agreement was being prepared between the Academy and the inventor besides contracting on marketing some distinctive Egyptian inventions such as plastic arial, model for output of automatically machine used for cutting wood (Archet), and electrostatic water neutralizer.
14. Four Egyptian inventions were being included in Business and Technology AI Mansoura Incubators such as the portable anaesthetic apparatus, a new method of sterilization of all types of materials by using dry heat ultra violet rays and moist heat in one and the same, a new method of sterilization by using a mini autoclave of a simplified operating procedure, and an artificial pancreas for controlling the blood glucose level in patients of insulin-dependant diabetes mellitus.

15. Three distinctive patented Egyptian invention abstracts were being included in the International Foundation for Inventors Association web site.
16. 28 copies of hydro geological maps of the National Centre of Nuclear Security and Radiation Control of Atomic Energy; National Institute of Astrological and Geophysical Researches, Centre of Engineering Researches and Consultations, and Institute of Surveying Researches.

The Academy adopted and supported several Egyptian inventions and innovations, besides supporting all ideas that could be turned in to technological projects via celebration on 30th September 2003, towards the distinctive patented Egyptian inventors of 2002, which was supported by Dr. Mofeed Shehab, Minister of High Education and Scientific Research. This was done for the purpose of encouraging inventors and innovators. As a result, a number of marketed distinctive Egyptian inventions were selected (seven inventions). Others were expected to reach the market by first carrying out feasibility studies, in order to reach quantitative stage, and commercialization. A number of inventions have been sent to a specified committee to manufacture their prototypes. It was clear that the Academy had an effective role in developing technology via certain mechanisms in order to take the distinctive Egyptian inventions into quantitative stage and commercialization. This was to encourage creativeness, innovations and inventions. As a result, a lot of benefits will return to applied bodies in addition to their achievements in technological development– a way forward to social and economical development.

Part III

S&T Policy Implications in Socio-Economic Sectors

Chapter 19

Assessing the role of local institutions in adoption of innovations for sustainable agriculture in Kenya

Lutta Muhammad and Paul G.A. Omanga

Abstract

The paper presents findings of a study conducted in South Western Kenya with the objectives of assessing the impact of small scale irrigation and construction of terraces and analyzing factors influencing in its adoption. Methodology involved a survey of 298 members of the Homa Bay Agriculture and Environment Program (HB-AEP) of the Catholic Diocese, using rapid appraisal techniques and group discussions. Data analysis involved extraction of descriptive statistics and regression procedure.

Farmers adopted terracing first, raising grain production by 12 per cent, then bucket irrigation. Mean length of terraces on cropland was 77 meters per farm. Two systems, (bucket and T-pump) were adopted by 22 per cent for irrigating kitchen gardens and raised beds. Factors affecting adoption and impact were land tenure, and micro finance. The micro-finance system allowed access to local and external capital, contributing to entrepreneurial acumen and household welfare. It suggests that a policy on the Village Bank development should be adopted.

Keywords: Group-formation, Kenya, Micro-finance, Service delivery, Sustainable farming practices.

Introduction

Post-independence public investments in education, land and other institutional reforms over the 1960-80 period yielded impressive dividends in terms of agricultural modernization and increased productivity of smallholder systems. These gains were evident in coffee, tea, pyrethrum, cotton, sugar cane, tobacco, horticulture and in the food crops sub-sectors. The impacts of these achievements were more pronounced in the high potential agro-ecological zones than in the semi-arid regions. Moreover, population growth rates during the period were high, resulting into intense pressure on land and consequential threats to the environment. The downward trend in the world market prices for Kenya's commodities coupled with rising prices for imported farm inputs and high transportation costs also adversely affected the smallholder sector. In consequence, smallholder farming now faces declining soil fertility, increased erosion, reduced bio-diversity and declining productivity. Both public and voluntary sector strategies to improve the welfare of smallholders now incorporate sustainability and equity considerations.

The public sector provides agricultural research, education and extension services, economic infrastructure and a secure socio-economic environment. The private and voluntary sectors complement the roles of the government sector by providing extension services and other resources that are needed for smallholder development.

The Kenya Agricultural Research Institute (KARI) is the premier national agricultural research organization. The Institute has the legal mandate to carry out research in fields specified in the fifth schedule of the Science and Technology Act, Cap 250, which includes agriculture, veterinary sciences, cooperation with other organizations and institutions of higher learning in training programmes and on matters of relevance to research, liaise with other research bodies worldwide carrying out similar functions, disseminate research findings, cooperate with the relevant ministries, the National Council for Science and Technology (NCST), and relevant research committees in matters pertaining to research policies and priorities, and do all things as appear to be necessary desirable or expedient to carry out its functions.

The Institute has 22 main research centres and 14 sub-centres located all over the country. The centres are allocated mandate areas within their locality for adaptive research and strategic research depending on their comparative advantages to carry out research. The Catholic Diocese of Homa Bay has, since 1987, been implementing agriculture and environment related projects within the area over which it has jurisdiction. This area covers Homa Bay, Rachuonyo, Suba, Migori and Kuria administrative Districts in the south-western Kenya. The area is characterized by high infant mortality, lack of appropriate farm tools, implements and other inputs, poor access to markets, environmental degradation and food shortages. In recognition of the multiple dimensions of the food security problem, the diocese has adopted an integrated approach embracing three programme thrusts, namely, Maternal Child Health and Community Based Health Care (MCH/CBHC); Water and Sanitation; and, Agriculture and Environment Programme (HB-AEP). The (HB-AEP) Programme, scheduled to start in July 1997 and end in July 2000, was designed to address sustainable agriculture and relat d problems. The office of the Diocesan Integrated Development Services (DIDS) was the implementing agency. The Diocese conducted a study of HB-AEP, between August 31st and September 6th 2000, with the purpose of obtaining and documenting base-line information on the program activities, beneficiaries and collaboration, for possible use in planning and evaluation of the next phase of HB-AEP. In this document, the terms of reference, procedures and methods and the main findings of the HB-AEP study are presented (Catholic Diocese of Homa Bay, 1997).

Physiography

Jaetzold and Schmidt (1982) estimated that total land of the said area is 5,714 km^2, out of which, 4,533 km^2 is agricultural. There is considerable variation in topography, agro-climatic and soil resource endowments as well as density of human settlement across the region. There are plains in the centre of the region and alluvial valleys away from steep slopes especially to the east. Near the lake shore line are steep slopes, *e.g.*, the Gembe and Gwasi hills. Under influence of these physiographic features, contrasts in rainfall due to local circulation are tremendous. Average annual amounts range between 700-1800 mm. The first and second rainy seasons start in March and November, respectively. Towards the lakeshore line, the second season is too short and weak to reliably support a reasonable crop. There is also a middle season and its expected time of onset is mid August. The region is characterized by high temperatures and humidity. Evapo-transpiration rates are high (2000-2200 mm/ year) due to the strong winds experienced. Most of the soils of the region are overlaid with relatively acid parent rock. Only small areas with basaltic rocks occur. Soils of the hill slopes are relatively deep and fertile. The southern and northern parts of the region have shallow soils, which are not fertile. The eastern areas of the region have soils of moderate to high fertility.

Farming in the region

The smallholder farm surveys that were carried out in Agro-ecological Zones[1] (AEZ) 2 and 3 in the area (Jaetzold and Schmidt, 1982) indicated that average farm size was 3.4 ha. About half of this acreage was devoted to crop production, leaving the rest for homesteads development and grazing. Intensification of farming had, by the early 80s, become a matter of growing importance. Country wide, the government set about putting in place several measures. For example, controls on trade in farm inputs and farm produce were relaxed (Nyoro, 1996). Agricultural education and extension services were strengthened through implementation of the National Extension Project (NEP 1 and 2). Adaptive research activities were intensified. As a result of the effort directed at easing the problems which constrain productivity in the smallholder sector, many useful technologies have been developed and tested within the On-farm Research Programs (OFR). Improved varieties of maize, sorghum, millet,

[1] Temperature zones LH, U/LM designate Lower highlands (15-18°C), Upper Midlands, (18-21°C) and Lower Midlands (21-24°C). Mean annual temperatures are enclosed in parentheses. Temperature zones are combined with humidity zones 2 (sub-humid), 3 (semi-humid), 4 (transitional), 5 (semi-arid) and 6 (arid), respectively.

cowpeas, pigeon peas, beans, cassava and sweet potatoes as well as appropriate husbandry practices have been developed and evaluated widely by smallholders in the region. While a number of farmers reported using improved seeds, evidence of the use of fertilizers and crop protection agents was limited. Yields were low and most of the produce harvested was consumed within the region. Diseases were reported as hindering introduction of improved livestock breeds. There was general lack of suitable animal drawn mechanization. Low rates of farmer adoption have tended to reduce the impacts that these technologies might have had on smallholder household foods and income security objectives. These have translated into low productivity, post harvest losses, low income and inadequate nutrition leading to low yields and unsustainable land use.

Even when technologies are available, adoption may be constrained by lack of appropriate grass-root level institutions which can foster and promote adoption. Constraints on public sector extension services have reduced access of farmers to the knowledge they need. The poor state of transport and other infrastructure in the rural areas, and high transaction costs involved in the distribution of farm inputs through trade or other means have adversely affected the development of effective markets for farm inputs (fertilizers, drugs, seeds, chemicals) as well as farm produce. This, in turn, has tended to constrain the private sector participation. Thus, HB-AEP in collaboration with other stakeholders set out to address the following constraints on smallholder development:

1. Development of appropriate grass-root level organizational structures,
2. Information on appropriate land use practices,
3. Suitable institutional arrangements to facilitate adoption of improved practices,
4. Agricultural technology (*e.g.*, suitable varieties),
5. Complimentary technology (*e.g.*, irrigation, oil presses etc.), and
6. Mobilization of resources needed for adoption of appropriate land use practices through micro-finance.

HB-AEP interventions

Three projects, which were concerned with agriculture and the environment, were implemented during 1987 and 1997. These were, the Community Afforestation Programme/Sustainable Farm Management Programme (CAP/SUFAMA), Mobilization Against Desertification (MAD) and South Nyanza Agro-forestry Project (SNAF). The staff of Homa Bay Diocese and MISEREOR, a donor, conducted an evaluation in May 1994. This evaluation revealed that whereas the first two programmes had a limited impact on the communities, SNAF had empowered the communities to identify their needs and to address them through diverse activities other than those initially identified. Discussions between the local communities and DIDS indicated a preference for an integrated programme covering the entire Diocese in which the approach that most empowered the community would be used.

A model community structure that has been developed and successfully operated by communities in 5 Divisions which were covered by SNAF was adopted for replication over the entire programme area. This community structure was Division-based. 'Farmers committees' representing 10 groups in each Division had been set up. Each of these groups was to be a cohesive unit made up of between 10 and 15 farmers who were well known to each other. Group members were expected to have understood the rules of conduct. These structures demonstrated effectiveness in mobilizing community resources, supporting the adoption of skills and appropriate technologies. The 'Farmers Committees' were responsible for the day-to-day management of the programme at the grassroots. The community needs which had been identified during Participatory Rural Appraisals (PRAs) in July and August 1995 formed the basis of the interventions in the Homa Bay Diocese Agriculture and Environment Programme (HB-AEP). HB-AEP replicated this system in the other 17 Divisions while expanding the range of interventions without compromising the existing mechanisms of mobilizing local resources and ensuring credit recovery. HB-AEP interventions were organized around the following activities.

Table 19.1: HB-AEP interventions

Homa Bay Agriculture and Environment Program (HB-AEP) interventions		
Activities	*Interventions*	*Demonstration/trials*
Kitchen gardens	Improved storage	On Farm trials
Fruit trees	Storage pest control	Demonstration of improved agricultural practices
Manure management	Processing and utilization	
Integrated pest management	Micro-finance	
Livestock production	Community mobilization	
Fodder establishment	Group formation	
Upgrading zebus	Group training	
Feeding dairy cows	Savings mobilization	
	Loan approval, disbursement and recovery	

In South Western Kenya, high population growth rates have given rise to pressure on land, declining soil fertility, increased erosion, reduced bio-diversity and declining productivity. Problems associated with inadequate moisture supplies at critical crop growth stages and particularly so in semi arid environments which characterize much of the South Western Kenya (SWK) region can be mitigated through rainwater harvesting techniques by construction of water conservation structures or smallholder irrigation. Terracing would prevent soil erosion and effect water harvesting. Micro-irrigation has, where adopted, rewarded implementing farmers by allowing more flexibility. Moreover, policy measures such as reduction of controls on trade in the farm inputs and farm produce (Nyoro, 1996) and improved agricultural extension services through the National Extension Project (NEP 1 and 2) should have supported increased adoption of numerous technologies for soil conservation, rainwater harvesting and micro-irrigation which were developed and evaluated widely by smallholders in the region. Yet yields remained low (Jaetzold and Schmidt, 1982). Some of the changes and adaptation in farming systems that took place over a period of three years in South Western Kenya following institutional interventions by the Homa-Bay Agriculture and Environment Program and possible explanations for the changes are given below.

South Western Kenya

The South Western Kenya area covers the catchment area for the Gucha, Mogusi and Sondu rivers and extends over Homa Bay, Rachuonyo, Suba, Migori and Kuria administrative Districts. These rivers chart a south westerly course, draining into Lake Victoria. Thus, the area is part of the greater Nile Basin. Most of the soils, however, are suitable for raising reasonable crops under irrigation. Sources of water that can be used for irrigation include surface run off, rivers, ponds and the lake. Average farm size is about 3.4 hectares. About half of this acreage is under crops. Lack of appropriate farm tools, implements and other inputs, and poor access to markets, contribute to food shortages, and low income. Demand for vegetables by the expansion of the rural and urban market in the area is rarely met. In consequence, diets of members of rural households are generally deficient in vitamins and minerals and for this reason, skin diseases are common especially among the fishing and pastoral communities. Prevalence of many diseases is high. Factors such as lack of appropriate grass-root level institutions (Carney, 1996) can constrain adoption of improved farming techniques while non-availability of financial resources needed to support adoption can undermine the efforts of extension agents (Beynon, 1996).

One outcome of the public policy reforms of the Government of Kenya of 1986-96 was the reduced participation in the provision of services on the one hand, while on the other hand, the role of the NGOs as service providers increased. The Agriculture and Environment Program of the Catholic Diocese of Homa Bay (AEP-HB) interventions included mobilization of farmers into groups, construction of soil conservation structures, and the bucket and T-pump technologies for irrigation (Catholic Diocese of Homa Bay 1998-2000). By August 2000, groups affiliated to HB-AEP had recruited 3662 individual members. The HB-AEP sponsored savings and credit scheme was in place.

Methods

Field methodology for this study comprised four activities, *viz.* (*i*) a sample survey of 298 out of the 3600 members of the HB-AEP sponsored groups using Rapid Rural Appraisal Approaches, (*ii*) focus group discussions with representative groups, (*iii*) informal discussions with key resource persons within institutions which collaborate with HB-AEP in the area, and (*iv*) group discussions with HB-AEP field staff. The challenges of data handling and interpretation for this kind of study not withstanding, summary statistics were extracted and regression analyses performed.

Field methodology involved a survey of members of the HB-AEP affiliated groups following the rapid appraisal approaches (Chambers, 1992). A proportionate random sampling procedure was used (Moser and Kalton, 1979). A sample of 300, out of over 3000 households, or 10 per cent of all the members belonging to all groups was selected. Due to the need to represent agro-ecological as well as socio-economic variation in the area, a stratification criterion was devised in such a way that although all districts within the Diocese were covered, distribution within districts was weighted by respective shares of members. Questionnaires and check lists were developed. Enumerators received training and briefing covering the general principles of surveys in social studies, the purpose and objectives of the survey, structure, question wording and sequence. In the session preceding the field pre-testing, participants were exposed to techniques for approaching interviewees, securing interviews and managing time. Farmers were assisted by enumerators to estimate the length of soil conservation measures that had been put in place. Roles of the enumerators and supervisors were spelt out and agreed. The questionnaire was pre-tested within the local farming community prior to formal administration.

Following Shariff and Visara (1993) and (Kumar, 1993), Focus Group Discussions (FGDs) addressing collective assessment of benefits of AEP sponsored interventions were held. Adoption costs, information needs, sustainability, and cost of membership were discussed. Activities of other institutions involved in agricultural development in the area were discussed, laying emphasis on prospects for collaboration between organizations which promote development in the area. Summary statistics were extracted from survey data. Modelling relationships between observations was facilitated by simple cross-tabulations, descriptive statistics and the fitting of the data to ordinary least squares (OLS) and logistic regression procedures.

Results and discussion

Land ownership averaged 3.24 ha, and of this, between 49 to 69 per cent was under crops. Adoption of the ox-plough technology, used mainly for ploughing, stood at 23 per cent. Popular organic farming methods were boma manure and composting, among others. Up to 95 per cent of the respondents were using at least one of the organic farming practices.

Access to information for farming

There were at least 15 sources of information. The most frequently mentioned sources were HB-AEP (88 per cent), the Ministry of Agriculture (6 per cent) and CARE-Kenya (2 per cent). Verbal exchange between the farmer and extension agent during contact (59 per cent), demonstrations (13 per cent), and tours (12 per cent) were commonly used. Less frequently mentioned were seminars (5 per cent), contact with 'technicians' (4 per cent), and farm visits (4 per cent). HB-AEP used all the methods of conveying information to farmers. The Ministry of Agriculture, Netherlands, C-MAD and DANIDA also used demonstrations. As the study on Agricultural Knowledge and Information Systems (ETC 1999, Table 6.) showed, HB-AEP had linkages with all the 25 agencies that were active in disseminating information in the area, being the only agency with such comprehensive linkage apart from the Ministry of Agriculture.

Some farmers stated that they rarely were in contact with the extension agent (35 per cent), a fair proportion (34 per cent) reported that they were in contact at intervals of one month or less. A further 15 per cent, while not giving a quantitative indication of frequency of contact, stated that they were in contact with extension agents.

Savings and credit

Micro finance was an important factor in adoption of smallholder irrigation systems and construction of terraces. A total of 234 farmers in the sample indicated that they were saving with their respective Village Banks (VBs) monthly. Of these, at least 61 had benefited from the loan facilities offered by their VBs. The money borrowed was used to finance general business, crop and livestock production, general farm and general household expenditure. Most respondents (39 per cent) used loan money to set up, improve or to operate general business such as cattle, cereals, clothing, dress making and general trade. This was followed by crop production (16 per cent) where the money borrowed was spent on purchase of seeds and general crop husbandry. Third place (15 per cent) was taken up by expenditure on livestock. This involved purchases of cows and goats and investment in dairy, poultry and the setting up of zero grazing units. The general farm (13 per cent) and household (5 per cent) expenditure were remarked as 4th and 5th, respectively. The main items of general farm expenditure were purchases of a wheelbarrow and oxen. General household expenditure was incurred on purchase of bicycles and payment of school fees. Other items of expenditure accounted for 12 per cent of the loanees.

Major problems were difficulties with processing of loan applications (53 per cent), delays from the date of application to disbursement, high interest rates and the short grace period. Two of the VBs were in the process of putting in place the required infrastructure. Members seemed to be learning, and adjusting to obligations, responsibilities and privileges associated with belonging to the VBs. The savings base of VBs was not robust enough to support the credit requirements of all the members. These problems were more frequent among members of recently established VBs located in areas of lower agricultural potential. Overall, however, the VB system was able to facilitate mobilization of local financial resources, and, indirectly, access to capital from outside the area. There was evidence of contribution to the local entrepreneurial acumen and general household welfare.

Soil and water conservation

The strategy for promoting sustainable land use by smallholders that was adopted by HB-AEP embraced soil conservation, tree planting, kitchen gardening, raised beds and double dug beds. As most of these practices are labour intensive, their adoption would promote on-farm employment and possibly, generation of farm income through sales of vegetables and tree products. Survey results as well as the synthesis of Focus Group Discussions (FGDs) suggest that farmer response to these interventions was encouraging.

A total of 48 km (mean 175 meters per farm) of trash lines and terraces for soil and water conservation had been laid. Many farms had check dams, retention ditches, cover cropping and mulching in place. Terracing was the most popular soil conservation measure to be adopted by farmers (20 per cent), followed by grass strips (12 per cent). Third in popularity was mixed cropping, check dams and retention ditches (each accounting for 10 per cent). These were followed by tree planting, alley cropping, cover cropping and trash lines (7-8 per cent). Some farmers carried out mulching as a soil conservation measure (3 per cent). Regression results for determinants of adoption or length of terraces are presented in Table 19.1.

The coefficients of the "borrowing from Village Bank" and "crop land size" variables bear positive and negative signs, respectively. The implied positive association between borrowing variable and increase in the length of terraces laid is understandable. As already noted, funds borrowed were used for investment in farm and non-farm activities and this appears to have influenced adoption of terracing either directly or indirectly. The 't' statistic for this variable, however, is not significant. The cropland variable has a negative slope indicating that farmers with large acreages tend to invest less in the construction of terraces.

Irrigation

Production of vegetables in the area is low and many families do not have access to sufficient quantities of this essential dietary component. To reliably raise reasonable vegetable crops farmers need to supplement rainfall with irrigation. HB-AEP adopted two low cost approaches to the promotion of smallholder irrigation (the bucket and T-pump systems). Both are simple technologies, which, if adopted on a wide scale, would increase productivity of the rural labour force. The bucket method was adopted by the highest number of smallholders. Adoption of the T-pump technology did not feature prominently in the results of the survey. The reasons for the low rate of adoption of the T-pump were that the cost of the technology is high, and that operating the pump is tedious. Regression results for determinants of adoption of bucket irrigation are presented in Table 19.2.

Irrigation was used for raising small horticultural crops on kitchen gardens, double dug seed beds, and raised seed beds.

The signs on the coefficients of the intensity of participation at group meetings, gender of household head, and borrowing from Village Bank variables indicated a positive association with adoption of the bucket irrigation intervention. Intensity of participation was self assessed by members as the frequency at which they raised issues or offered suggestions, responses or views at their group meetings.

Closely related to the irrigation intervention were setting up and managing kitchen gardens, raised seed beds and double dug beds. These three were intended to help farmers raise fruits and vegetables to help improve nutrition especially for women and children. The three interventions could also help farmers and especially the more vulnerable members of the households obtain cash income through sale of fruits and vegetables that may be surplus to their subsistence needs. Results of the regression analysis (Table 19.2) indicate that the adoption of the kitchen garden intervention may have been influenced by the size of crop land and the length of terraces. As for the raised beds intervention, length of terraces was shown to be positively associated with extent of adoption. Gender of the household head and intensity of participation in meetings were shown to influence adoption of the double dug beds intervention positively. The kitchen garden, raised beds and double dug beds were shown to have been positively influenced by the presence of irrigation on the farms.

Table 19.2: Results of regression analysis

Dependent variable	*Variables in the regression equation*	*Coefficients*	*S.E.*	*Sig.*	*Asympotic ratio*
Length of terraces					
	Constant	0.010	1.241	0.994	0.008
	Intensity of participation at group meetings	0.181	0.215	0.399	0.842
	Gender of household head	0.591	0.533	0.267	1.109
	Permanency of land tenure	-0.345	0.291	0.237	-1.186
	Borrowing from Village Bank	0.037	0.588	0.950	0.063
Bucket irrigation					
	Constant	0.010	1.241	0.994	0.008
	Intensity of participation at group meetings	0.181	0.215	0.399	1.188
	Gender of household head	0.591	0.533	0.267	0.902
	Permanency of land tenure	-0.345	0.291	0.237	-0.843
	Borrowing from Village Bank	0.037	0.588	0.950	0.063
Kitchen gardens					
	Constant	-4.246	2.737	0.121	-1.551
	Borrowing from Village Bank	-0.213	1.134	0.851	-0.188
	Crop land size (ha)	1.712	1.041	0.100	1.645
	Terrace length (m/farm)	0.002	0.004	0.627	0.500
	If irrigation adopted	2.304	1.282	0.072	1.797
Raised beds					
	Constant	41.739	322.19	0.897	0.130
	Borrowing from Village Bank	-0.053	2.161	0.980	-0.025
	Terrace length (m/farm)	0.002	0.009	0.848	0.222
	If irrigation adopted	2.604	1.618	0.108	1.609
	Gender of household head	-0.857	2.398	0.721	-0.357
	Participation intensity (at group meetings)	-10.772	80.543	0.894	-0.134

Contd...

Table 19.2–Contd...

Dependent variable	Variables in the regression equation	Coefficients	S.E.	Sig.	Asympotic ratio
Double dug beds					
	Constant	-2.833	1.491	0.057	-1.900
	Borrowing from Village Bank	-1.410	0.735	0.055	-1.918
	If irrigation adopted	1.178	0.703	0.094	1.676
	Gender of household head	1.139	0.752	0.130	1.515
	Intensity of participation at group meetings	0.246	0.311	0.430	0.791
Grain production					
	Constant	-12.565	585.99	0.983	-0.0210
	Borrowing from Village Bank	31.232	240.58	0.898	0.1300
	Crop land size (ha)	-10.206	115.85	0.930	-0.0880
	Gender of household head	-50.685	236.49	0.832	-0.2140
	Terrace length (m/ha)	2.200	0.969	0.031	2.2700
	Adoption of organic fertilizers	207.415	226.74	0.368	0.9150

Crop production

Out of more than 18 foods, horticultural and other cash crops identified, maize was ranked as the first crop by 55 per cent of all respondents. Second, third and fourth in importance were cassava, sorghum and beans, respectively. This order of importance of crops, however, varied from one district to another. Sorghum received high ranking in Homa Bay and Suba districts while cassava was the second most important crop in Migori and Rachuonyo districts. Beans featured in Kuria and Migori Districts only. Low status of legumes suggests that diets lack proteins and need to be supplemented with milk, meat, eggs, fish or other animal derived food. This would be an important motivation for investing effort in promoting production and consumption of grain legumes by smallholders in the area.

Overall, average grain production before 1997 was 1,600 kg/farm. Farmer estimates indicated that average grain production had increased to 2,033 kg/farm in 2000. Median production levels were 1,165 kg before 1997 and 1,380 at the time of the survey. Factors that may have contributed to raising grain production are further analyzed in Table 19.2.

The results of regression analysis show that the terrace length coefficient is positive and significant indicating that it may have contributed to increased grain production in the region. The other variables, which were shown to influence increased grain production, were borrowing from the village bank and adoption of the organic fertilizer intervention.

Conclusions and recommendations

Many farmers were saving with their respective VBs and some had obtained credit from these. This seems to have resulted from the flexibility that this facility offers. Farmers who borrowed were able to service a variety of needs from this facility. The VB system is a potentially important focal point for collaboration with other stakeholders especially those in the private sector and NGO communities. The VB system still needs to receive external support in the form of governance, managerial skills and modest infusions of loanable resources until sustainability is attained. Participation by collaborating agencies should be encouraged. Ways of simplifying the procedures for processing applications and disbursement should be sought. Improved access to credit facilities to enable farmers to overcome the capital constraint will contribute to increased adoption of the technology and increased fruit and vegetable production in the area.

Acknowledgement

The work reported in this paper formed part of the assessment of the Homa Bay Agriculture and Environment Project that provided resources for field work. We wish to acknowledge their contribution and that of the participating farmers and institutions. The authors wish to record their debt of gratitude to the Catholic Relief Services (CRS) who provided the grant for the assessment.

References

Beynon, J., 1996. Financing agricultural research and extension for smallholder farmers in sub-saharan Africa. Natural Resource Perspectives. ODI. London.

Carney, D., 1996. Formal farmers organizations in the agricultural technology system: current roles and future challenges. Natural Resource Perspectives. ODI. London.

Catholic Diocese of Homa Bay, 1998-2000. Agriculture and Environment Programme AEP. Quarterly Progress Reports. (P.O. Box 326, Homa Bay, Kenya)

Chambers, R., 1992. Rural Appraisal: rapid, relaxed and participatory. Discussion paper, Institute of Development Studies, University of Sussex, Brighton, U.K.

Jaetzold, R. and Schmidt, H., 1982. Farm Management Handbook of Kenya. Vol II A, West Kenya. Ministry of Agriculture, Nairobi.

Kumar, K., 1993. An Overview of Rapid Appraisal Methods in Development Settings. World Bank. Washington D.C.

Moser, C.A., and Kalton, G., 1979. Survey Methods in Social Investigation. Second edition. Heinemann Educational Books, London.

Nyoro, J.K., 1996. Impact of market reform on seed development, multiplication and distribution. In proceedings of the Egerton University Policy Analysis Matrix Conference, September 25th 1996, Egerton, Kenya.

Shariff, A., and Visara, P., 1993. The role of focus group interviews in assessing the primary health and family planning programme in India. Rapid Rural Appraisal Notes No. 17. pp 34-39. International Institute for Environment and Development, 3 Endsleigh Street, London WC1H 0DD.

Chapter 20

Tracer study of science and technology (S&T) graduates passed out from the universities in Sri Lanka during 1998–2003

P.R.M.P. Dilrukshi Ranathunge and
Seetha I. Wickremasinghe

Abstract

A mailing survey was conducted to study the state of employment of graduates passed out from the science, agriculture and engineering faculties of the national universities in Sri Lanka during the period 1998-2003. The study was conducted in four steps, respectively, in 2001, 2003, 2004 and 2005. This paper highlights the salient results of the survey.

The employment rates recorded in the four years of the surveys were, respectively, 78 per cent in 2001, 61 per cent in 2003, 72 per cent in 2004, and 93 per cent in 2005. The number of graduates pursuing further studies increased from 3 per cent in 2001 to 9 per cent in 2003 and declined to 1.5 per cent in 2005. The unemployment rate declined from 19.7 per cent in 2004 to 4.5 per cent in 2005. Among all the S&T graduates, the highest employment rate was recorded for the B.Sc. engineering graduates followed by B.Sc. special science, B.Sc. agriculture, and B.Sc. general science. The rate of the employability of B.Sc. agriculture graduates increased from 70 per cent in 2004 to 94 per cent in 2005. Similarly, the employment rate of B.Sc. general science graduates increased to 60 per cent in 2004 and to 91 per cent in 2005. The highest proportion of graduates was employed in the public sector (government and semi-government). The employment in public sector showed a gradual decrease from 60 per cent in 1999 to 38 per cent in 2004 but increased to 64 per cent in 2005. The private sector employment also increased from 38 per cent in 1999 to 57 per cent in 2004, but declined to 34 per cent in 2005. The waiting time for a graduate to be employed was analyzed. The results showed that 50 per cent graduates secured their first employment within six months of graduation. The major obstacles faced by the S&T graduates while finding a job were indicated to be the lack of work experience, and the absence of employment opportunities in the field of their specialty.

Keywords: Graduate employment, Science and technology (S&T) graduates, Sri Lanka, Sri Lankan universities, Tracer studies.

Introduction

Manpower and employment planning is an integral part of the socio-economic development a country. Development of human resources is a function of well knit education and training system. With the advancement of technology, it is not possible to achieve an economic development without a contribution coming from the knowledge of science and technology. To achieve these goals S&T educational institutions in the country impart training and education at different levels, *viz.* degree, diploma and certificate. The universities are one of the major institutions that supply S&T manpower qualified at graduate level in Sri Lanka. According to the UGC statistics 2004, the Government of Sri Lanka spent USD Million 415.58 (Recurrent USD Million 6.37) on education in the year 2003 (2.22 per cent of GDP). Out of that, USD Million 67.16 (Recurrent USD Million 55.87) was expended on the University education, which was 0.42 per cent of the Gross National Product-GNP (University Statistics, 2004). For the effective utilization of this considerable amount of money being spent on the university education, a well planned approach is necessary. The graduates produced by the universities should be utilized well for the economic development of the country. For this purpose, monitoring of the regular labour market is also important.

In the labour market monitoring system, the employment planning and the manpower planning are two essential steps. For this purpose, labour market trends need to be assessed and monitored to minimize simultaneous shortages or surpluses in the given skill categories. Therefore, the first essential step in employment planning is obviously to estimate the magnitude and trends in the graduate employment. The National Science Foundation, Sri Lanka conducts a regular tracer study of the S&T graduates passed out from the local universities. In the present study, the analysis is focused on gaining an insight into the current employment status of graduates passed out from different areas of S&T in the country, waiting time of the graduates for permanent employment, absorption pattern and rates for various sectors in the labour market, and the demand for different fields of studies. The results of this study are likely to be useful for policy makers in taking policy initiatives to minimize the gaps due to the shortage or over supply of S&T graduates. It will also assist the university educationists to develop university curricula to suit the current needs in the country.

Methodology

The target group of the study was the graduates passed out from the faculties of agriculture, engineering and science in 13 universities in the country during the period of 1998-2004. The addresses of the graduates were obtained from the respective universities. A structured questionnaire was posted to all the addresses obtained from the universities with a covering letter indicating the importance of the study. This study was conducted in four steps. The first step in 2001 was to survey the graduates passed out in years 1998 and 1999. The second step in 2003 was to survey the graduates passed out in the years 2000 and 2001 followed by the third step in 2004 to study for the graduates passed out in 2002. In the fourth step, a subsequent study was conducted in 2005 for the graduates passed out in 2003. The employment rate of the graduates was measured in different subject areas in surveyed years respectively 2001, 2003, 2004 and 2005. After one month of posting the questionnaire to graduates a non-respondent survey was conducted for the graduates who did not respond to the first round of the survey. The letters of reminders with a questionnaire were sent again to all the non-respondents. After second round of the survey, 30 per cent of non-respondents were selected, randomly, and direct interviews were conducted. The collected data were entered using visual basic (VB) interface into the database prepared by MS Access. The data analysis was conducted using the statistical Software Package for Social Sciences (SPSS).

Results and discussion

The study was carried out for 14,270 number of graduates passed out during 1998-2003 from the faculties of science, agriculture, and engineering in all the universities of Sri Lanka, namely, University of Colombo, University of Peradeniya, University of Kelaniya, University of Morotuwa, University of Ruhuna, University of Sri Jayawardenepura, the Open University of Sri Lanka, University of Jaffna, Rajarata University of Sri Lanka, Sabaragamuwa University of Sri Lanka, South Eastern University of Sri Lanka, Eastern University of Sri Lanka and Wayaba University of Sri Lanka. The Newly established faculties in some universities had not produced any graduates during the survey period 2001-2005. These were the Faculty of Agriculture in the Rajarata University and the Faculty of Engineering in the University of Ruhuna. The overall number of graduates surveyed consisted of 8,216 (58 per cent) graduates in science, 4,034 (28 per cent) graduates of engineering, and 2,020 (14 per cent) graduates in agriculture.

The total number of 5,928 graduates responded to the survey achieving 42 per cent response rate. Table 20.1 indicates the percentage of the graduates responded according to their field of the degree. The response rate of the graduates was closely similar to the average number of graduates passed out in each degree field in surveyed years and therefore, it is assumed that the respondent sample adequately represented the surveyed graduate population.

Table 20.1: The response rate of graduates in different disciplines (in percent)

Degree	*2001*	*2003*	*2004*	*2005*
B.Sc. Agriculture	16.8	22	13	15
B.Sc. Engineering	24.2	19	31	22
B.Sc. Science Special	5.6	8	10	15
B.Sc. Science General	53.4	51	46	48
Total	100	100	100	100

Basic characteristics of the graduates

Distribution of graduates by gender

The overall responded sample comprised comparatively higher number of males (Figure 20.1). The sample of engineering graduates was dominated by male (85 per cent male and 15 per cent female) as compared to other disciplines. The gender participation in the agriculture and science streams was almost equal in number (47 per cent female in agriculture and 50 per cent female in science).

Employment status of graduates at the time of the survey

In the questionnaire, graduates were asked to mark their current employment status. They were given five choices to select their current position as 'engaged in employment only'; 'engaged in further education/training only'; 'employed and engaged in further education'; 'self- employed' and 'unemployed'. Graduates who worked as paid employees at the time of the survey were considered as employed. This group might be involved in full time/part time work either in temporary or permanent basis. The graduates pursuing post graduate degree/diploma were

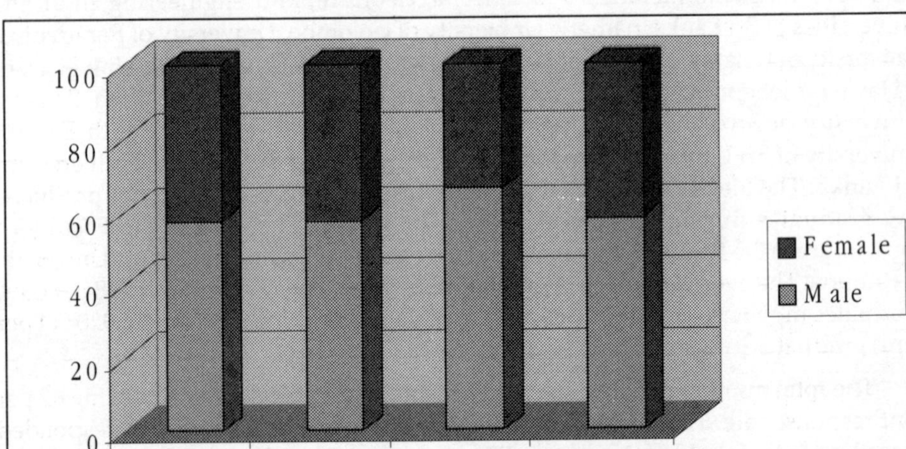

Figure 20.1: Response rate of graduates according to the gender during 2001–2005

considered as 'graduates engaged in further education'. The graduates who were pursuing postgraduate degree/diploma while employed were taken as 'employed and engaged in further education'. The graduates available and looking for employment, and who were unemployed at the time of the survey were considered as 'unemployed graduates'. The graduates involved in further educations were not considered as unemployed as they were involved in further education.

The analysis of the data indicates that the employment rate of the graduates decreased gradually from 89 per cent in 1999 to 61 per cent in 2003, and again increased to 93 per cent in 2005. Similarly, the unemployment rate increased from 6 per cent in 1999 to 31 per cent in 2003, and dropped again to 4.5 per cent in 2005 (Table 20.2). The number of graduates involved in self-employments and further education shows an increase in the years that recorded low rate of employment (Figure 20.2). The rate of graduates involved in further studies decreased from 5.5 per cent in 2004 to 1.5 per cent in 2005 disrupting the steady increase of graduates involved in full-time postgraduate studies (Figure 20.2).

Table 20.2: Employment rate of S&T graduates 1999-2005

Surveyed year	*1999**	*2001*	*2003*	*2004*	*2005*
Year of graduation	1997	1998, 1999	2000, 2001	2002	2003
Employed (per cent)	89.1	78.0	61.0	71.5	92.9
Engaged in further education only (per cent)	4.5	3.0	3.6	5.5	1.5
Self-employed (per cent)	**	1.0	2.3	3.3	1.1
Unemployed (per cent)	5.7	14.6	30.5	19.7	4.5
Not Specified (per cent)	0.7	3.4	2.6	-	-

Sources: Tracer study of S&T graduates 1999, Ministry of Science and Technology in Sri Lanka and National Science Foundation of Sri Lanka 2001-2005.

* Survey of the Ministry of Science and Technology for the graduates passes out in 1997.

** Not analyzed as a separate component in 1999.

The employed graduates were categorized according to the sectors of their employment as 'government', 'semi-government', 'private' and other organization (*e.g.* Non Government Organizations-NGO or foreign/local funded projects etc). The highest proportion of graduates was employed in the public sector (government and semi-government). The employment in the public sector showed gradual decrease from 60 per cent in 1999 to 38 per cent in 2004. However, this again increased to 64 per cent in 2005. With the decreasing employment in the public sector, the private sector employment increased from 38 per cent in 1999 to 57 per cent in 2004, which again declined to 34 per cent in 2005.

The highest number of unemployment recorded in 2003 may be partly due to the then government policy implemented by suspending recruitments to the government service. This has reduced the employment opportunities in the public sector in 2003,

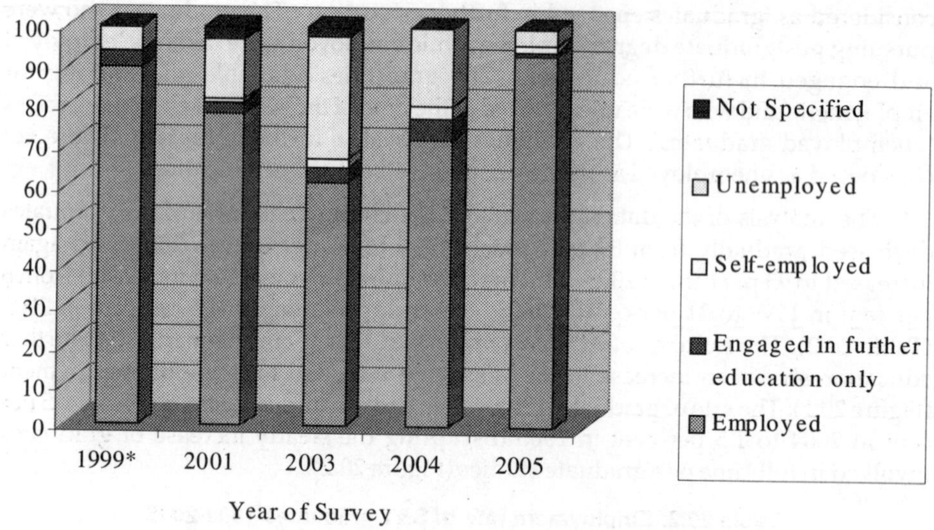

Figure 20.2: Trends in the employment rate of S&T graduates in the period of 1999–2005

thus, elevating unemployment of graduates in the country. However, the suspension was lifted in the later part of 2004 allowing the government sector organizations to recruit employees, which may have lowered the unemployment rate of graduates in 2004 (Table 20.3).

Table 20.3: Absorption rate of graduates into the different sectors during 1999–2005

In association with	*In association with*				
	1999	*2001*	*2003*	*2004*	*2005*
Government	41.7	30.9	23.2	11.5	47.9
Semi-government	17.9	23.7	30.0	26.0	16.0
Private	38.4	37.8	43.8	56.6	33.7
Any other	1.8	2.5	3.0	5.9	2.4
Not specified	0.4	5.1	0.0	0.0	0.0
	100	100	100	100	100

Sources: Tracer study of S&T graduates 1999, Ministry of Science and Technology in Sri Lanka and National Science Foundation of Sri Lanka 2001-2005

The analysis shows that the employment rate of the S&T graduates depends mostly on the government sector employments, despite the fact that there are large number of private sector organizations in the country that are involved in the work related to S&T. However, it is noted that when there is a low absorption of the S&T labour force in to the government sector, the employment rate in the private sector increases, as shown in 2004. Similarly, when there are opportunities in the government sector, as happened in 1999 and 2005. The S&T graduates tend to move from their positions in the private sector to the government sector employments even for low salaries due to the job security and prestige (Table 20.3).

Employment status and degree course

When considering the degree course and the employment status, the employment rate was highest for the graduates who followed B.Sc. engineering, followed by B.Sc. special science, B.Sc. agriculture and B.Sc. general science degrees. The highest percentage of graduates involved in further education were graduates who read B.Sc. special science degree followed by B.Sc. agriculture, B.Sc. general science and B.Sc. engineering. Therefore, the graduates who followed the degree in the field of engineering seemed to have the lowest rate of involvement in further education.

Unemployment rate is comparatively higher for the graduates who read B.Sc. general science degree followed by the graduates with B.Sc. agriculture, B.Sc. special science and was lowest for the graduates who followed B.Sc. engineering. The B.Sc. agriculture degrees were offered by seven (7) universities in the country. In these universities graduates have opportunities to specialized different subject areas. in general, the agriculture graduates had specialized in the fields such as agriculture biology, agriculture economics, agriculture engineering, agriculture extension, animal science, crop science, food science and technology and soil science. The demand for the graduates of different disciplines of agriculture varied in the labour market in different years (Figure 20.3).

All the study series revealed that graduates specialized in the area of agriculture extension, food science and technology, and animal science have the highest employment rate. The graduates who specialized in the areas of agriculture biology, crop science and soil science seem to have low employment rates. However, this trend showed changes in 2005. In this case, the actual demand for graduates of different disciplines was very difficult to envisage since the recruitments were done

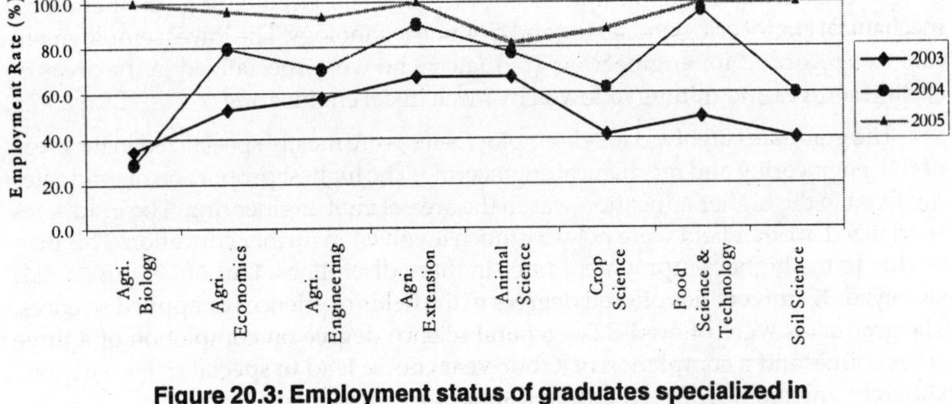

Figure 20.3: Employment status of graduates specialized in different fields in the area of agriculture during 2003–2005

in large numbers without giving much consideration to the actual demand in the organizations for different areas of specializations.

When considering the proportion of agriculture graduates involved in self-employments, the responses received were only from the graduates specialized in the area of agriculture engineering and animal science. The graduates specialized in all the areas mentioned above were equally involved (10-15 per cent) in further education. This rate, however, was low in 2005 compared to other years.

Table 20.4: Employment rates of S&T graduates of different disciplines

Field	*Surveyed year*				
	1999	*2001*	*2003*	*2004*	*2005*
B.Sc. Agriculture	88	48	90	70	94
B.Sc. Engineering	94	96	80	87	96
B.Sc. Special Science	84	90	68	79	95
B.Sc. Science General	87	71	57	60	91
Overall	88	78	61	72	93

Sources: Tracer study of S&T graduates 1999, Ministry of Science and Technology in Sri Lanka and National Science Foundation of Sri Lanka 2001-2005

The B.Sc. engineering and technology graduates are produced by four universities in Sri Lanka, *viz.* University of Peradeniya, University of Morotuwa, University of Ruhuna and the Open University of Sri Lanka. The Faculty of Engineering in the University of Ruhuna was recently established and, therefore, no students were awarded degrees during the period of the survey. These graduates have opportunities to specialize in the fields, namely, civil engineering, mechanical engineering, computer science, mineral and mining engineering, textile and clothing technology, production engineering, chemical and process engineering, material science, electrical, electronics and telecommunication. The analysis recorded 100 per cent employment rate among the graduates specialized in the areas of chemical and processing engineering, computer science, electrical and electronic, material engineering, mathematics, mechanical engineering and textile and clothing technology. The lowest employment rate was recorded for engineering graduates who were specialized in the areas of earth resources and mining followed by mechanical engineering.

The graduates involved in self-employments were mostly specialized in the areas of civil engineering and mechanical engineering. The highest proportion of graduates involved in the higher education was in the area of civil engineering. The graduates specialized in other field were not very much involved in further education. This may be due to the higher employment rates in these disciplines. Out of 13 universities surveyed, 12 universities offered degrees in the fields of sciences or applied sciences. The graduates were offered B.Sc. general science degree on completion of a three years course and a completion of a four-year course lead to special science degree. Different universities offered different subject areas for specializations.

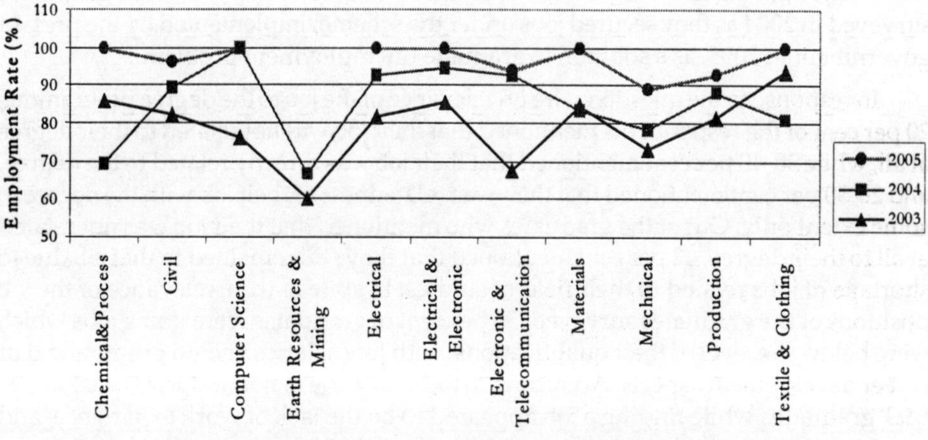

Figure 20.4: Employment status of engineering graduates of different disciplines in survey during 2003–2005

When compared across the surveyed years, the highest employment rate for science graduates was constantly recorded for the graduates specialized in the fields of statistics, industrial management and computer science (Figure 20.5). The lowest employment rates were recorded for the graduates who specialized in the areas of geology followed by zoology. However, a record of high employment rate was seen for the geology graduates in 2005. A reason for this could be the Tsunami disaster, which occurred in December 2004. As a result, frequent landslides were experienced in the country that raised the importance of awareness in distance investigation and arrangement activities relevant to the area of geology.

The waiting time for a graduate to be employed was analyzed. The results showed that 50 per cent graduates secured their first employment within six months of graduation. The 79 per cent of graduates were able to secure their first paid employment

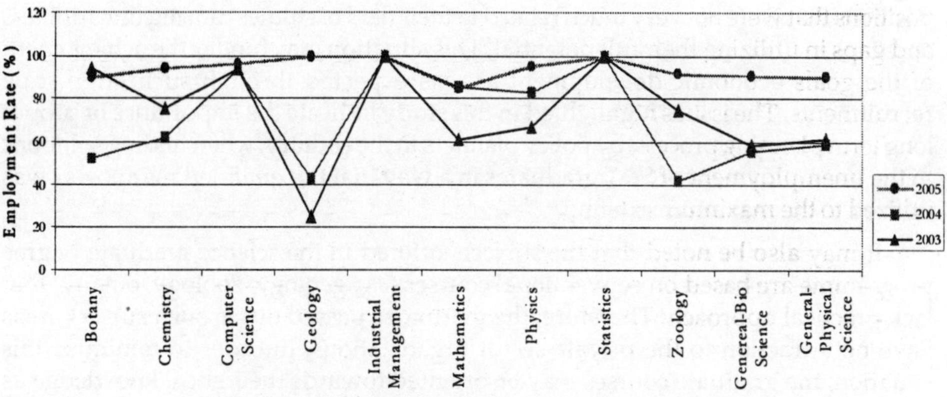

Figure 20.5: Employment status of science graduates of different disciplines in surveys during 2003–2005

within one year period. However, this criterion cannot be applied to the graduates surveyed in 2004 as they secured jobs under the scheme, implemented by the present government in 2005, as a solution to graduate unemployment problem.

In response to the questionnaire on relevance of the job to the degree programme, 20 per cent of the respondents mentioned that their job was not related to their degree at all, while 30-40 per cent mentioned that their job was directly related to the degree, and 20-30 per cent mentioned that their was relatedness of their job with the degree to some extent only. Out of the graduates who mentioned that their job was not related at all to their degree, 42 per cent mentioned that they still remained in that job due to shortage of jobs related to their field of studies. In spite of the irrelevance of the job positions of the graduates surveyed, 36 per cent of graduates were doing jobs which were below the level of their qualifications, with low salaries and no promotional or career development aspects. According to the study, the major obstacles faced by the S&T graduates while finding a job appeared to be the lack of work experience, and the absence of employment opportunities in the field of their specialty.

Conclusions

The fluctuation in employment rates of S&T graduates showed a clear relationship to the government policies, adapted from time to time, to remedy the graduate unemployment in the country. Various governments had taken several measures to overcome the unemployment problem of graduates. The most common step taken was to recruit all unemployed graduates into the government sector employments. In 1999, the government at that time recruited around 9,000 to 15,000 graduates into the government employments. Similarly, at the beginning of 2005, the present government recruited 43,967 graduates into the government sector employment. As a result, the highest employment rates were recorded in 1999 and 2005. Though there is a decline in unemployment at present it cannot be considered as a permanent solution to the graduate unemployment in the country. There is no clear strategy to recruit graduates who will pass out from the universities in future. Mass recruitment also creates problems in assigning jobs to suitable persons. Number of graduates, who responded to the survey, complained that they were recruited into positions that were not very much related to their field of studies causing unhappiness and gaps in utilizing their full potential. This situation may hinder the achievement of the goals economic development as was expected through such large scale recruitments. The issues highlighted in this study indicate the importance of proper long term planning process, by policy planners in the country, when finding solutions to the unemployment of S&T graduates in a way that the qualified manpower was utilized to the maximum extent.

It may also be noted that the subjects offered in the science graduate degree programme are based on conventional courses (*e.g.* geology, zoology, botany) that lack practical approach. Therefore, the graduates passed out in such subject areas have no attraction to the private sector organizations. In order to minimize this situation, the graduate courses may be oriented towards theoretical knowledge as well as practical knowledge. Creating link programmes with the industrial sector may open up avenues to gain working experience. Introduction of new programmes into science stream for improving knowledge in information technology, along with basic aspects in management, administration, law, and economics will also increase employment opportunities for S&T graduates when competing for the labour market in the private sector industries. Such initiatives will no doubt enhance the contribution of science and technology towards economic development of the country.

Acknowledgment

The financial assistance by Asian Development Bank and the Science and Technology Personnel Development Project of the Ministry of Science and Technology of Sri Lanka is gratefully acknowledged. The contribution of the former Scientific Officers Mr. A.P Hettiarachchi and Mr. R. Pathirage to the initial studies is very much appreciated.

References

MOST, 1999. Tracer Study of University S&T graduates passed out from universities in Sri Lanka during 1997. Technical Report, Science and Technology Personnel Development Project, Ministry of Science and Technology, Sri Lanka.

NSF, 2001. Tracer Study of S&T graduates passed out from universities in Sri Lanka during 1998/1999. Technical Report, The National Science Foundation, Maitland Place, Colombo 7, Sri Lanka.

NSF, 2003. Tracer Study of S&T graduates passed out from universities in Sri Lanka during 2000/2001. Technical Report, The National Science Foundation, Maitland Place, Colombo 7, Sri Lanka.

NSF, 2004. Tracer Study of S&T graduates passed out from universities in Sri Lanka during 2002. Technical Report, The National Science Foundation, Maitland Place, Colombo 7, Sri Lanka.

NSF, 2005. Tracer Study of University S&T graduates passed out from universities in Sri Lanka during 2003. Technical Report, The National Science Foundation, Maitland Place, Colombo 7, Sri Lanka.

UGC, 2006. University Statistics 2005, University Grants Commission, 20, Ward Place, Colombo 7, Sri Lanka.

Chapter 21

Food security in Ghana: The development and diffusion of appropriate technologies

Sylvester Gyamfi

Abstract

Ghana, like many other African countries, is well endowed with natural resources, yet it is not food self-sufficient. The economy revolves around subsistence agriculture, which accounts for about 36 per cent of the Gross Domestic Product (GDP) and 60 per cent of the work force. In this context, the government of Ghana has initiated many projects and specific poverty reduction strategies such as, youth empowerment and skills training programme, and micro-finance programmes to ensure food security and economic and financial wealth. The paper shares information about some of such initiatives and activities.

Keywords: Appropriate technologies, Diffusion of technology, Food security, Ghana.

Introduction

The right to food is a fundamental human right. It implies that the individual households must have reliable supply of food to maintain a good health. This necessity of life has eluded over 800 million people world wide in spite of the fact that it is technically feasible to attain food security for all (FAO, 1996). Ghana, like many other African countries, is well endowed with natural resources yet it is not food self-sufficient. The economy revolves around subsistence agriculture, which accounts for about 36 per cent of the Gross Domestic Product (GDP) and 60 per cent of the work force. Ineffective production techniques, low yielding crop varieties and inadequate supplies of irrigation water have been identified as major constraints to achieve food security. The combined effect of these factors coupled with bush fires resulted in severe famine in the country in 1983. The concerted efforts were, therefore, needed to forestall such occurrences in the future and to significantly improve food security in the country.

Action by various governments

For all intents and purposes, the Governments over the years have shown an interest in improving the food situation in the country. The country, under the leadership of the National Liberation Council (NLC), for example, introduced the 'Operation Feed Yourself Programme' where individuals were encouraged to make their own backyard gardens. The National Democratic Congress (NDC) Government pledged its support to the World Food Summit Plan of Action which included coordinated efforts at ensuring food security at the individual, household, national, regional and global levels, eradicate poverty to improve access to food and ensuring peaceful and stable environment, a fundamental condition for the attainment of sustainable food security. The ruling government, the New Patriotic Party (NPP) Government commissioned the Food and Agricultural Organisation (FAO) and the World Food Programme (WFP) to assess the food supply situation in Northern Ghana in 2002 (FAO, March 2002).

Activities undertaken to improve food security in Ghana

Ghana is basically an agrarian country. Improvement in agriculture has implications for food security and, hence, the nutritional requirements of the population. This highlights the issue of moving away from the traditional methods to modernizing agriculture through the application of science and technology. The activities for the improvement of food security have concentrated on (*i*) increased production of selected food crops for food and industrial use, (*ii*) production of fish through aquaculture, (*iii*) livestock production, and (*iv*) prevention and management of diseases.

The Ministry of Agriculture has linked up with the research institutions and is funding various research activities. The dissemination of research findings on some of the major food crops in Ghana such as maize, rice, sorghum, millet, and cassava, is a good starting point.

The application of science and technology has featured prominently in the production of breeder seeds for the production of foundation seeds through various multiplication projects (Grains and Legumes Development Board, 2006). For instance, in the case of cassava, the application of science and technology was carried out through the 'Participatory Cassava Breeding Project' and the 'Cassava Planting Material Multiplication Project'. Under the Participatory Cassava Breeding Project (a farmer/scientist participatory project) the cassava varieties with resistance to major diseases are developed from seeds from 18 cassava families. In the case of the Cassava Planting Material Multiplication Project, planting materials of improved varieties of cassava referred to as breeder stock are multiplied on a large scale. Teams of crop protectionists inspect the planting materials to ensure that they are healthy before they are given for distribution to farmers (Moses E., 2003). As part of the transfer of technology or dissemination of research findings to farmers, field trips are organized to research-managed demonstration fields to serve as tool for the dissemination of improved crop production technology.

The application of science and technology has resulted in the development of new hybrids of maize, sorghum, cowpea, soyabean and other crops. These have been given local names as: maize–*mamaba*, sorghum–*kapaala*, cowpea–*akpaabala*, and soyabean–*salintuya*.

A comparison of crop yield indicates that yield is far lower under traditional management practices than under improved technology. For cowpea, for example, under traditional management practices, the national yield per hectare is 0.3 metric tonnes whereas under improved technology the yield per hectare is 1.8–2.0 metric tonnes (Daily Graphic, September 2006).

Agriculture in Ghana is mainly rain-fed. Inadequate supply of water has been a hindrance to achieving food security. To solve this problem, in 2002, the African Development Bank through the Food and Agriculture Organization's (FAO) special programme for food security (SPFS) committed US $ 1 million for a three-year food security programme in Ghana (FAO, 2006). The key areas of intervention were the effective use of low-cost irrigation techniques, diversifying from crops to livestock and fish, and the introduction of high value horticultural crops. Twenty-three Chinese irrigation experts were deployed to work hand-in-hand with the Ministry of Agriculture extension services to rehabilitate dams in the project areas, thus, ensuring a steady supply of water to farmers. Through the dissemination of efficient irrigation procedures, yields for newly introduced high value horticultural crops such as rice, maize and onions increased three fold, representing a considerable increase in farmer income and access to food in the project areas. Currently, urine is being used, after 3-4 months sterilization, as liquid manure to fertilise maize, sorghum, plantain, pawpaw and other crop fields on pilot basis. Farmers have been exposed to the urine technology, which is being supported by the German Research Ministry (Daily Graphic, October 2006).

Release of improved varieties of crops for cultivation

After years of research by the national scientific institutions connected with food security, the National Varietal Release Committee, the only body responsible for releasing newly developed and improved crop varieties in the country, announced the release of the improved varieties into the public domain. This was done after going through rigorous procedures and processes including field visits, measurements and comparison of size of field with yield.

Prevention and management of crop diseases

The International Society for Plant Pathology Task Force on Global Food Security meeting in Bangkok, Thailand in 1999 accepted to train farmers to acquire simple knowledge in controlling diseases that attack their major staples to increase yield and improve on their food security. In Ghana, this was consciously factored into all crop protection projects. It takes the form of farmer/scientist participatory activities.

Extension officers from the Ministry of Food and Agriculture educate farmers on the effective way to prevent and combat diseases. The plant protection and regulatory services are also playing a major role in the fight against diseases, all aimed at improving food security in Ghana. Seed testing and seed health laboratories have been established as part of the seed quality control programme.

Fish production and preservation

The Ministry of Fisheries is at the forefront of ensuring that the country achieves fish self-sufficiency. It is, therefore, encouraging the construction and stocking of fishponds to improve dietary protein intake as well as generate employment. The Ministry, in collaboration with research institutions, both local and foreign, and private individuals are supporting activities that transfer and disseminate productivity enhancing technologies and information across the country. Field days are organized for the fish farmers at the well-established fish farms as a tool for technology transfer. Observation and participatory approaches are adopted during such trips to enable fish farmers learn and apply the knowledge acquired on their farms.

Farmer income levels are considerably reduced in times of bumper harvest due to low prices. Additionally, a considerable proportion of their catch gets rotten due to inadequate storage facilities and inappropriate preservation methods. To stem the recurrence of this phenomenon, appropriate technological innovations have been developed in place of the traditional methods. One such innovation is the Chorkor Smoker, which maximizes the energy generated for smoking fish.

Institutions behind innovations

At the forefront of the technological innovations are the various agricultural faculties of the country's universities, both public and private, the various agricultural research institutes such as the Savanna Agricultural Research Institute (SARI), the Food and Crop Research Institutes of the Centre for Scientific and Industrial Research (CSIR), non-governmental organizations such as Catholic Relief Services (CRS), Techno Serve and foreign partners such as the FAO, WFP, and German Research Ministry.

Other complimentary activities being implemented to ensure food security

It has been realized that nutritional intake of the most vulnerable population could deteriorate if programmes are not put in place to ensure financial access of all people to food. Therefore, specific poverty reduction strategies targeting the most vulnerable population groups are being implemented (Policy Framework, November 2005). The Government has embarked upon the Youth Empowerment and Skills Training Programme. It is also implementing various micro-finance programmes to ensure wealth creation and ultimately to reduce poverty and improve financial or economic access to food by all Ghanaians. The Government has also instituted the first Friday of December every year as Farmers Day to honour the country's farmers. These measures are aimed at increasing food production to ensure that the country is food self-sufficient.

References

Daily Graphic, 2006. Low Productivity Constraint in Agricultural Development, The Graphic Communications Group Limited, Friday September 29, p 40.

Daily Graphic, 2006. Valley View University Uses Urine as Manure, The Graphic Communications Group Limited, Thursday October 5, p 43.

FAO, 1996. Report on the World Food Summit, November 13-17, Rome

FAO, 2002. Special Report: FAO/WFP Crop and Food Supply Assessment Mission to Northern Ghana, 13 March,. pp 1-23 (http://www.fao.org/docrep/005/y6325e/y6325e00.htm).

FAO, 2006. The Special Programme for Food Security, SPFS at Work: Ghana (http://www.fao.org/tc/spfs/ghana_en.asp)

GLDB, 2006. Half Year Performance Report, Grains and Legumes Development Board, Ghana, pp 1-16.

NDPC, 2005. Growth and Poverty Reduction Strategy (GPRS II) 2006-2009, Policy Framework, National Development Planning Commission, Volume 1, November 2005, pp 41-45.

Moses E., 2003. Activity 4: Farmer Training in Simple Disease Management on Cassava in Ghana, In: Report for International Society for Plant Pathology Task Force on Global Food Security, February 2003 Meeting., Crops Research Institute, Kumasi, Ghana. P.O.Box, 3785, (E -mail: e.moses@cropsresearch.org)

Chapter 22

Measuring the technical efficiency level of sauce processing industry in Malaysia

Abu Kasim Bin Ali

Abstract

A brief perspective of the industry is first presented, followed by empirical results from a survey of fifty-one sauce processors all over the Peninsular Malaysia. This study found that the average technical efficiency (TE) index is 0.27 or 0.29 depending on the estimation method used. There were large efficiency variations among firms, ranging from 0.10 to 1.0. Productive efficiency can be improved by narrowing the efficiency gap between the 'average' and the 'best practiced' firms. Inefficient resource use such as under utilisation of labour and machines are common in this industry. This paper will also present a brief discussion on profile analysis and major problems of the industry.

Keywords: Food processing, Malaysia, Sauce processing, Technical efficiency.

Introduction

The market size of the Malaysian processed food and drinks increased from RM 5,600 million in 1981 to about RM 12,400 million in 2,000, representing an average annual growth rate of 4.2 per cent. However, during this period the contribution of food processing to the overall manufacturing output declined from 16 percent to about 9 per cent. This showed that the industry had been growing rather slowly relative to other manufacturing sub-sectors.

Growth of any industrial sectors has always been associated with productivity or efficiency. Efficient use of resources will contribute positively to economic growth. In Malaysia, the labour as well as the overall productivity has improved mainly through mechanization, especially for firms that fall under the Industrial Coordination Act (ICA) which require manufacturing license to produce. These firms are large, and more capital intensive compared to the small and medium (SME) food processors. The average labour productivity of the large firms is higher than the small one. However, this does not necessarily mean that the large firms are more efficient. As pointed out by Farrell (1957), the average labour productivity is inadequate as a measure of efficiency, since it ignores all the other inputs. The role of efficiency may be viewed as an integral component of any policy to stimulate production and to promote resource conservation. Hence, it is important to know how an industry can increase its output by simply increasing its efficiency without absorbing further resources.

The sauce industry is one of the sub-sectors that have a significant Bumiputra participation. There are more than 200 of them and majorities are SMEs catering for the surrounding consumers. The large companies such as Nestle, Maggi, Cheong Chan Hup Kee's Kimball dominate the supermarket, hypermarket and retail outlets all over the country. The popular sauces sold locally are soya sauce (kicap), chilly, tomato, oyster and peanut.

The domestic market size in 2000 was about RM 820 million. It has shown remarkable improvement in terms of export performance. Its exports had increased from RM 16 million in 1985 to RM 126 million in 2000. With imports of only RM 70 million in 2000, it registered a RM 56 million positive balance of trade. In view of the popularity of the products, its per capita consumption is unlikely to decline. Hence, the increase in export can be attributed to higher domestic production. Major export destinations are Singapore, Hong Kong, Britain and Australia (Abu Kasim and Hamdzah, 2003)

Based on this background, the primary objective of this study was to measure the technical efficiency of the sauce processors in the Peninsular Malaysia. Other objectives include, understanding the status and performance of the industry players especially that of Bumiputra entrepreneurs.

Concept and measurement of efficiency

The study of efficiency relates to observation that firms producing homogenous products have different factor intensities and varying average factor productivities. A test of efficiency will provide a satisfactory explanation of this phenomenon. A technically efficient firm is defined as one, which produces on its outer-bound production function to obtain the maximum possible output that is feasible using current technology (Kalirajan and Tse, 1989). Figure 22.1 illustrates the basic model used to measure technical efficiency (TE) and allocative efficiency (AE) in the case of one variable input and one output. The curve TPP_m shows the maximum possible total output as input X increases while curve TPP_a shows the input response on an average firm. All points lying below TPP_m are technically inefficient because they give less output at a given input.

The profit maximization criterion suggests that a producer will choose to utilize level X_1 where marginal value product (MVP) is equal Px. Thus, a firm that uses X_2 and produces Y_2 is both technically and allocatively inefficient. On the other hand, a firm that uses X_2 and produces Y_3 is technically efficient but allocatively inefficient. Referring to Figure 22.1, the efficiencies can be measured as follows:

Technical Efficiency (TE) = OY_2/OY_3

Allocative Efficiency (AE) = OY_3/OY_1

Economic Efficiency (EE) = TE x AE (as defined by Ferguson, 1966)

$$= (OY_2/OY_3)(OY_3/OY_1)$$

$$= OY_2/OY_1$$

A production function is the relationship that describes the 'maximum possible' output for a given combination of inputs. However, the function estimated by the Ordinary Least Square (OLS) method shows an 'average' response (TPP_a in Figure 22.1) and does not qualify for the theoretical definition of a production function (Ali and Chaudhry 1990). Therefore, it is the production frontier that corresponds to the theoretical notion of the production function.

Efficiency measure

In establishing Technical Efficiency (TE) standard, the deterministic frontier function as proposed by Aigner and Chu (1968) is first specified as

$$Y = f(X)\, e^u \qquad u \leq 0 \qquad (1)$$

In the Cobb-Douglas form, equation (1) is

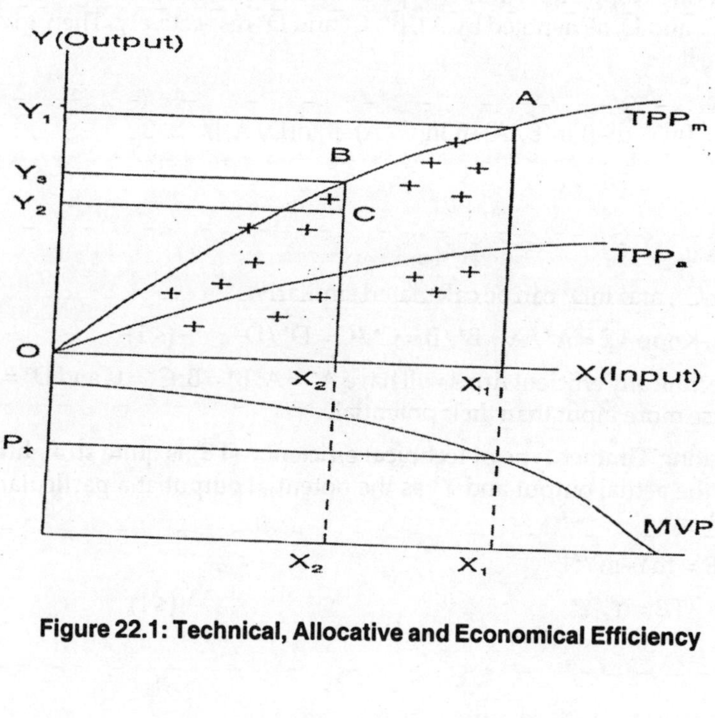

Figure 22.1: Technical, Allocative and Economical Efficiency

$$\ln Y = \alpha + \sum_{i=1}^{n} \beta_i \ln X_i + u \text{ ....} \qquad (2)$$

where,

Y: Output

α: Intercept

β_i: Estimated coefficients

X_i: Set of input variables

u: One- sided residual

To estimate equation (2) OLS is applied, and the best linear unbiased estimates of the β_i coefficients are obtained. The intercept estimate is then shifted upward such that all residuals are either zero or negative {i.e. an amount equal to the largest positive residual (u_{max})}. This procedure will generate a Best Linear Unbiased (BLU) estimate of β_i and a biased but consistent estimate of α (Green, 1980)

Firm specific TE can be measured both through the ratio between actual and potential output as well as through that of inputs. Timmer (1971) used the ratio of actual output to potential output, given the level of input use on that particular firm to measure TE. It thus, indicates how much extra output could be realized if a firm is operating on the frontier. On the other hand, Kopp (1981) measure a firm efficiency by comparing the actual levels of input use to the levels, which could be used if a firm, is located on the frontier, given the actual output of the firm and given the same ratio of input usage.

To generate Kopp and Timmer type of TE measures, let the optimal usage of input A, B, C, and D, be denoted by A*, B*, C* and D* respectively. Then, given Y for each firm

$$\ln A^* = [\ln Y - \alpha^1 - \beta_2 \ln(B/A) - \beta_3 \ln(C/A) - \beta_4 \ln(D/A)] / \sum_{i=1}^{4} \beta_i \qquad (3)$$

where,

$\alpha^1 = \alpha + u_{max}$

ln B*, lnC*, and lnD* can be calculated similarly.

Thus, Kopp TE = $A^*/A = B^*/B = C^*/C = D^*/D \qquad (\leq 1) \qquad (4)$

Only technically efficient firms will have $A^* = A$; $B^* = B$; $C^* = C$ and $D^* = D$, other firms will use more input than their potential.

Calculating Timmer type of technical efficiency (TE) is quite straightforward. Given Y as the actual output and Y* as the potential output at a particular level of input use, then,

$\text{LnTE} = \ln Y - \ln Y^*$

OR $\quad TE = Y/Y^* \qquad (\leq 1) \qquad (5)$

Only the technically efficient firms will meet the constraint Y* = Y. All other firms will have an output less than their potential. Repeating the above calculation procedure for each firm will generate both the Timmer and Kopp type of TE indexes.

Data Used

The listings of sauce processing firms were obtained from the secondary sources, especially, the government agencies that provide assistance to the industry such as Malaysia Agricultural Research and Development Institute (MARDI), banks and the Majlis Amanah Rakyat (MARA). Personal interviews using structured questionnaires were conducted on 51 sauce processors throughout the Peninsular Malaysia based on stratified random sampling. This study aggregates the various sauces (chilli, tomato, soy etc.) produced by the same firm as one output 'sauce'. For the purpose of estimating and generating TE indexes, all variables are expressed in value terms (RM).

Empirical results and discussion

Specification and estimation

The value of output is determined by the level of four input costs, *viz.* direct labour, machine, building and administrative expenses. Using the cross sectional data collected from each firm, the 'average' Cobb-Douglas is

$$\underset{(2.304)}{\ln Y = 2.44369} + \underset{(3.762)}{.58552 \ln LW} + \underset{(1.122)}{.15212 \ln MCH} + \underset{(1.595)}{.17974 \ln BLD} + \underset{(2.371)}{.16096 ADM} \quad (6)$$

$R2 = 0.77$

$u_{max} = 1.5514$

$$\sum_{i=1}^{4} \beta_i = 1.07834$$

t ratios in parentheses

where,

Y: Ex-factory value of all finished products

LW Total production wages or direct labour

MCH: Machine cost (depreciation, rental, maintenance and utilities)

BLD: Rentals of building and land, depreciation, and maintenances

ADM: Administrative expenses (wages, telephone, stationeries etc).

After the appropriate shift in the intercept, and denoting Y* as the maximum value of output obtainable from the given level of input usage, the frontier function becomes:

$$\ln Y^* = 3.99509 + .58552 \ln LW + .15212 \ln MCH + .17974 \ln BLD + .16096 ADM \quad (7)$$

OR

$$Y = 54.33073 LW^{.58552} MCH^{.15212} BLD^{.17974} ADM^{.16096} \quad (8)$$

The OLS coefficient for labour (LW) is highest relative to the coefficients of the other three variables, indicating that labour is the most important input that determines the output level of the small and medium scale Industry (SMI's) sauce manufacturers. In other words, it is a labour intensive sector.

Using procedures explained in equation (4) and (5), generated the technical efficiency (TE) indexes of each firm as summarised in Table 22.1 below.

Table 22.1: Technical efficiency (TE) of a sample of 51 sauce factories

Efficiency ranking of firms	*Timmer TE*	*Kopp TE*
1	1.00	1.00
2	0.93	0.93
3	0.77	0.78
4	0.75	0.76
5	0.68	0.70
–	–	–
–	–	–
51	0.10	0.11
Average	0.27	0.29

If the production function in equation (6) exhibits constant return to scale, the Timmer and Kopp measures would be identical (Russell and Young, 1983). In this study, mild increasing return to scale are found, that is $\Sigma \beta_i = 1.07834 \geq 1$. The sample firms have the average TE efficiency level of 0.27 and 0.29 based on Timmer and Kopp measures of efficiency, respectively. These results show that sauce SMI's as a whole are technically inefficient. On the other hand, a study by Kalirajan and Tse (1989) on mainly large food manufacturing industries in Malaysia found that their mean TE to be 0.734. This study shows that there is a wide variation in TE among the firms sampled. The wide variation indicates that SMIs dealing with sauces have much room for efficiency improvement. Improved efficiency together with better products quality will go a long way in the development of this food sector. A recent study conducted by Embi *et al.* (2003) found that sauces cluster ranked highest compared to 13 other clusters studied in terms of development potential and competitiveness.

Low efficiency levels found on a few sauce factories studied are not necessarily related to production capability. There are cases where productions capacities are not fully utilized due to slowdown in demand. The model used cannot differentiate whether it is bona fide inefficiency or just artificial inefficiency due to unexpected circumstances.

Profile analysis and major problems

Legal status and years of operation

Most of the factories surveyed were SME establishments. Forty nine per cent were registered as sole proprietorships, followed by partnerships (23.5 per cent), private limited companies (23.5 per cent) and co-operatives (3.9 per cent). Semi-government agencies, namely, Farmers Association (FA) and the Federal Land Consolidation and Rehabilitation Authority (FELCRA) established the co-operative factories. About 70 per cent of the factories were located in the rural areas or small towns in which the processors had started their businesses with a small factory on their own land. Eventually, the processors had expanded the factory size to meet production and market requirements.

Nearly half (47 per cent) of the establishments were new players with only five years or lesser period of involvement in business. Thirty six per cent of the respondents have been in operation between six to ten years while about 20 per cent have more than 10 years experience.

Investment

The paid-up capital was used to indicate the investment level of respondents. Again, majority of the respondents are SME processors with 31.9 per cent of them having less than RM10, 000 of paid-up capital. Merely 6.5 per cent of them have paid-up capital of RM 500, 000 or more and hence were not considered as SMI based on the official definition of the Malaysian Industrial Development Authority (MIDA). The mean paid-up capital was RM 95,705, while investment of RM 50, 000 (the mode) was reported by five processors.

Employment

More than half of the respondents employed less than 10 permanent workers. Family members who helped the business were included since most of the processors surveyed were owner operator establishments. Family members normally play an important role in running the business. Only one respondent (2 per cent) had more than 75 permanent workers that can be considered as large scale. Those employing a number varying from 10 to 24 workers represented 37.3 per cent of the respondents.

Major problems

The findings showed that limited capital funds and market competition were the major problems faced by processors. Machine break-down and the ability to satisfy the Food Law Standards were considered as the least problematic issues by a majority of respondents. Although the sauces produced by SMEs are generally believed to be below the standards specified in the Food Regulation (1995), the reasons attributed for same are given as the following factors:

1. Lack of understanding on the importance of meeting the required standards as a measure to ensure consistent product quality,
2. The Food Act (1983) governing the regulations is not strictly enforced, and if the offenders are prosecuted, the fine imposed is usually too little to deter the processors from committing future offences, and
3. In meeting the standards, some extra cost on quality control equipment, material and wages are required. Subsequently, this will add to production cost and the product may have to be sold at a price higher than the competitors who do not meet the required standards.

Shortage of labour is also a major problem, especially, for those operating in the west coast of the Peninsular Malaysia. About 21 per cent of the respondents had commented that getting good workers and retaining them is their biggest problem. Very often food processors are unable to compete for labour with the other sectors of the economy, especially, the electronic and service sectors, for the typical reasons, *viz.* better working environment, prestige, and also due to the availability of cleaner jobs in the non-agro-based sectors. Better incentives including the offering of minority shares to the employees should be explored to overcome the labour problems.

Conclusion

Sauce processing is a growing industry as indicated by the increase in domestic market size and exports earning. The low TE indices as well as large efficiency variation point to a development stage that is yet to reach maturity. Hence, there is ample room for efficiency improvement.

The idling of operation in some firms is partly due to the tight capital fund situation and the limited market for their finished products. Efficiency improvement is possible through improved mechanization and automation to ease labour problems, which require more capital injection in the form of loans or grants from the relevant agencies. Market access can also help the industry in minimizing underutilization of resources that will lead to better efficiency level. The SMIs need support particularly in marketing and technology aspects. In marketing, an aggressive advertising campaign and promotion of products would be possible through funding by FAMA using a common brand. This may lead to better market penetration especially to the level of super and hypermarkets. However, to maintain and improve the market position, a continuous product development is vital, which is only possible through conducting contract research in collaboration with the research institutions because R&D is generally beyond the capacity and capability of small and medium scale industries.

References

Abu Kasim, A., and Hamdzah, A.R., 2003. Peluang perniagaan dalam insustri sos, rempah dan kondimen. Paper presented at Food Processing Business Development Forum, 23 October 2003, Sungai Petani, Penang, Malaysia. Organizer: MARDI.

Aigner, D. J., and Chu, S. F., 1968. On estimating the industry production function. American Economic Review, 58 (4), pp. 826-39.

Ali, M., and Chaudhry, M. A., 1990. Inter-regional farm efficiency in Pakistan's Punjab: A frontier production function study. Journal of Agricultural Economics, 41 (1), pp. 62-74.

Embi, Y. *et al.*, 2003. Value adding of locally produced food crops. Task force report for the Ministry of Agriculture, (Pusat Pentadbiran Kerajaan Persekutuan, Putrajaya, Malaysia).

Farrell, M. J., 1957. The measurement of productive efficiency. Journal of Royal Statistical Society, Series A (General), 120, pp. 253–90.

Ferguson, C. E., 1966. Microeconomic theory. Homewood:Irwin.

Green, W. H., 1990. Maximum likelihood estimates of econometric frontier function. Journal of Econometrics, 13, pp. 27-56.

Kalirajan, K. P., and Tse, Y. K., 1989. Technical efficiency measures for the Malaysian food manufacturing industry. Journal of Developing Economies, xxvii (2), pp. 174-184

Kopp, R. J., 1981. The measurement of productive efficiency: A reconsideration. Quarterly Journal of Economics, 97, pp. 477-503.

Russell, N. P., and Young, T., 1983. Frontier production functions and the measurement of technical efficiency. Journal of Agricultural Economics, 34, pp. 139-150.

Timmer, C. P., 1971. Using a probabilistic frontier production function to measure technical efficiency. Journal of Political Economy, 79 (4), pp. 779-94.

Note: (1 US dollar = 3.39 RM as at June 2007)

Chapter 23

Importance of S&T in providing shelter

M.W. Leelaratne

Abstract

Shelter is a basic need of the people. Almost each country of the developing world is facing the problem of providing shelter to their people. The root cause of this crisis is that the development of new habitats in these countries is still running on the traditional practices through which it seems impossible to provide affordable shelters to their exponentially growing population.

The paper conveys some recent advances of science and technology in relation to design and construction of shelter that are being used in Sri Lanka. These technologies are developed by the National Engineering Research and Development Centre (NERDC). The Centre aims to provide affordable shelters by using alternative building materials, minimizing the usage of traditional raw material like timber, sand, rubble, etc. It shares the significant achievements made in introducing new technologies and the challenges in good governance in providing shelter.

Keywords: *In situ* construction, New technology, Pre-cast building, Shelter, Sri Lanka, Traditional technology.

Introduction

Shelter is a basic need of people. Hence, it is the duty of all responsible governments to provide suitable and affordable shelter for all citizens. Shelter can be defined as a structure or building that provides cover from the weather or protection against danger.

Every family aspires to have its own shelter, because it is the basic infrastructure that keeps the family together and gives protection (Figure 23.1).

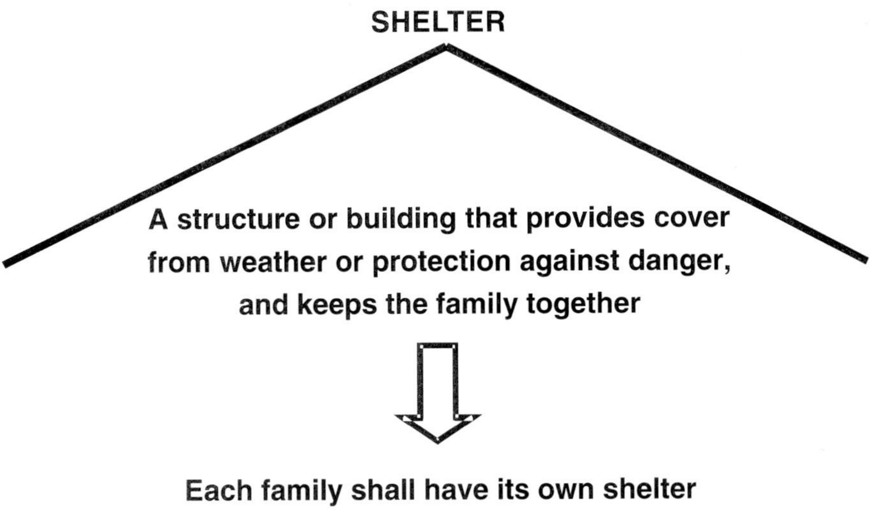

Figure 23.1: Defining shelter

With the increase of population and resources becoming scarce, the task of providing an affordable shelter has, however, become a major challenge. The situation is alarmingly acute with respect to those dependent on meagre income sources for their living, not to mention the fate of those below the poverty line. According to 2001 Census (Central Bank Report, Sri Lanka, 2005), the shortage of houses in Sri Lanka has been estimated at 200,000 units, and another 100,000 or more sub-standard units needed improvement. To make the situation worst, Tsunami disaster occurred in December 2004, destroyed more than 70,000 housing units and partially damaged in excess of another 30,000 units. According to the Central Bank Report (2005), the annual demand for houses in Sri Lanka exceeds 100,000 units.

Science and technology (S&T) can play a major role in meeting such challenges at demanding situations. This paper highlights some recent advances in using science and technology in relation to design and construction of shelter for the people in Sri Lanka.

Challenges in providing affordable shelter

Every family would like to have its own shelter. However, owning a shelter is a question of affordability for many, especially for the low or fixed income groups. Affluent families can afford a shelter of their choice, and are being served by many housing developers, very often at a premium price. However, there is hardly any developer, except the State, for proving shelter for the low-income families. In addition, increasing demand for housing by the upper class, for example, Condominium Houses, Villas, Housing Complexes, etc. and the development related infra-structure construction projects have created a big demand for construction materials and resources, resulting in a supply-demand gap. Many of the present day challenges are the result of the cumulative effect of worsening the supply–demand gap and not introducing new technologies, which can meet these challenges. The situation has worsened in the recent past due to environmental impacts associated, when trying to meet the increasing demands, especially for raw materials. Figure 23.2 shows the interactive effect of some of the important issues needing attention in providing

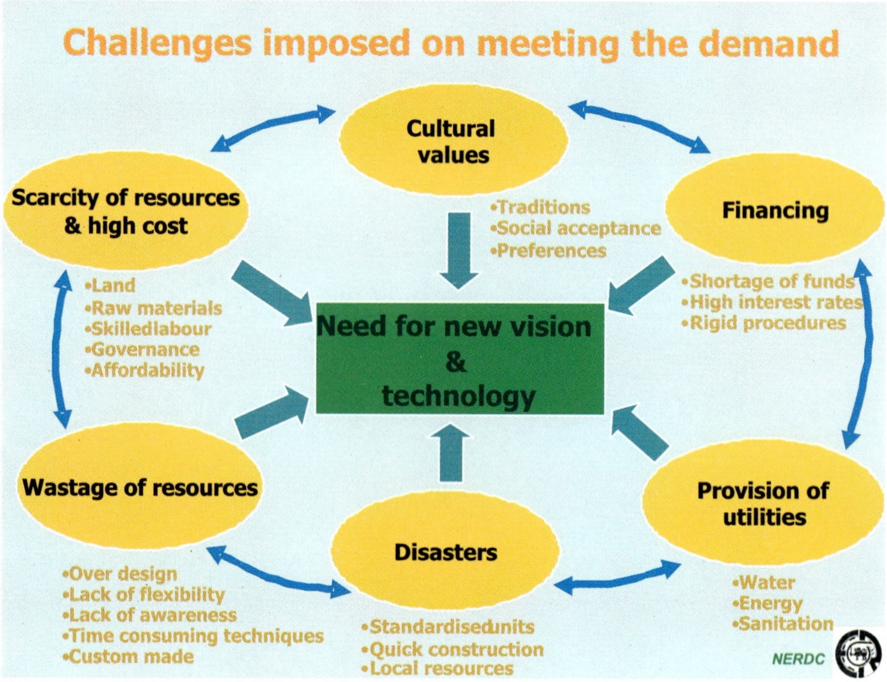

Figure 23.2: Challenges in providing shelter

Price escalation of resources

Two major challenges in providing an affordable shelter have been the scarcity of raw materials and the resulting price escalations. Figure 23.3 shows how the prices of basic construction materials have been increasing in the last several years.

As Figure 23.3 shows that the prices of almost all basic construction materials have been increasing dramatically over the past 15 years with sand (10 times), bricks (6 times) and timber (4 times) leading the trend. Such price escalations are beyond the affordable limits of a low or fixed income earner, whose income does not increase in the similar rate.

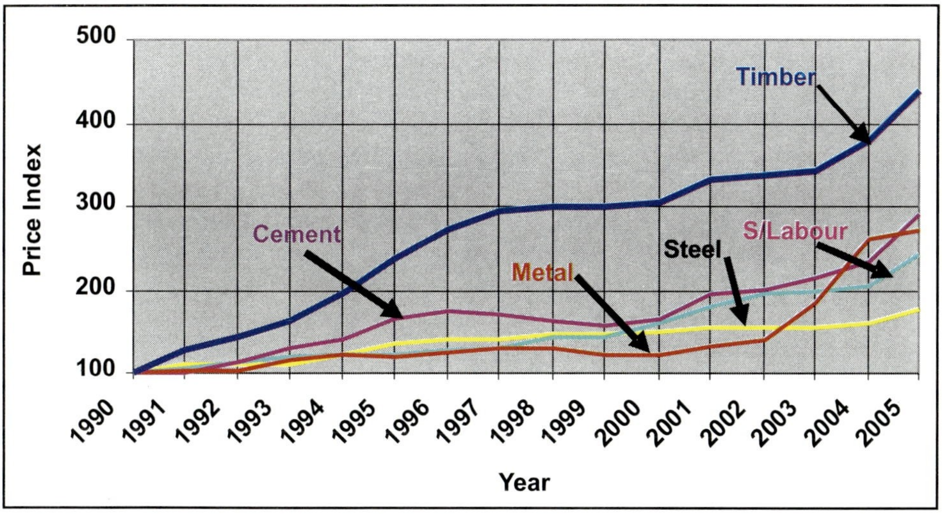

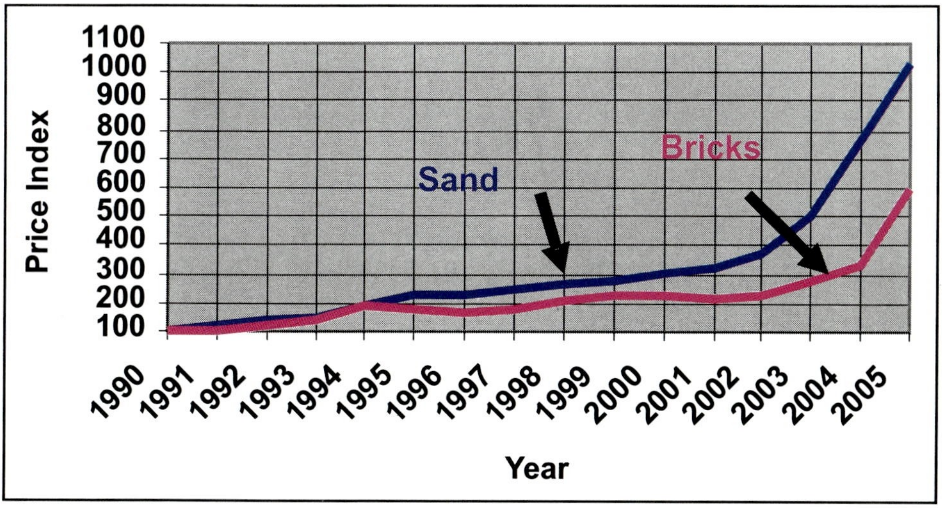

Figure 23.3: Price indices of basic construction materials

***Source*: Construction Statistics, Institute for Construction Training and Development (ICTAD), 2006**

Scarcity and dearth of skilled labour

Cost of skilled labour has also gone up considerably during the past 4 to 5 years. The main reason for this has been the short supply of skilled labour and the increasing number of skilled labour seeking foreign employment. Also, it should be noted that the traditional construction technology, which is still being used in most of the construction work, demands skilled labour. In recent times the component of labour in construction cost has been rising rapidly.

Wastage in traditional construction methods

Up to now, houses have been basically custom made, except the flats, to satisfy individual needs. This process involves the use of non-standardized house plans, building components, and traditional technologies which are inherently inefficient and sometimes wasteful too. Inability to use standardized components, methods and techniques, and the need to construct in situ with traditional technology has made the construction of houses labour intensive, time consuming, and very dear to the ordinary people. In addition, traditional technology depends more on highly skilled expensive labour rather than on innovative, technologically proven construction methods which do not require high skills. Undue wastage of construction materials, for example sand, timber, bricks, cement, etc, which are already in short supply, associated with traditional technology is a major issue of national importance for the sustainability of the house construction. Introduction of new technology is important to address these challenges.

Meeting the challenges through innovative technology

Bringing the humble shelter within the affordability limits of the ordinary citizen calls for new technology, which could maintain the cost of house construction within the reach of the poor. These challenges are met through the introduction of innovative building technologies developed through R&D. The focus of such research has been aimed at developing alternative building materials, minimizing or eliminating the usage of timber, sand, rubble, etc., optimizing designs, introducing pre-fabricated building components, faster construction methods and techniques, reducing the need for secondary materials, etc. The National Engineering Research and Development Centre did some pioneering research work in addressing these issues and was able to develop some novel technologies, which could bring the shelter closer to the people at an affordable price. Given below are the areas focused for R&D in trying to bring shelter within the affordable limits.

1. Introduction of pre-cast standardised building components to house construction
2. Minimising *in-situ* construction
3. Minimising usage of sand
4. Introduction of new and cheaper raw materials
5. Size reduction of components by new and better structural designs
6. Minimising the use of indirect materials
7. Minimizing skill labour requirement by newer construction methods
8. Reduction of wastage by centralised mass manufacturing of building components.

Figure 23.4 shows some of the important new and innovative technologies introduced after carrying out R&D.

Technological change through innovation - R&D		
Old technology	*Proposed new technology*	*Advantages*
• Rubble foundation	Plinth beam foundation for walls	Less excavations, Less material
• *In-situ* concrete column foundation	Pre-cast concrete pocket foundation	Quick construction, Less skilled labour
• *In-situ* columns	Pre-stressed/cast columns	Mass production, Less skilled labour, Standardized components, Less material, Quick construction, No form work
• Brick/Cement-sand block wall	Slip-form wall	New cheaper materials, No sand, Re-usable form work, No plastering, Quick construction, Less skilled labour
	Cement stabilised soil block	New cheaper materials, Material widely available, Less transport cost
• *In-situ* slab	2″ thick Composite slab with pre-stressed beams	Less material, Less/No form work, Fast construction, Mass production – beams, Standardised components
• Timber door/ window frames	Pre-cast concrete door/window frames	Alternate cheap material, Mass production, Standardised components, Less skilled labour, More durable, No lintel required above
• Roof with timber frame	Roof with pre-stressed/cast concrete framework	Alternate cheap material, Less skill labour, Standardised components, Mass production, More durable

Figure 23.4: New and innovative technologies introduced

Significant achievements made by introducing new technology

Since the proper introduction of this new technological vision, more than two decades ago, benefits have been accruing to the society and its people in numerous ways. Given below are some of the major achievements that have benefited the society at large:

1. Over 400 new licensees (contractors) have been established on this new technology
2. Nearly 100 pre-cast yards for manufacturing building components have been established
3. Nearly 20 pre-stressed yards for manufacturing pre-stressed building components have been established
4. Over 5,000 new employment opportunities have been created
5. Buildings/Shelters having floor area of more than 60,000 m^2 has been constructed in the last decade using the new technology
6. Currently, on the average, pre-cast building components required for 25 houses are made per day
7. Alternative construction materials sans sand have been gaining wide popularity
8. Over all sand savings from 40 per cent–70 per cent
9. Over all resource saving up to 40 per cent
10. Over all cost savings up to 40 per cent

Demand for this new technology is increasing, especially from low-income groups. This trend is an indication that new technological vision has created an environment and a matching system, which ensures the provision of shelter to the low-income groups at an affordable price. Figure 23.5 shows some photographs in the application of new technology.

New challenges needing attention

The Tsunami disaster occurred in December 2004 and the other natural and man-made disasters, which were not taken seriously up to now, have opened up new areas for special consideration, in formulating policies and mechanisms in meeting future shelters needs, especially, in disaster situations. Some key issues are given below:

1. Soil investigations in Tsunami affected, land slide and flood prone lands
2. Technology for meeting shelter needs in disaster situations
3. Newer structural designs and construction techniques for withstanding disasters
4. Newer financing mechanisms for low/fixed income groups

Good governance in providing shelter

Governance in the provision of shelter has been addressed through the establishment of a conducive organizational and a regulatory framework

Figure 23.5: Photographs showing application of new technology

Column foundation

Column footing connected by plinth neam

Contd...

Figure 23.5–Contd...

Using cement-soil block

Concrete door/window frames

Contd...

Figure 23.5–Contd...

Concrete roof framework

Slip form wall

Contd...

Figure 23.5–Contd...

Floor slab using pre-cast beams

(Figure 23.6). It may be noted that in the face of the new challenges created through natural disasters, the scope of the good governance and regulatory framework will have to be reviewed and revised appropriately.

Conclusion

Demand for shelter increases with the increase in population. The requirements for shelter vary depending mainly on the affordability of people. Affluent class has very little problems in owning a shelter of their wish because there are many housing developers fulfilling their needs. It is the poor and the low/fixed income groups that are not catered properly to the provision of shelter. Continuous use of traditional technology in the construction of shelter has failed to make the present day shelter affordable to many. This was due to the absence of solutions, in the traditional technology, to rapid increase of the cost of construction materials, scarcity of basic construction materials, new regulations covering environment and usage of land and raw materials, scarcity of labour and time consuming construction methods and techniques. Hence, need for new technological vision is imperative to address the problem of shelter.

Good Governance in providing shelter

Organisational Framework

Ministry of Housing and Construction

National Housing Development Authority — **Main facilitator for fulfilling nations housing needs (low income groups)**
- Review
- Planning
- Ownership
- Motivation
- Management
- Infrastructure

Other institutions directly involved in the provision of shelter

- Ministry of Fisheries and Aquatic Resources
- Plantation Human Development Trust
- Private property developers -
- Individual house builders

Housing Development regulatory bodies

- **Urban Development Authority**
  - Issue permits for urban housing projects
- **National Physical Planning Department**
  - Integration of social, physical & environmental Issues
  - housing designs to suit environment.
  - new township development
- **Central Environmental Authority**
  - Environmental Protection license (EPL)
  - EIA
- **Coast Conservation Department**
  - regulates developments in the coastal zone
- **Land Reclamation and Development Corporation**
  - reclaim land earmarked for development

Institutions Providing Utility Services

- **Ceylon Electricity Board**
- **National Water Supply and Drainage Board**

Figure 23.6: Good governance and regulatory framework in the provision of shelter

The National Engineering Research and Development Centre (NERDC) took a pioneering effort in early 1980's, with the able guidance of its former Chairman, the late *Deshabandu, Vidya Jothi* Dr. ANS Kulasinghe, in addressing this problem. With his vision and guidance, the NERDC's dedicated engineers and scientists have been able to develop new technologies to bring the shelter within the affordable limits of ordinary people in Sri Lanka.

Chapter 24

Popularization of science and technology in eastern zone of Tanzania

Juma Hamsini Seph

Abstract

Popularization of science and technology aims at promoting the application of scientific and technological products to have impact on the social and economic development of the people in Tanzania. This is achieved through the use of the radio programmes, science and technology (S&T) forums, publication of newsletters and stickers to the S&T stakeholders as well as to the general public. The United Nations Educational Scientific and Cultural Organization (UNESCO) funded a project, as part of the technical assistance programme aimed at popularizing science and technology and its effective use to the society in Tanzania. The project was expected to bring about a proper understanding of science and technology by popularizing scientific and technological research findings to end users and the general public. The paper shares the results of the survey conducted under the said project. The findings of the study indicate that the contribution of S&T has had more positive impacts than negative. The only shortcoming observed was that most of the new technologies disseminated depended on the external development partners' financial support. When the project ended, the local government did not take initiatives for extending the best-practiced technologies. Therefore, scaling up the technologies to other areas had been difficult.

Keywords: Popularization of science and technology, R&D institutions, Research extension, Science and technology, Tanzania, Technology transfer.

Introduction

Science advances and technological changes are becoming increasingly important in developing a knowledge-based economy. Realizing this, the Tanzanian government continues to support S&T development that promotes productivity driven growth and provides for competitive advantage. It continues to invest heavily in Research and development (R&D) activities as an effort to strengthen the technological capability of the nation.

Knowledge has always played a role in the progress of nations but knowledge today is about speedy communication and the spread of information and data. Everything that anyone needs to know in order to make a decision must be available immediately. S&T information such as emergence and applications of new technologies, technology trends, human resource capability, R&D activities, research findings and scientific articles and journals must be made available quickly, cheaply and efficiently to policy makers, researchers as well as to the business communities and the public at large. Information Communication Technology (ICT) advancement and the Internet have helped tremendously in pushing these information and knowledge faster and more efficiently but their use in the developing countries including Tanzania, is at a minimal level due to lack of capacity to utilize them.

Today, effective applications of science and technology are vital to various production sectors so as to ensure sustainable national development in the era of increasing competitiveness in the global market. The government through the Ministry of Higher Education, Science and Technology has established the mechanisms for popularisation of science and technology since its establishment in early 1990's. These mechanisms aim at promoting the application of scientific and technological products to have an impact on people's social and economic development. In popularizing S&T, the ministry has been using the radio programmes, S&T forums, publication of newsletters and stickers to the S&T stakeholders as well as to the general public. It is envisaged that after accomplishing this project, a high percentage of Tanzania population will understand the potentials of science and technology in bringing about social-economic development. Therefore, the primary aim of the survey is to gather information that will facilitate the improvement of the modes of the dissemination of the research and development products that have socio-economic impact on society. Thus, it enables easy access to the scientific and technological information by the public.

This report contains a review of the useful scientific and technological research findings to the end-users in the Eastern Zone. Analysis and conclusion of study on popularization of science and technology in Tanzania and key recommendations are provided. The Department of Science and Technology in the Ministry of Science Technology and Higher Education (MSTHE) conducted a study over a period of six months between October 2005 and March 2006, as agreed with the funding agent, United Nations Educational Scientific Organization (UNESCO). The UNESCO has funded this project, as part of the technical assistance programme aimed at popularizing science and technology and its effective use to the society in Tanzania.

Objective of the study

General objective

This project was expected to bring about a proper understanding of science and technology by popularizing scientific and technological research findings to the end users and the general public. The project created public awareness on the prevailing technical changes brought about by S&T development. Currently, such an understanding of S&T issues among the general public seems to have been blocked by different factors that affect the process of sharing of scientific and technological information. Such factors include poor teaching methodology in science and technology education at all levels, lack of awareness on the relevance of S&T development to daily lives of the general public, science and technology education is limited only for the selected few, due to non- availability of proper facilities for teaching science and technology subjects *i.e.* ill-equipped laboratories at secondary and higher education levels, poor infrastructure for collection and dissemination of S&T information, lack of commitment among top decision makers in spear-heading science and technology development in the country, lack of exposure to science and technology during the past, which resulted in inadequate grasp of science and technology, later on in life (especially politicians), and poor government support on S&T activities due to budgetary constraints. Other factors were not dealt because they were beyond the scope of this project.

Specific objective

Specifically, the project involved the strengthening of the current popularization programmes, which are carried out by the Ministry. Relevant research findings could be applied by the end-users in order to bring about social and economic development of the country.

The study team

The study was carried out by a team coordinated by the Director of Science and Technology Department, Mr. Titus Mteleka. The team comprised of two members, namely, Ms. Dhahia R. Mbaga (Study Team Leader), and Mr. Juma Hamsini (Member).

Methodology of study

The team was guided by the terms of reference (TOR), which outlined broad aspects related to the subject of the project. Therefore, the team concentrated on collection of information regarding useful scientific and technological research findings that are useful to end-users in the Eastern zone, and establishing a database of scientific and technological research findings that had socio-economic impact in the country at large.

In order to collect relevant information on the project, the team used the data collection methods such as interviews, structured and unstructured questionnaires, and physical observation for collecting primary data. In addition, secondary data was retrieved from documents in research and development institutions in the Eastern zone. Two groups of respondents were involved in the survey conducted from

September to December 2005. These were namely, key Informants Interviews and the survey questionnaires.

The first research instrument used was the guided questions. The interviews were conducted with 16 key informants. These included the Chief Executive Officers, Project Coordinators, Researchers and Extension staffs in the selected research and development (R&D) institutions per region in the Zone. The institutions involved were as follows:

1. Kibaha Sugarcane research Institute–Coast region,
2. Tanzania Industrial Research Development Organization–Dar-Es-Salaam Region,
3. Mikocheni Agriculture Research Institute–Dar-es-Salaam Region,
4. Sokoine University of Agriculture–Morogoro Region, and
5. Ifakara Health Research Institute–Morogoro Region.

The second research instrument was a survey questionnaire, which included structured and unstructured questions. The structured questions were qualitative covering issues on personal profile, information on access to new technologies, preference of media for accessing new technology information, linkage with R&D institutions, obstacles in accessing the new technologies, and appropriate modes for dissemination of the research findings etc. Due to meager financial resources allocated for the field work, the sample size of the respondent ranged from eight to ten depending on the total number of people benefiting from scientific and technological research findings/products. The survey covered 101 users of scientific and technological research *i.e.* individuals, villagers and community members.

The villages and community members covered three regions, namely, Morogoro region, Coast region, and Dar-Es-Salaam region. These included the Mbasa village, Kilombero Ruaha Village, Kilombero Kidodi Village, Ifakara Village, Msolwa A, and Msolwa Ujamaa in the Morogoro region; Chanika Village, Magindu Village, Dundani Mkuranga, and Kiparanganda Mkuranga in the Coast region; and Mushroom growers in the Dar-Es-Salaam region. In order to have an equal representation of the types of respondents who are the end users of the research findings, the selection of sample was based on three variables, *viz.* gender, leadership role, and working relationship with research and development institution.

Illustration of sample selection

The sample included 4-5 key informants. The survey questionnaires for villagers and other end users of science and technology included 8 to 10 research and development institutions' products. There were 4-5 males, and 4-5 females.

Procedure for selection of respondents

Guided by the three variables, selection of respondents was done by each member of the team using information from the Chief Executives and Project Coordinators of the R&D institutions and their extension staff. Each selected respondent was contacted in person. Thereafter, each respondent was informed about the objective and purpose of the study. Then the study team members administered the structured questions to the key informants and the survey questionnaires to the end users of the R&D institutions products. An approximate time-duration was spent with the each respondent.

Complete confidentiality in responses was assured to each individual by non-registering of his/her name in the questionnaire. Before administering of the questionnaire a respondent was asked to express his/her willingness to participate or not. The informed consent form was appended as Appendix II. The questionnaire was administered only to those who expressed willingness to participate.

Organization of the report

In order to ease the reading of this report, an attempt has been made as far as possible, to separate the observations and findings, assessment, recommendations, and finally, the conclusions.

Study findings and observations

Useful scientific and technological research findings to the end-users in the eastern zone

According to the TOR, the study team was required to collect the information regarding the scientific and technological research findings that are useful to the end- users in the Eastern zone (Dar-Es-Salaam, Morogoro and Coast Region). The team developed two sets of research instruments, in order to collect reliable data, which assisted the team in making an overall assessment of the useful scientific and technological research findings/products to the end users in the Eastern zone. For the S&T research institutions, the structured questions were administered to the Chief Executives, Projects Coordinators and Extension Staff. These key informants provided data on level of research conducted, sources of funding and facilities, partnership and technical support, and significance of the research and development institutions products to socio-economic development. Whereas, the end users provided information on educational level of the end users, access to the new technologies information, mode preferred to get information on new technologies, application of the new technologies for income generation, significance of various institutions on technology transfer, collaboration with the R&D institutions, visits of the Extension staff, availability of demonstration plots and/or sites and obstacles that hinder adoption of new technologies.

Key informants

Level of the research conducted

The respondents from the R&D institutions mentioned that the level of research conducted was both basic and applied research.

Sources of funding and available facilities

For the R&D institutions covered in the study namely, Kibaha Sugarcane research Institute, Coast region, Tanzania Industrial Research Development Organization, Dar-Es-Salaam region, Mikocheni Agriculture Research Institute, Dar-es-Salaam region, Sokoine University of Agriculture (SUA), Morogoro region and Ifakara Health Research Institute, Morogoro region indicated that most of the sources of funding for operational activities were, mainly, from government allocation, local funding and institutions own generated funds (Figure 24.1). The external development funds were used, mainly, for conducting various researches and improving infrastructure for conducting research within the R&D institutions, depending on the agreement made with the particular development partner.

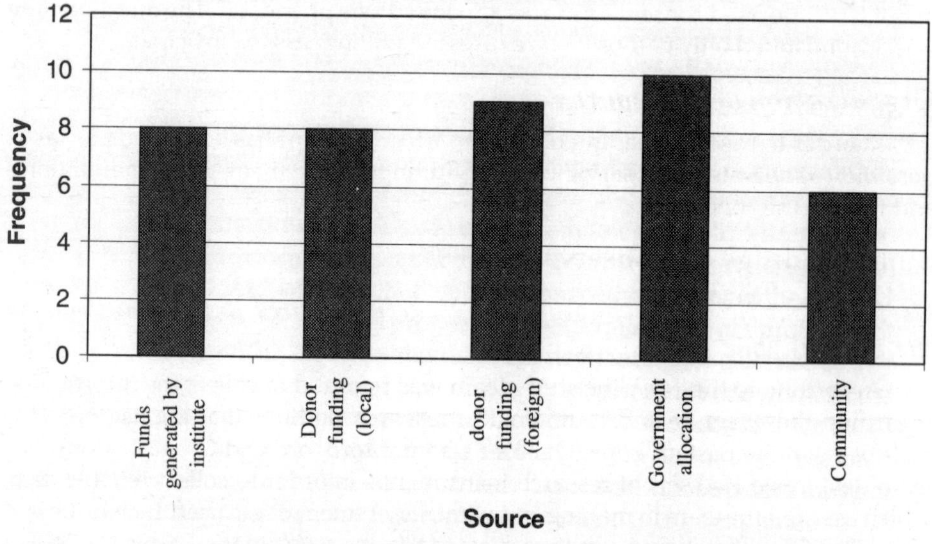

Figure 24.1: Source of funds received by R&D institutes

Regarding the facilities of R&D institutions, the status was good as all institutions had good working environment with good networking to outside world due to access to Internet and availability of a number of journals as indicated in Figure 24.2. Further, the findings of the study indicated that there were a number of extension sites or demonstration sites per each R&D institution that implying that it was a popular mode of dissemination of the research findings/products to the end users. However, on the research and reports publications produced by institutions, the number was few, which indicated that this was a less popular mode of dissemination of the research findings/products.

Partnership and technical support

From the discussion with the Chief Executives and the Project Coordinators, it was pointed out that there were various partners in the implementation of the projects. These included:

1. External development partners that provided financial support when project proposals were accepted and did mid term and terminal evaluation of the implemented projects,

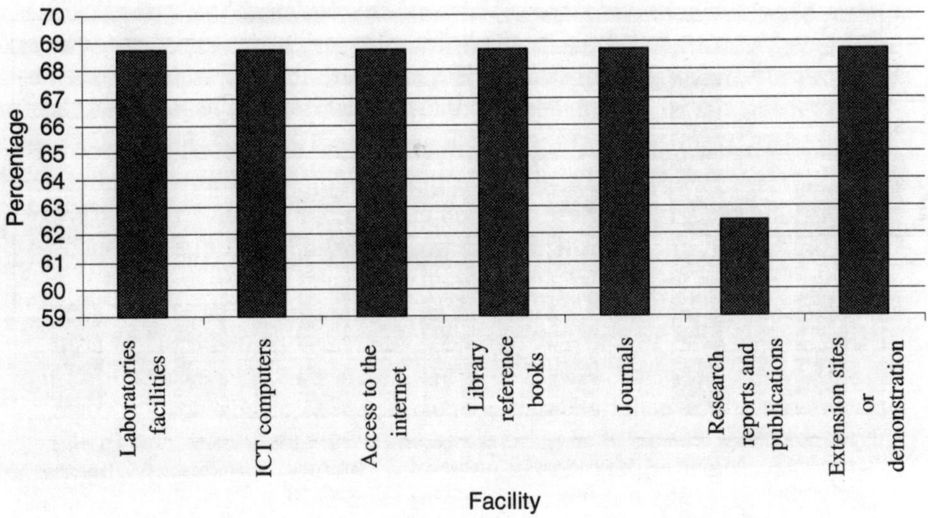

Figure 24.2: Facilities available in R&D institutes

2. Collaborative research with other institutions locally, a good example was the implementation of the Tanzania Agricultural Research Project (TARP) II project at SUA, which involved the researchers from Mikocheni Agricultural Research Institute in supervising the implementation of Mixed Cropping Project in Mkuranga district, and
3. Extension staff working in various villages where the project was implemented had been instrumental in transferring the technologies to the end users and supervising its application.

Significance of the research and development institutions products to socio-economic development

The majority of respondents informed that the type of research conducted in their respective institutions were mainly the basic research with some applied research (Figure 24.3). This could be one reason to have an increased number of new scientific formulae, S&T products and processes etc., in the R&D institutions and the universities that were surveyed.

End users

Education level of end users

The literacy level varied from primary education to university level in the three regions *i.e.* Dar-Es-Salaam, Morogoro and Pwani. The majority did have primary school education as indicated in Figure 24.4. When compared by gender, the variation of education level in primary and secondary levels for the three regions among female and male was not significant. A small proportion of the people interviewed in the three regions of Dar-Es-Salaam, Morogoro and Pwani had University education but these were mainly retirees (men).

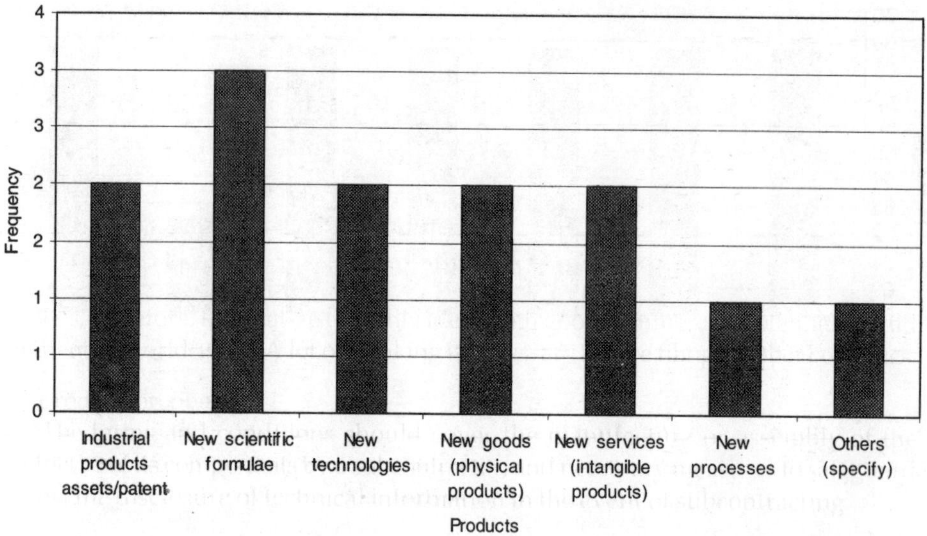

Figure 24.3: Distribution of various research products and processes

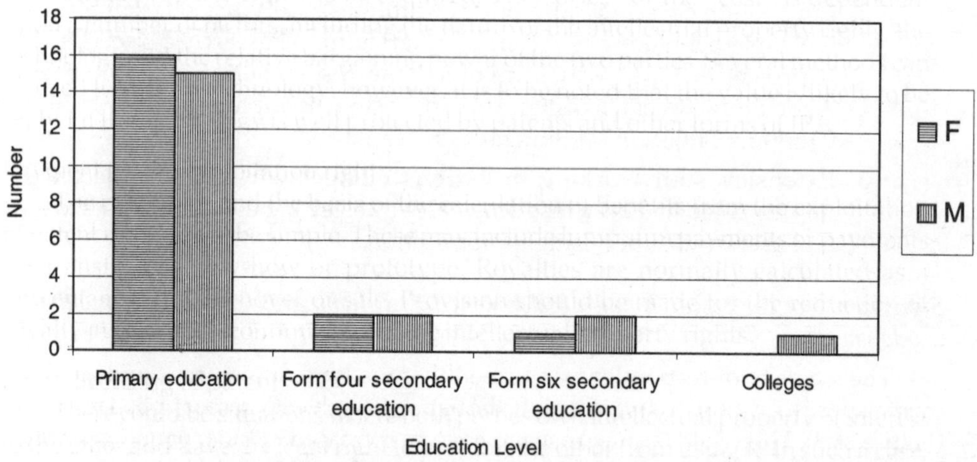

Figure 24.4: Education level of the end-users by gender

Access to the new technologies information

The majority of the respondents expressed that seminars and workshops, extension staff visits, and demonstration plots were authentic sources for new technology information (Figure 24.5). This tally well with the education level of the end users interviewed. The less educated individuals relied on audiovisuals in internalizing the new technologies, whereas, better educated respondents reported to have greater reliance on print materials (newspapers, newsletters, posters, journals and scientific publications). TV dishes were used mostly by the end users living in urban areas and the end users living in rural areas with a better education level.

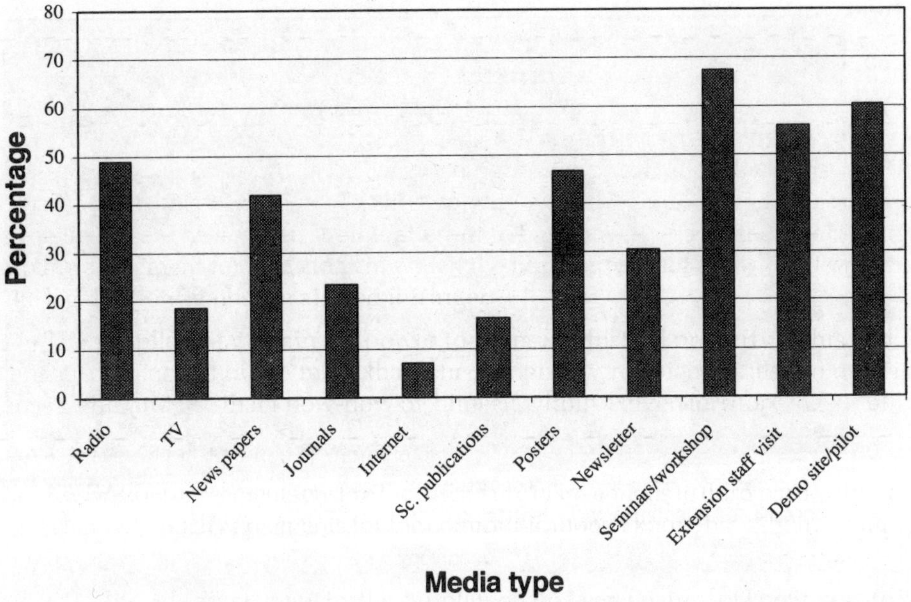

Figure 24.5: Sources of information of new technologies available for the end users

Application of the new technologies for income generation

A majority of the respondents were using food and agriculture technologies (>80 per cent) and reported positive livelihood impacts in terms of cash income, food availability, reduced work load for farming and improved shelters. Other technologies that were important were energy technologies that reduced workload for women and industrial technologies that were used for processing food products and, therefore, increased the food security (Figure 24.6).

Significance of various institutions on technology transfer

A majority of respondents reported access to new technologies for health, efficient use of energy, increase in agriculture production, post harvesting storage, water management, environmental management were from the R&D institutions followed by the universities (Figure 24.7). If one considers the collaboration of R&D institutions with end users in transferring technologies, Pwani region was leading for strong collaboration between the R&D institutions and end users in transferring new technologies for socio-economic development among the three regions (Figure 24.8). University and R&D institutions in three regions (Dar-es-salaam and Pwani) have played instrumental role in developing various technologies mentioned above and transferring them to end users, Figure 24.9 gives the distribution of extension sites among the institutions.

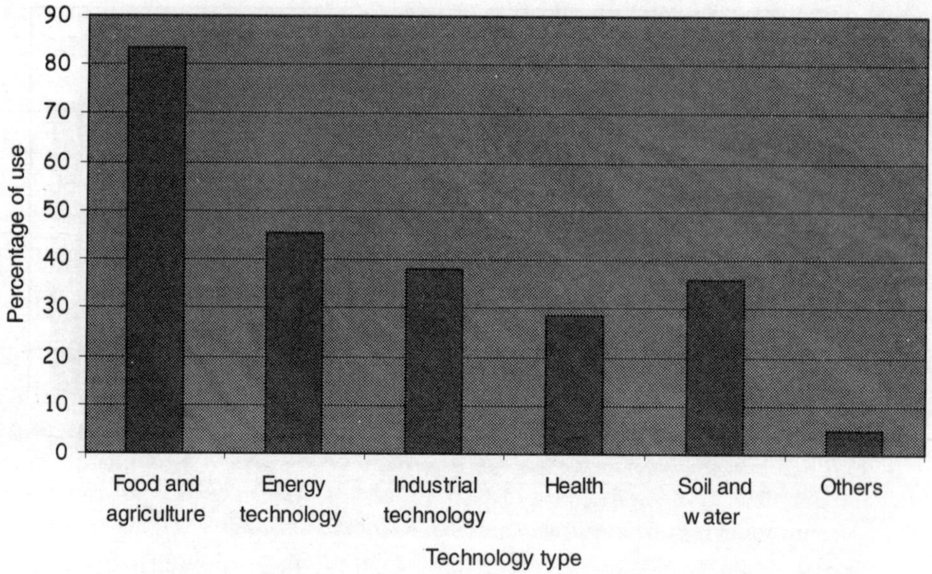

Figure 24.6: Areas of new technology application for income generation

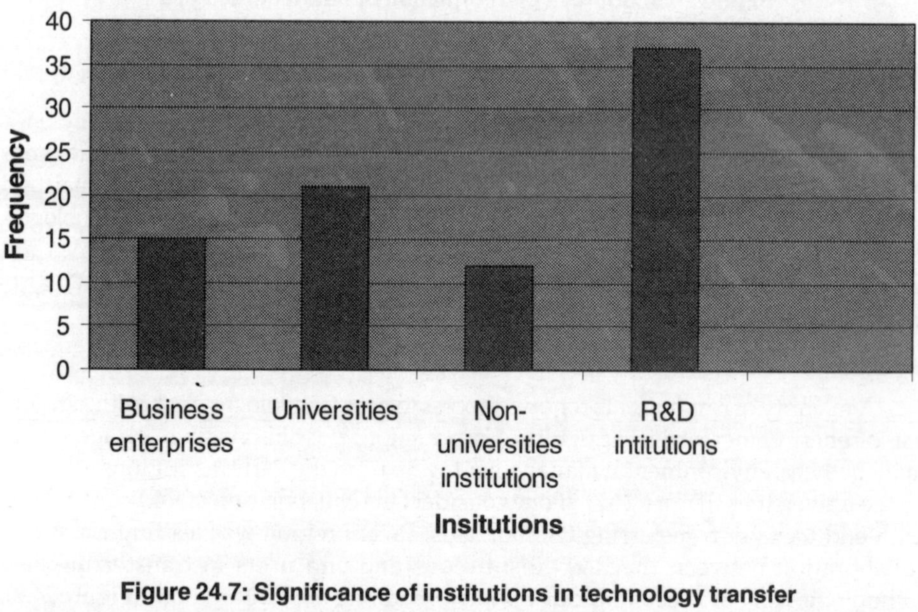

Figure 24.7: Significance of institutions in technology transfer

Obstacles for adoption of the new technologies

Various reasons have constrained the end users in using the introduced technologies. These included lack of information on technologies (60 per cent), lack of technical know-how (45 per cent), education status (36 per cent), poor linkages (30

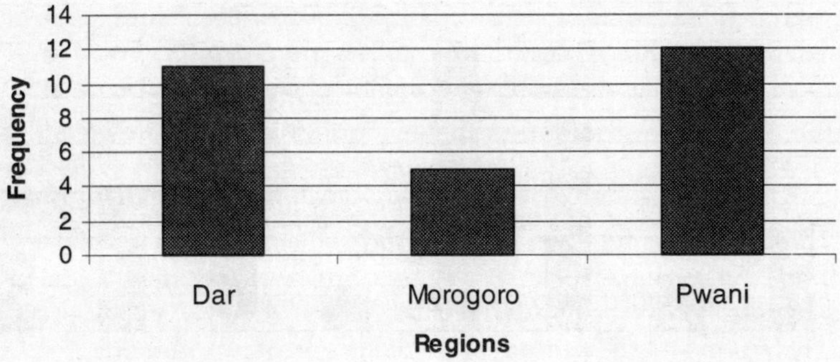

Figure 24.8: Collaboration with R&D Institutions in Accessing New Technologies

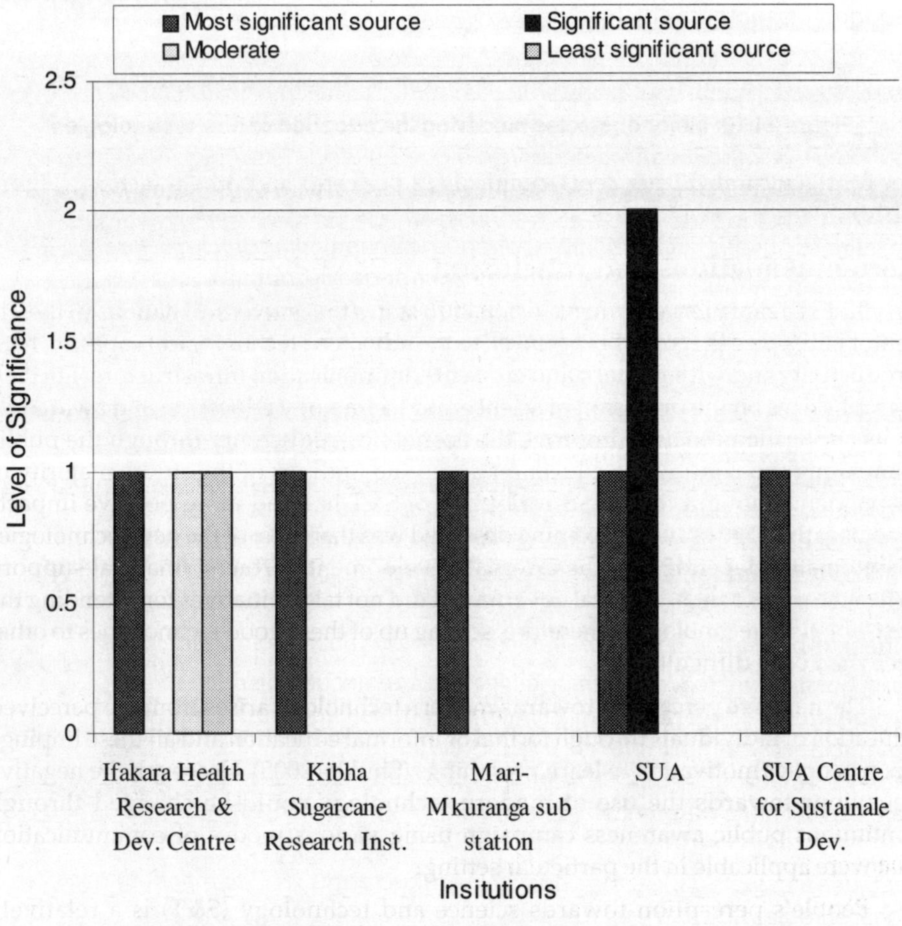

Figure 24.9: Distribution of extension sites belonging to R&D institutions

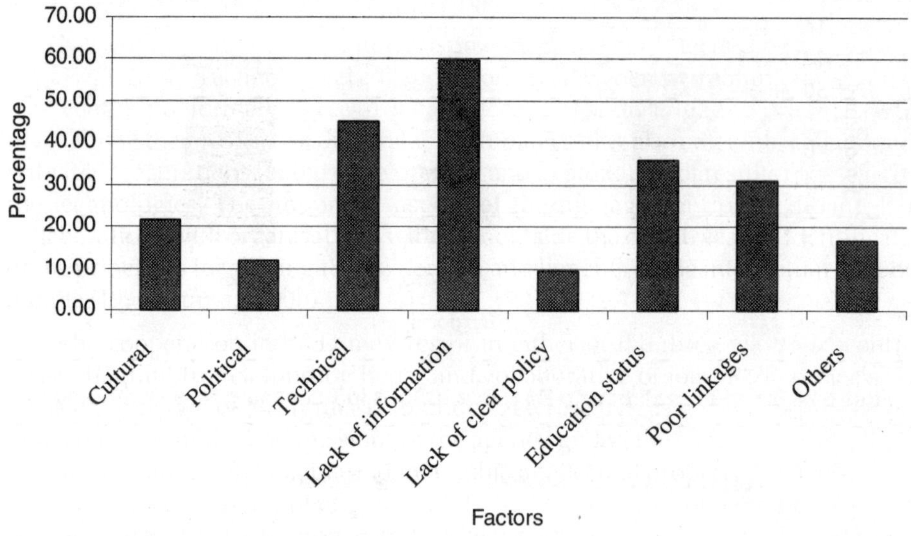

Figure 24.10: Major obstacles hindering the adoption of new technologies

per cent), cultural (22 per cent), political (12 per cent) and others factors. (Figure 24.10)

Recommendations and conclusion

The Tanzania Development Vision 2025 states that successful nations in the 21st century are those that can compete well in the advanced technological capacity, high productivity and with modern and efficient communication infrastructure. Further, it emphasizes on the provision of science and technology education and awareness of its applications so as to improve the economic productivity through the public campaigns and continuous education. From the findings of the study, majority of respondents thought that the contribution of S&T has had more positive impacts than negative. The only shortcoming observed was that most of the new technologies disseminated depended on the external development partners' financial support. When the project ends, the local government did not take initiatives for extending the best-practiced technologies, therefore, scaling up of these good technologies to other areas had been difficult.

The negative perception towards modern technology arises from the perceived education of individuals through formal or informal education and all these impinge upon people's motivation to learn new things (Shukla, 2005). However, the negative behaviour towards the use of modern technology could be changed through continuous public awareness campaign using various modes of communication that were applicable in the particular setting.

People's perception towards science and technology (S&T) is a relatively unexplored area of study and the existing surveys carried out on the subject are only indicative but not representative in nature. To understand the people's perception about scientific and technological issues, their awareness, and how closely the public use science and technology, the Department of Science and Technology (DST) of the Ministry of Higher Education, Science and Technology (MHEST) was commissioned by UNESCO to undertake a study titled "Popularization of science and technology in the Eastern zone of Tanzania". The study has found that the S&T stakeholders need to be strategic in selecting the modes of dissemination of the research findings products/processes to their end users. Similarly, the Department of Science and Technology may consider the following measures in strengthening its popularization programmes on S&T to the general public.

1. Conduct the study on the educational level of their end users as the study indicated that the access to information to technologies vary according to the educational levels and gender,
2. Compliment the radio programmes with other modes of dissemination of S&T information such as promoting the conducting of regular seminars and workshops by R&D institutions to popularize their research findings products/process to the end users, establishment of S&T TV Programmes, dissemination of newsletters with information on the new technologies, and promotion of the establishment of demonstration sites or extension sites for new technologies developed by universities and R&D institutions in various parts of the country,
3. Encourage the universities and R&D institutions to collaborate with the local government and the extension staff on moving from the demonstration sites to an up scaling mode without losing the value added (reaching a

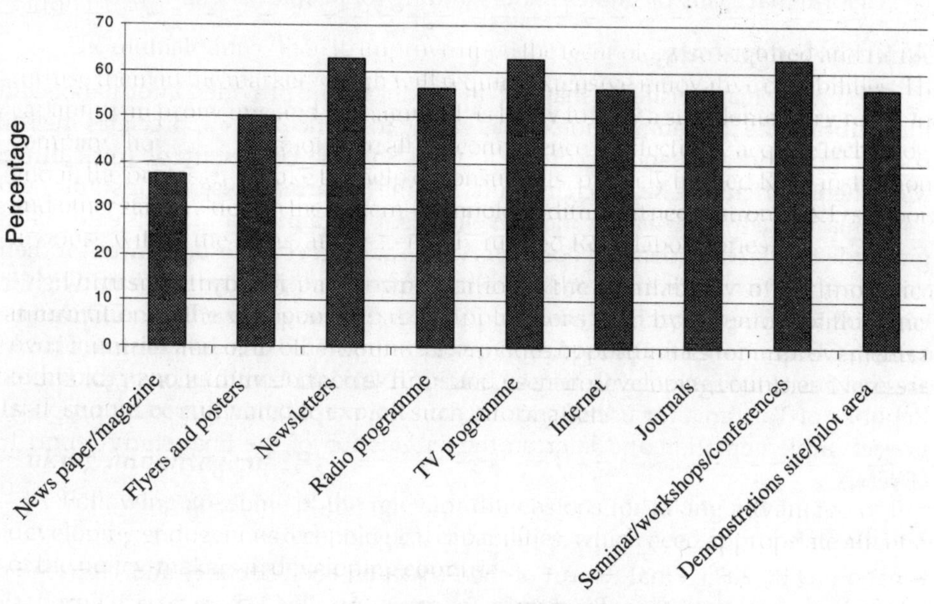

Figure 24.11: Means of dissemination

higher number of end users, having a higher impact to rural as well as urban areas and improve the national economies),

4. Encourage the local government, universities and R&D institutions on provision of incentives, improved working facilities to extension staff disseminating the new scientific technologies to end users, especially those living in remote areas in the country (United Nations Centre for Regional Development (UNCRD) 2002, S&T Sub-Master plan (2003-2018),
5. Encourage the universities and R&D institutions to provide continuous technical support to extension staff as these have been found out in the study as important vehicle for introduction and dissemination of the new technologies to the end users (Kinabo *et al.*, 2005), and
6. Finally, the researchers whether from R&D institutions, or universities should be encouraged to view the extension staff and end users as partners and collaborators so that the feedback obtained from them would be instrumental in the appropriate dissemination of the technologies.

Despite the fact that this study seems to be the first of its kind in Tanzania, the findings and the national science and technology policy for Tanzania (1996), and the guidance on popularization of S&T permit the Director of Science and Technology to offer some recommendations for policy makers and S&T stakeholders on the best modes for dissemination of research findings/products. The introduction of new technologies must as a condition necessitate the introduction of some form of feedback mechanism from end users so as to monitor the effectiveness of communication between researchers, extension staff and the end users (Rutatora *et al.*, 2004). It must be in a form that easily facilitates understanding for people.

Acknowledgments

The study team wishes to thank all individuals who provided information and ideas during the implementation of this study. In particular, it appreciates the full cooperation and assistance accorded to it by Sokoine University of Agriculture, Mikocheni Agriculture Research Institute, Kibaha Sugarcane Research Institute, and Tanzania Industrial Research and Development Organization. It also extends its gratitude to all respondents of both interviews and survey questionnaires for their voluntary participation that made this task successful. Finally, the team wish to continue to express sincerely thanks to Director/Representative UNESCO Dar-es-salaam Cluster Office for providing financial support for this assignment. Thanks are also due to the Secretary General, UNESCO National Commission of the United Republic of Tanzania for useful guidance on the preparation of project proposal, project implementation and later on the submission of the final study report to UNESCO.

References

Kinabo L.D.B., 2005. Final Report of the Project on Food Security and Household Income for Smallholder Farmers in Tanzania, Applied Research with Emphasis on Women under the Tanzania Agricultural Research Project Phase Two (TARP II) -SUA Project 2005.

Mwandosya, M.J. *et al.*, 1996. Towards a New Millennium Perspectives on Tanzania's Vision 2025, Proceedings of a Retreat Workshop held at Mkonge Hotel, Tanga, 16th–20th September 1996, The Centre for Energy Environment, Science and Technology (CEEST).

Rutatora, D.F. *et al.*, 2004. Perspectives and Approaches for Sustainable Rural Development in Africa, Proceedings of the International Conference held at the Institute of Continuing Education, SUA, Morogoro, Tanzania

Shukla, R., 2005. India Science Report: Science Education, Human Resources and Public Attitude towards science and Technology. National Council of Applied Economic Research, New Delhi, India.

UNCRD, 2002. Social Development Issues in Africa, Funded by the United Nations Development Account, 2nd Edition, UNCRD Office, Nairobi, Kenya. Edited by Kidanu A. *et al.*

URT, 1999. The Tanzania Development Vision 2025, The President's Office, Planning Commission, Dar-Es-Salaam, Tanzania.

URT. 2003. Science and Technology Sub-Master Plan (2003-2018), Directorate of Policy and Planning, Ministry of Science, Technology and Higher Education, Dar-es-salaam, Tanzania.

URT, 1996. The National Science and Technology Policy for Tanzania, Ministry of Science, Technology and Higher Education, Dar-es-salaam, Tanzania.

Contributors

Abu Kasim Ali, Deputy Director, Economics and Technology Management Research Centre, Malaysian Agricultural Research and Development Institute (MARDI), P.O. Box 12301, General Post Office, 50774, Kuala Lumpur, Malaysia. (E-mail: abukasim@mardi.my)

Mr. Ali completed Master of Agriculture Science, majoring in agricultural business and economics in 1982 from A&M University, West Texas, USA. He worked on technology foresight at the Policy Research in Engineering, Science and Technology (PREST), University of Manchester, UK. He is responsible to lead and conduct research in the field of resource economics and technology management including S&T policy, technological innovation process, impact assessment and S&T indicators. He has contributed to several tasks and activities related to S&T policy and industrial technology development, especially, within the agriculture and agro-based sectors.

Dilli Raj Joshi, Unit Chief, Science and Technology Policy Research Unit, Nepal Academy of Science and Technology (NAST), P.O. Box 3323, Khumaltar, Lalitpur, Kathmandu, Nepal. (E-mail: info@nast.org.np)

Mr. Joshi completed M Sc (Development Economics) in 1996 from University of Strathclyde, UK and MSc (Statistics) in 1989 from Garhwal University, India. He is responsible for designing, implementing, and monitoring project activities, and preparing guidelines for technology transfer and for carrying out comprehensive industrial studies. He contributes in the overall programme planning, budget distribution, and review of NAST in-house programmes. He has contributed in amendment of Science and Technology Policy and formulation of a 20 years perspective plan on Science and Technology of Nepal.

Dudi Hidayat, Head of Division for S&T Policy Research, Centre for S&T Development Studies, Indonesian Institutes for Sciences, Jl. Gatot Subroto 10, Jakarta 12710, Indonesia. (E-mail: dudi.hidayat@lipi.go.id)

Dr. Dudi Hidayat obtained MSc (Science and technology policy) in 1999 from Science Policy Research Unit, University of Sussex, UK. He is a researcher since 1995 at the Centre for Analysis of Science and Technology Development. He has been conducting a number of studies concerning Indonesian National Innovation System and related S&T policy in Indonesia. Presently, he is conducting R&D survey in the government sector and higher education institutions, in association with the Ministry of Research and Technology.

Ernesto Fernandez Polcuch, Programme Specialist, UNESCO Institute for Statistics (UIS), C.P. 6128, Succ. Centre-Ville, Montréal (Québec) H3C 3J7, Canada. (E-mail: e.fernandez-polcuch@uis.unesco.org)

Mr. Ernesto Fernandez Polcuch did his MSc (Science, technology and society) from National University of Quilmes, Argentina. He leads the Science and Technology statistics team at UIS where he takes care of the production of worldwide R&D statistics survey, as well as the implementation of training activities and the development of new methodologies in the area of S&T statistics. In his previous position as Technical Secretary of the Iberoamerican Network on S&T Indicators (RICYT), he contributed significantly to the establishment of the network and its success in producing S&T indicators for Latin America and the Caribbean.

Ettu Obassi, Scientific Officer, Raw Material Research and Development Council (RMRDC), Plot 427, Aguiyi-Ironsi Street, Maitama District, PMB 232 GARKI, Abuja-FCT-Nigeria. (E-mail: ettustic@yahoo.com)

Mr. Obassi completed MSc (Mining/Mineral processing engineering) in 1995 from Moscow State Mining University. His current responsibility is to assist in identifying and solving problems of raw material exploitation, development and utilization for potential small and medium scale industries in country. He has been involved in preparing the young students into junior engineers, technologists and scientists for national and international competitions. He also takes care of the implementation of federal government policies on capacity utilization of local raw material and the development of process technology for ideal industries.

Hendrik Christoffel Marais, Head of Secretariat, National Advisory Council of Innovation, Department of Science and Technology until the end of March 2007. He is currently a consultant involved in programme evaluation, and research capacity building at a private South African university. PO Box 13054, HATFIELD, Pretoria, South Africa (Email: Marais.stnetwork@gmail.com)

Dr. Marais holds PhD (Experimental social psychology) in 1975 from the University of Port Elizabeth. His work includes strategic leadership functions of the Council, including identification of priorities, initiating evidence-generating STI policy projects and the conversion of such studies into policy. He has published widely in social psychology, communications, research methodology and STI policy, including perspectives on science policy in South Africa (2000). His recent research interests focus on research management, science, technology and innovation policy, and research methodology.

Simon Mpele, Manager, Policy investigations, National Advisory Council of Innovation, Department of Science and Technology, CSIR Campus, Building 53, Private Bag X894, Meiring Naude Road, Pretoria 0001, South Africa. (E-mail: Simon.Mpele@dst.gov.za)

Mr. Simon Mpele holds an MBA degree in 2004 from University of Potchefstroom, South Africa, and an MA degree (Economics) in 1998 from University of Nottingham, UK. He is a portfolio manager of Policy Investigations at the secretariat of the National Advisory Council on Innovation (NACI). He has earlier worked for Department of Science and Technology, managing bilateral relations between South Africa and other countries. He has also worked for the Department of Trade and Industry (DTI), and in a World Bank project for statistics on living Standards and development in South Africa.

Janet Ibrahim Youseef, President, Innovation and Invention Development Agency (IIDA), Academy of Scientific Research and Technology (ASRT), 101 Kasr Al Aini Street, Cairo, Egypt. (E-mail: asrt@asrt.sci.eg)

Mrs. Janet completed BSc (Chemical engineering) in 1976 from Cairo University. She is holding the chair of president in IIDA because of her vast experience in technical research in housing and construction. She has also been attending the seminars and training courses related to the various issues of Intellectual Property.

Juma Hamsini Seph, Agricultural Engineer, Ministry of Higher Education, Science and Technology, P.O. Box 2645, Dar-es-Salaam, Tanzania. (E-mail: msthe@msthe.go.tz, Hamsinij2002@yahoo.co.uk)

Mr. Juma Hamsini Seph accomplished his BSc (Agricultural engineering) in 2000 from Sokoine University of Agriculture, Morogoro, Tanzania. His current responsibilities are to conduct situational analysis and prepare and execute the development programmes for innovators and inventors. It also includes coordination with scientific institutions for socio-economic development, the establishment of Science Parks and Centres, and application and acquisition of appropriate technology in the country. He has the experience in IT management and S&T related issues. He worked as a team member to draft the cyber law and its regulation in the country and is now working as a secretary of Institute of Technology Task Force, a proposed IT university in the country.

Kamaruhzama Bin Mat Zin, Principal Assistant Director, Malaysian S&T Information Centre (MASTIC), Level 4, Block C5, Parcel C, Federal Government Administrative Centre, 62662 Putrajaya, Malaysia. (E-mail: kamarul@mastic.gov.my)

Mr. Kamaruhzama Bin Mat Zin completed his MSc (Information system in management) in 1994 from Sheffield Halaam University, UK. He is designated to collect statistical information related to Science and Technology (S&T) indicators. His prime duties are designing and conducting surveys in the country *e.g.* R&D survey, innovation survey, and survey on public awareness of science and technology. He analyses the data and prepares reports, and also undertakes bibliometric studies.

Lutta W. Muhamad, Senior Research Officer, Socio-economics and Biometrics Division, Kenya Agricultural Research Institute, P.O. Box, 57811 Nairobi, Kenya. (E-mail: LWMuhammad@kari.org, luttam2002@yahoo.com)

Dr. Lutta Muhamad acquired PhD (Economics) in 1997 from University of New England, Australia. The Socio-economic and Biometric Division is involved in the formulation and implementation of national S&T policies for the management of the agriculture industry. His research work is mainly aimed for analysis of farming systems, impact assessment and technology transfer. He has the expertise in the agronomic and livestock research for dryland farming systems in Kenya.

M. Bandyopadhyay, Senior Expert and Administrative Officer, Centre for Science and Technology of the Non-aligned and other Developing Countries (NAM S&T Centre), Core 6A, 2nd floor, India Habitat Centre, Lodhi Road, New Delhi-110003, India. (E-mail: bandodst@yahoo.com, namstct@gmail.com)

Mr. Bandyopadhyay completed Master of Technology in Food Technology and Biochemical Engineering in 1970 from Jadavpur University, Calcutta, India. He has expertise in S&T policy management, international cooperation, technology transfer, technological innovation and academic-industry linkages. Presently, he manages the technical and administrative affairs of the NAM S&T Centre. Earlier, he worked in the industry and in the Department of Science and Technology, Government of India where he was in-charge of several programmes, including international S&T cooperation, technology business incubators, S&T parks, S&T entrepreneurship development and rural technology. He has also worked in the University of Oxford as a senior visiting fellow and has been a UN consultant on technology for rural industrialization.

M. W. Leelaratne, General Manager, National Engineering Research and Development Centre, 2P/17B, IDB Industrial Estate, Ekala, Jaela, Sri Lanka. (E-mail: leelaratne@nerdc.lk)

Eng. Manikku Wadu Leelaratne obtained MSc (Integrated environmental management) in 2002 from University of Bath, UK, and MBA in 1994 from Postgraduate Institute of Management, University of Sri Jayewardenepura, Sri Lanka. His present duties include planning, execution, monitoring, and reviewing the R&D portfolio of the Centre as well as managing its annual operational budgets, and formulation and implementation of its policies. He has conducted R&D on harnessing renewable sources of energy and executed energy management and conservation activities in the industrial, commercial and domestic sectors. He has been a UN Consultant for energy management programmes.

Naresh Kumar, Scientist-G and Head, Research and Development Planning Division, Council of Scientific and Industrial Research (CSIR), Rafi Marg, New Delhi–110001, India. (E-mail: headrdpd@csir.res.in, nareshsahajpal@yahoo.co.in)

Dr. Naresh Kumar acquired Doctorate degree in analytical chemistry in 1978 from Panjab University in India. He has been involved in intellectual property capture, protection and management, technology transfer, R&D planning and management, and human resource development. He is also heading the Human Resource Development Centre, Ghaziabad of CSIR. He has experience of teaching in the S&T related issues in various institutions in the country. He was awarded by CSIR technology prize (1984) for biological science and CSIR shield (2002) for process technology. He was nominated by the Government of India to attend WIPO Academy session in Geneva.

P.R.M.P. Dilrukshi Ranathunge, Scientific Officer, Science and Technology Policy Research Division, National Science Foundation, 47/5, Maitland Place, Colombo-07, Sri Lanka. (E-mail: dilrukshi@nsf.ac.lk)

Dr. P. R. M. P. Dilrukshi received PhD (Zoological sciences) in 2004 from the University of Peradeniya, Sri Lanka. She coordinated the National Research and Development Survey 2004. She is the chief investigator of the surveys on the tracer study of S&T graduates passed out in the universities in Sri Lanka (2002-05), employers perceptions on science and technology graduate employees, and women scientists in Sri Lanka. She also coordinates the National Survey of Research and Development in Sri Lanka and analyzed the National S&T indicators.

Praveen Arora, Scientist F, National Science and Technology Management Information System (NSTMIS) Division, Department of Science and Technology, Technology Bhavan, New Mehrauli Road, New Delhi–110 016, India. (E-mail: parora@nic.in)

Mr. Arora completed MSc (Agricultural statistics) in 1986 from Indian Agriculture Research Institute, New Delhi and is now pursuing his PhD in Science policy from Jawaharlal Nehru University, New Delhi, India. He has been involved in the field of S&T statistics, indicators, and S&T management and policy, and pursuing active research in the area of national innovation systems in the knowledge intensive sectors. He has been managing various surveys and projects for the quantification of S&T resources, development of new S&T indicators, S&T policy and planning, and transfer of technology in the country.

Rajeswari Sarala Raina, Senior Research Fellow, Centre for Policy Research, Dharma Marg, Chanakyapuri, New Delhi – 110021, India. (E-mail: rajeswari_raina@yahoo.com, rraina@cprindia.org)

Dr. Raina is a researcher in the social studies of science, specializing in agricultural and rural innovation. She has qualifications in the agricultural sciences (B.Sc. and M.Sc.) and economics (Ph.D. in 1994 from Kerala University, Centre for Development Studies, India). Recent publications analyse features of and lessons from successful innovation systems, contest the notion of political neutrality of science, emphasise the need to distinguish between institutional and organizational change in the agricultural sciences, and highlight the role of professional associations in enabling innovation in developing countries. Current collaborative research includes work on new insights into civil society led pro-poor innovation using specific cases of solar and bio-mass based rural energy, and reforms to enable financing of rural innovation and rural enterprise development.

Seetha I. Wickremasinghe, Head, Science and Technology Policy Research Division, National Science Foundation, 47/5, Maitland Place, Colombo-07, Sri Lanka. (E-mail: seethaiw@nsf.ac.lk)

Dr. Seetha I. Wickremasinghe has BSc Special (Hons.) degree in Botany and MSc in Agricultural Biology from University of Peradeniya, Sri Lanka. She completed PhD (Science Policy) in 2004 at the Centre for Studies in Science Policy, Jawaharlal Nehru University, New Delhi, India. She presently supervises policy research projects, leads the research team in conducting the National Research and Development Surveys and is responsible for maintaining the Science and Technology Management Information System at the NSF. Her current interests include S&T policy, national innovation system, women in science, and research management. She is the Assistant Editor of the Journal of the National Science Foundation (JNSF) of Sri Lanka.

Seyed Mohsen Masoumzadeh, Head, Division of Innovation and Entrepreneurship, Iranian Research Organization for Science and Technology (IROST), Research Institute for Technological Development Studies, No. 71 Snghelab Ave., Shahid Mousavi Street, Tehran 15819, Iran. (E-mail: masoum@irost.com)

Dr. Seyed Mohsen Masoumzadeh has Doctorate (Technological change in developing countries) in 1996 from University of Bradford, UK. He is working in the Research Institute for Technological Development, a multidisciplinary research non-profit organization under the IROST. He is also contributing in the Innovation and Entrepreneurship program in the Institute. His research experience is in technological innovation in Iranian manufacturing industries and valuation of intellectual property rights. The current areas of interest are technological development, innovation, entrepreneurship and national innovation system.

Sylvester Gyamfi, Assistant Chief Statistician, Ghana Statistical Service, Head Office, P.O. Box 1098, Accra, Ghana. (E-mail: gyamfi@yahoo.com, sgyamfi1958@yahoo.com)

Mr. Sylvester Gyamfi obtained MA Degree (Population studies) in 2002 from University of Ghana and a PG certificate course in Statistics in 1997 from University Southampton, UK. Mr. Gyamfi has supervised, coordinated and monitored various statistical projects of Ghana Statistical service which has the mandate to collect, analyse and disseminate statistical information on all aspects of the Ghanian economy to guide policy decision making and planning. He has also been instrumental in training the Regional Planning and Coordinating Units (RPCUs) in Ghana.

Tariq Bashir, Deputy Chief (Science), Pakistan Council for Science and Technology, Shahraj-e-jamhuriat G 5/2, Islamabad, Pakistan. (E-mail: tbashir@pcst.org.pk, drtariqbashir@yahoo.co.uk)

Dr. Tariq Bashir received PhD (Entomology) in 2000 from Natural Resources Institute, University of Greenwich, UK. He has diversified experience of scientific research, agricultural extension, science and technology policy, planning and development, and policy research. He has been involved in planning of R&D, evaluation of scientific research, promotion of science and research in the country, and in studies relating to the preparation of Science and Technology Indicators of Pakistan. Additionally, he is bearing the responsibility as Director of Pakistan Technology Board. He also worked briefly with Science, Innovation and Technology Team of North West Development Agency, Warrington, UK.

Thidar, Lecturer, Department of Technical and Vocational Education, Yangon, Myanmar. (E-mail: most25@myanmar.com, kmzin@Z6.com)

Dr. Thidar completed PhD (Engineering Mathematics) in 2005 from Mandalay Technological University, Myanmar. She is involved in the program of national education promotion for human resources development, which is being implemented by the states to build a modern nation. She carried out her research work on White Noise Analysis.

Vinod Kumar Gupta, Scientist, National Institute of Science, Technology and Development Studies, Pusa gate, K.S. Krishnan marg, New Delhi – 110012, India. (E-mail: vkgdwarka@gmail.com, vkgupta@nistads.res.in)

Dr. Gupta received MSc (Mathematics) in 1972, and obtained PhD in the area of intellectual property rights from the University of Mysore in 2006. His areas of current research interests include science and technology policy reforms in India, management of intellectual property rights in R&D organisations, and WTO related issues for science and technology. His areas of specialisation include scientometric analysis of patents, international affairs in science and technology, S&T management including technology transfer, consultancy, venture capital, start-up companies, dual use technology and export controls, and human resource development. He is presently on the editorial board of the Journal of Intellectual Property Rights being published from India.

Venâncio Massingue, Minister of Science and Technology, Government of Mozambique, Av. Patrice Lumumba 770, Maputo, Mozambique. (E-mail: Venancio.massingue@uem.mz, venancio@bandla.uem.mz)

Prof. Dr. Eng. Venâncio Massingue holds the doctoral degree in Information and Communication Technology Application and Management from Technology University of Delft, Netherlands. He also studied for the degree of Licenciatura in Electro-Technical Engineering at the Eduardo Mondlane University (UEM), Mozambique, where presently he holds a Faculty position. He is a member of the Mozambican Informatics Policy Task Force for development of a national information policy. He is currently the President of the UNESCO Regional Informatics for Africa (RINAF).

Vito N. Quevedo Rodríguez, Director, Innovation and Technology, Ministry of Science, Technology and Environment, Prado y San Jose, National Capitol, Havana, Cuba. (E-mail: vito@citma.cu)

Dr. Vito Quevedo got Doctor in Sciences (Energy) in 1980 from Energy Institute of Moscow, Rusia University, Cuba. He has a vast research experience in diversified areas of S&T and in participating and managing various projects, workshops and training courses related to the subject. Chairing as the Director of Technology and Innovation, Dr. Vito manages the focal point of the net of Indicators of Science and Technology (RICYT). He has been the President of the National Group of Technological Cooperation of Energy and had an active participation in the organization, management and implementation of the Cuban System of Science and Technological Innovation.